THE CHEMISTRY AND METABOLISM OF DRUGS AND TOXINS

AN INTRODUCTION TO XENOBIOCHEMISTRY

THE CHEMISTRY AND METABOLISM OF DRUGS AND TOXINS

AN INTRODUCTION TO XENOBIOCHEMISTRY

by

MICHAEL BRIGGS DSc, PhD, FRIC, FIBiol
Director of Biochemistry
Alfred Hospital, Prahran Melbourne 3181
Australia

and

MAXINE BRIGGS MB, ChB, DPH
Assistant Medical Superintendent
Alfred Hospital, Prahran, Melbourne 3181
Australia

William Heinemann Medical Books Ltd.
London

First published 1974

ISBN 0 433 04225 7

Printed in Great Britain by
William Clowes & Sons Limited
London, Colchester and Beccles

CONTENTS

PREFACE

There is an area of knowledge that lies between Medicine, Pharmacology, Microbiology, Biochemistry and Organic Chemistry. For want of a better name we have called it "Xenobiochemistry" (Greek *xenos* : strange). This subject covers all chemical aspects of the interaction between an organism and foreign compounds. In the following book we are concerned only with human xenobiochemistry, but others, clearly, will have more interest in veterinary xenobiochemistry, insect xenobiochemistry, etc.

From the human viewpoint, foreign compounds are not normally found in the human body. They may be toxins of microorganisms, venoms introduced by snakes or spiders, drugs, or industrial chemicals of many different kinds. A special group are normal body constituents present in abnormal (i.e. foreign) concentration, such as an overdose of adrenaline, acetylcholine, etc.

Xenobiochemistry does not significantly overlap with any of the established disciplines mentioned in the first paragraph above. Medicine is concerned with the treatment and prevention of disease due to foreign compounds, Pharmacology with the distribution, efficacy and side-effects of therapeutic substances, while Microbiology is involved only with those aspects involving microorganisms. The Organic Chemist is concerned only with the structures and synthesis of foreign compounds.

Theoretically, it is Biochemistry that should be concerned equally with the metabolism and effects of both foreign and non-foreign compounds. Unfortunately, Biochemistry is already a huge subject (many would feel too huge!), that encompasses an enormous range of subject matter. Foreign compounds have largely been pushed to one side while the vast array of endogenous substances have been given special attention.

Research in Xenobiochemistry is therefore conducted by specialists of many kinds, in departments of universities, hospitals and institutes which bear a wide variety of names. We hope that by gathering together in this book all the various threads, the outlines of Xenobiochemistry will be made sharper and more interaction between the various component themes may be made easier.

It is impossible to contain all current information relevant to Xenobiochemistry in a single volume, and the present work is offered only as an introduction. Because the book is an *outline* of a field of knowledge, as well as to make it more readable, we have not tried to provide formal references to statements in the text, but have listed relevant reviews and major works at the end of each chapter.

In chapter 2 we were faced with the problem of how best to present the structural formulae of the many drugs we discuss. We chose to put most in a single, very long, table, rather than spread them through the text. We feel that many who may be unfamiliar with pharmacology will find it of interest to look through in a few pages the wide range of foreign chemical compounds used as drugs. The magnitude of the challenge to metabolism may be made more easy to grasp.

M.B.
M.B.
Melbourne, Australia

Chapter 1

METABOLISM OF FOREIGN COMPOUNDS

I. FOREIGN COMPOUNDS IN THE ENVIRONMENT

1. Pharmaceuticals

The human body is continually exposed to a wide range of substances which are not natural constituents of the body; indeed, many substances are the products of human industry and are never found in Nature. The human digestive tract has evolved to cope with food components, while the internal organs possess limited capacities to metabolize and clear foreign compounds accidentally ingested with foods, or released into the system by parasites. For the past century there has been an enormous increase in both numbers and types of foreign compounds ingested by human populations. We will be able to mention only a few of these, but more detailed information will be found in the literature listed at the end of the chapter.

Clearly, pharmaceuticals of most types are foreign compounds, though a few contain substances normally present in the human body. The chemistry of the various classes of pharmaceutical products will be discussed in the next chapter, here we take some commonly used substances and mention their actions and side-effects; both requiring biochemical explanations. Neither can be understood without knowing the metabolites formed and their actions on metabolism.

Perhaps the most widely used of all drugs is aspirin (acetylsalicylic acid), which is available without prescription in every country of the world. It is a moderately effective analgesic agent, of value in the treatment of headache, neuralgia and muscular pains. The commonest side-effect is gastro-intestinal discomfort, and gastric ulceration following chronic use is well documented. Aspirin appears to inhibit the formation and release of mucus by the stomach. Also, it is capable of inducing hypoglycaemia, particularly in children. In rats, the LD_{50} by the oral route is 0.2 mg/kg, while toxic symptoms occur in humans when serum salicylate levels exceed 35 mg/100 ml.

Above this level, inhibition of prothrombin formation results and coagulation defects may occur. Chronic use at therapeutic doses induces pancytopenia in some individuals, indicating a wide-ranging metabolic effect on many organs. About one person in 100 is unable to use aspirin owing to allergic reactions. As it is so freely available, aspirin has been used widely by suicides and attempted suicides. Hyperventilation is an early sign of an overdose and metabolic acidosis soon follows. Death results from oligaemic shock due to disturbed acid-base balance and dehydration. Prostaglandin biosynthesis is inhibited at all doses.

Another widely used class of pharmaceutical, entirely foreign to the body, is that of phenothiazine tranquillizers, such as chlorpromazine, thioridazine, imipramine, etc. These compounds have a complex action and depress activity in the brain stem without much altering activity in the spinal cord or cerebral cortex. The net result is reduced motor activity, without any significant hypnotic action. A very wide range of side-effects has been reported, affecting almost every important organ. Extra-pyramidal effects occur in some individuals, with Parkinson-like symptoms developing slowly. Other people show hyperkinetic side-effects, involving muscles of the face and neck. Irreversible dyskinesias have been recorded on chronic administration. Postural hypotension, skin eruptions, blood dyscrasias and jaundice have all been reported following the use of tranquillizers.

A large number of women throughout the world use oral contraceptives, which are based on synthetic analogs of the sex hormones. Usage has been estimated at about 20 million women. These products cannot be metabolized like the natural sex hormones owing to their chemical structures which possess substituents at regions of the molecules usually acted upon in the natural hormones. The use of these synthetic steroids is also associated with a wide range of side-effects, ranging from fatal thromboembolic disorders to chloasma.

Many other examples could be chosen from such widely used drugs as antibiotics, opiates, anti-malarials, etc. The metabolism of selected pharmaceuticals will be discussed in detail later in this chapter.

2. Industrial Chemicals

Many foreign compounds are used in industry as solvents, or as intermediates in manufacturing processes. Solvents such as benzene, carbon tetrachloride, trichloroethylene are all metabolized by the body only with great difficulty, yet are readily absorbed, both as liquids and vapours. Acute exposure can result in massive CNS depression, while chronic exposure produces respiratory and hepatic damage. Some workers become addicted to "sniffing" solvents from glues, paints, etc.

Aromatic nitro derivatives and amines are used widely in the manufacture of dyestuffs, explosives and pesticides. The association between exposure to some of these compounds, particularly β-naphthylamine, and bladder carcinoma is now well established.

Many of the products of the oil industry are hazardous and have widespread metabolic effects. Prolonged contact with any petroleum product is likely to produce dermatitis, while the previously mentioned addiction is common. Many of the additives included in petroleum products are also very toxic. Tetraethyl lead is particularly difficult to metabolize in the body, while tri-*o*-cresyl phosphate is a neurotoxin. In 1959 a large batch of cooking oil was accidentally contaminated by this latter compound and sold throughout Morocco where at least 10,000 people were poisoned. Over 1000 are bedridden due to permanent neurological damage, while an equal number required orthopedic procedures to correct severe deformities. Tri-*o*-cresyl phosphate has also been added to ginger extracts as a flavour.

A wide range of chemical detergents is now available: these are used for everything from washing kitchen utensils and clothing, to large scale industrial cleansing. Three major types may be distinguished chemically. The first are the anionic detergents, which are sodium salts of sulphated fatty alcohols, together with sodium sulphonates of long-chain alkyl derivatives of benzene. Chronic exposure may result in contact dermatitis, while ingestion of a large dose produces gastrointestinal irritation and diarrhoea. Next are the cationic detergents, which are alkyl or aryl quaternary ammonium compounds. Again they act as local irritants to the skin and death has been recorded following the accidental ingestion of only 3 g. The third class of detergents is non-ionic and include a range of chemically different substances, including polyethylene glycol alkyl or aryl ethers, alkyl or aryl polyether sulphates, polyoxythylene sorbitan monooleate, etc. These also are moderately toxic and skin irritants. Detergent residues are present in clothing after laundering and on kitchen utensils after they have been washed, so that most people living in civilized communities are continually exposed to low levels of all three classes of detergent.

Related industrial chemicals are the various types of bleaching or brightening agents. The commonest bleaches are hydrogen peroxide and chlorine, but in recent years there has been considerable interest in so-called "optical bleaches", which increase whiteness or brightness of laundered fabrics. Popularly used compounds include 4,4′-bis(4-*p*-methoxyanilino-6-morpholino-1,3,5-triazin-2-ylamino)stilbene-2,2′-disulphonate and related derivatives.

Numerous other examples could be mentioned in the modern expanding chemical industry.

3. Cosmetics

Perfumes and pigments have been used by both men and women since the dawn of recorded history. Until modern times, however, these products were plant and animal preparations, together with a few simple naturally-occurring mineral substances. There is now a vast chemical industry devoted to the synthesis of new agents for use in cosmetic products, and very large quantities of foreign compounds are in everyday popular use.

The alteration of hair colour is one major source of exposure to cosmetic chemicals. These range from peroxide bleaches to dyes containing aniline or metallic salt derivatives. Black hair dyes often contain *p*-phenylenediamine or *p*-toluylene-diamine and can cause skin reactions. These compounds penetrate the hair shaft, unlike metallic salt dyes, which are usually lead or bismuth compounds, and merely deposit a coloured film around the hair. Colour rinses are primarily organic acid dyes which may be removed by shampooing. While serious reactions to hair dyes seem rare, death has been recorded following the use of hair dye containing *p*-phenylenediamine; chronic use leads to low level skin, and, to a lesser extent, systemic symptoms of exposure.

Other foreign compounds derived from hair cosmetics include thioglycolates from depilatories, hair straighteners and cold-wave products, and polyvinylpyrrolidone from sprays and lacquers. Thioglycollates are potential skin sensitisors and produce gastro-intestinal upsets if taken orally. Polyvinylpyrrolidone and other resins are inhaled by users of hair sprays and may produce pulmonary infiltration, with granulation, inflammation and fibrosis. "Hair-spray thesaurosis" has been clearly described in a number of patients.

Large numbers of foreign compounds may be derived from substances present in lipsticks, mascara, nail polish, soaps, perfumes, and skin creams. A high percentage of the lipstick applied to the lips is orally ingested, while the capacity of the skin to transfer topically applied compounds into the blood is frequently under-estimated. Fatal intoxications in babies have been described following the absorption of constituents of marking-inks used to put laundry marks on napkins. Most cosmetics contain low concentrations of compounds known to induce severe metabolic reactions when used in large amounts. For example, death has been recorded following the ingestion of oil of citronella, while many perfumes have high lipid-solubility and tend to accumulate in nervous tissues.

4. Food Additives

Foods may be contaminated by all manner of different substances, but in recent years a wide range of synthetic chemical substances have been used on a wide-scale to improve the colour, flavour or keeping-qualities of foods. These additives now represent one of the major sources of foreign compounds in the human environment.

Some of these additives may be natural compounds, such as vitamins or inorganic phosphates, but are frequently used in a form differing significantly from the substances as they occur in natural foods (e.g. mixtures of natural and unnatural optical or geometrical isomers, or as salts or esters with some other compound).

Natural colouring agents for food, such as cochineal or saffron, have been largely superseded by azo and fluorescein dyes. Even the natural products are not free of dangers and outbreaks of salmonella have been traced to contaminated cochineal. One of the first food dyes was butter yellow (*p*-dimethylaminoazo-benzene), which was used to give winter-produced butters a stronger colour. It proved to be a potent liver carcinogen and its use was prohibited. Other commonly used azo dyes in foods are chrysoidine (1-*p*-sulphophenylazo-2-naphthol), Sudan I (1-phenylazo-2-naphthol), and tartrazine (5-hydroxyl-1-(*p*-sulphophenyl)-4-(*p*-sulphophenylazo) pyrazole-3-carboxylic acid) (for structure see Figures 1.23, 1.24, and 1.25).

Two synthetic compounds have found use as sweetening substitutes for sugars. They are cylamates (salts of cyclo-hexyl-sulphamic acid) and saccharin (*o*-sulphobenzimide).

Other deliberately added foreign compounds in foods include antioxidants, such as butylated hydroxyanisole, butylated hydroxytoluene or *N*,*N′*-diphenylphenylenediamine, preservatives, such as benzoic acid, sorbic acid, or *p*-hydroxy-benzoic acid, and flavour enhancers, such as monosodium glutamate.

5. Pesticides

Compounds toxic to unwanted plants, insects, molluscs, fungi, and even mammals are all in widespread use in agriculture or in public health programmes. The hazard to humans comes first from contact with relatively high concentrations of these compounds, as can occur in agricultural workers or persons involved in their manufacture; secondly, there is a more insidious hazard from low levels of pesticides throughout the human environment. Pesticides have been detected in the tissues of animals from the polar regions, from deserts, and from the depths of the sea. The compounds enter food-chains and can accummulate in the tissues of animals at the ends of these chains. They are present in detectable quantities in human tissues from all countries, although the highest amounts are seen in the developed countries where their use is the greatest.

Herbicides range from generally acting substances such as potassium chlorate, to selectively toxic compounds like 2,4-dichlorophenoxyacetic acid. The latter herbicide is of low toxicity, the lethal oral dose for an adult being about 50 g, but is often used mixed with 2,4,5-trichlorophenoxyacetic acid, which is far more persistent.

Insecticides probably represent the most widely used pesticides, not only for the extermination of insects in agriculture and the home, but also in the control of insect-borne diseases, such as malaria. Several quite chemically distinct classes of insecticides can be recognized. First there are chlorinated hydrocarbons, such as DDT (2,2-bis-(*p*-chlorophenyl)- 1,1,1-trichloroethane), aldrin (1,2,3,4,10,10-hexachloro- 1,4,4a,5,8,8a-hexahydro-1,4,5,8-endo, exo- dimethanonaphthalene) dieldrin (the epoxide of aldrin), and benzene hexachloride (1,2,3,4,5,6-hexachlorocyclohexane). Organophosphorus compounds represent a second class of insecticide and including substances such as parathion (*O,O*-diethyl-*O-p*-nitrophenylphosphorothionate), malthion (*O,O*-dimethyl-*s*-(1,2-dicarbethoxyethyl)-phosphorodithionate) and systox (*O,O*-diethyl-*O*-2- (ethylthioethyl)- phosphor-thionate). Carbamate insecticides include sevin (1-naphthyl-*N*-methylcarbamate) and zectran (4-dimethylamin-3,5-xylylmethyl carbamate), while certain natural products, such as pyrethrum (though this is often used in combination with piperonyl butoxide and bucarbolate) and nicotine are also used.

Many of these compounds also possess molluscocide activity, although pentachlorophenol is more specific in its actions. The latter compound also acts against fungi, but is less widely used than hexachlorobenzene.

The number of different compounds now in use as pesticides is probably approaching 1000 and world usage is in excess of one million tons per annum. A number of outbreaks of human disease have been traced back to the use of various pesticides. One such occurred in southern Turkey during the period 1955 to 1959 when over 600 cases of a vesicular and bullous condition resembling porphyria cutanea tarda were reported following the introduction of hexachlorobenzene to prevent fungal growth on wheat seedlings, some of which were later used for human consumption.

Substances aimed at the eradication of mammalian pests present a clear hazard to humans. Strychnine has been widely used as a rodenticide, and fluoroacetamide has been used to aid the elimination of unwanted deer. Warfarin (phenyl-acetylethyl-hydroxy-coumarin) is an anti-coagulant used as a human drug, but which finds its major use as a rodenticide.

6. Food Anutrients

We have already discussed the addition of foreign compounds to foods as colours, flavours, etc., and we have mentioned contamination of food by pesticides. The foods consumed by humans, however, contain substances devoid of nutritional value ("anutrients") derived from several other sources.

First there are naturally occurring anutrients present in particular foods. These include anthocyanins, which are the pigments of many fruits and vegetables; terpenes, which are present in the essential oils of many herbs and fruits; methylated purines, such as caffeine (1,3,7-trimethylxanthine), which is present in tea, coffee and coca, together with a variety of other substances, some of which are toxic and are discussed in the chapter on "Natural anti-metabolites" (p. 197).

Industrial chemicals of many different kinds may find their way into foods. Small particles of metal may be added by the use of faulty food processing machinery, while fragments of wrapping materials can also be present. The increasing use of plastic food containers also increases the chance of food contamination by dialkyltin compounds, such as diethyl tin chloride, which are used as stabilizers for plastics. The use of tetrafluorethylene polymers (Teflon) to coat "non-stick" pans can lead to the production of toxic fluorocarbons being released if the pan is overheated to above 300°C, but probably the commonest source of these substances ingested by humans is tobacco contaminated by grains of the plastic.

Widely used foods, especially fish, are likely to take up industrial chemicals. An outbreak of myoglobinuria occurring in people living near the Haff river was later traced to the accumulation of industrial resins by eels, which were later caught and eaten.

Finally, the method of food preparation can add foreign compounds. Meats barbecued over charcoal pick up traces of aryl hydrocarbons, but smoking fish or meat as a preserving method may add considerable amounts.

II. ABSORPTION AND EXCRETION OF FOREIGN COMPOUNDS

1. Routes of Absorption

We have seen that humans are subjected to exposure to numerous different types of foreign compounds from many sources. The compound can be presented to the body in several different ways. It may enter the lungs with inspired air; it may enter the gastro-intestinal tract with food or as a drug, or it may be absorbed through the skin from sprays, cosmetics or topical pharmaceuticals. Irrespective of the route of presentation, most foreign compounds are absorbed systemically, if only to a limited extent. It is extremely rare for a substance to come into contact with the body without some, at least, entering the blood and other tissues.

Once inside the body, almost every foreign substance is metabolized to some extent. The human ability to metabolize compounds which are unnatural to the body is inherited presumably from animal ancestors who had to deal with a wide variety of natural plant anutrients. Any compound foreign to the body is a potential poison and the outcome of the metabolism is either to break the compound down into fragments which enter normal metabolic pathways, or to alter the solubility of the compound so that it is excreted more readily. Foreign compound metabolism usually reduces the risk of toxic effects by getting rid of the compound as quickly as possible. Unfortunately, while metabolism of many foreign compounds produces substances of reduced toxicity, for some, metabolites are formed with *enhanced* toxic properties. Examples of the latter are as follows:

(i) The insecticide parathion is metabolized by desulphation to its oxygen analog, paradoxon, which is a potent inhibitor of cholinesterase, an enzyme upon which parathion has no effect.

(ii) Fluoroacetate is metabolized to fluorocitrate, which inhibits the enzyme, aconitase, and so blocks the metabolically important tricarboxylic acid cycle.

(iii) Carcinogens such as aryl hydrocarbons require metabolic activation to oxygen containing derivatives before they can induce cancers.

2. Factors Affecting Absorption

For compounds entering the body via the lungs, the major factors governing absorption are, first, the physical form of the compound (gas, liquid or solid) and secondly, its water-solubility. Insoluble solid particles inhaled into the lungs have only a low chance of systemic absorption, whereas soluble gases or liquid-droplets have a high chance. For example, coal dust (insoluble solid particles) or hair-spray resins (insoluble liquid-droplets) are poorly absorbed, but chlorine (soluble gas) or starch dust (soluble particles) are readily taken from the lungs into the blood.

Absorption through the skin is a more complex phenomenon and the presence of abrasions, cuts or other lesions may greatly assist direct transfer into the blood. Both water-soluble and lipid-soluble compounds can be absorbed through the skin, or may be held within the dermal and sub-dermal tissues. The skin is an extremely active organ capable of metabolizing many compounds, so that foreign compounds may be chemically altered within the skin prior to absorption.

Table 1.1. Effect of intestinal pH on drug absorption

Intestinal pH	Percentage of compound absorbed		
	Quinine	Salicylic acid	4-Hydroxypropiophenone
3.6–4.3	9	64	61
4.7–5.0	11	35	52
7.0–7.2	41	30	67
7.7–8.0	54	10	60

By far the most important route for the absorption of foreign compounds is via the gastro-intestinal tract and substances may be absorbed almost anywhere along its length, from the mouth (e.g. ethanol) to the colon (e.g. indoles produced by intestinal bacteria), although the most usual route is through the small intestine. The effect of small intestinal pH is often of primary importance to absorption and this is clearly shown in Table 1.1 of the effects of pH on the absorption of three drugs. It will be seen for quinine that absorption increases as the pH rises, whereas for salicyclic acid it is greatest at low pH. These are typical examples, whereas 4-hydroxypropiophenone is unusual in showing erratic changes in absorption at different pH.

Factors influencing the absorption of nutrients tend also to influence the absorption of foreign compounds through the intestines, so that, for example, intestinal infections can be of great influence.

3. Transport and Tissue Distribution

Foreign compounds in blood may be present in simple solution or, rarely, as droplets, but are often chemically combined with serum proteins. Lipid-soluble substances entering via the intestines are usually dissolved in the chylomicrons of dietary fat which are transported by both the lymphatics and the blood.

Water-soluble foreign compounds, or their water-soluble metabolites, are cleared by the kidneys for excretion in urine. Some also pass into the bile. Lipid-soluble foreign compounds and metabolites partition between blood and adipose tissues and may form a deep body compartment in the latter from which their removal may take months. Lipid-soluble compounds also enter the bile, but all biliary excretion is subject to an entero-hepatic cycle of continuous reabsorption and re-excretion, before escape into the faeces is possible.

Tissues other than adipose tissues can also bind and hold foreign compounds. For example, sulphonamides bind to collagen, while tetracyclines bind to teeth and bones.

Volatile foreign compounds present in blood tend to be lost via the lungs and this can be an important route of excretion. Other compounds bind to keratin and are gradually lost in hair and nails (e.g. heavy metals such as lead, mercury, arsenic).

4. Excretion of Foreign Compounds

Irrespective of the route of entry into the body, foreign compounds, or their metabolites, may be excreted by a wide variety of routes. Most foreign compounds are lost by two or three different mechanisms. The most important method of excretion for the majority of compounds is via the urine, but this requires water-solubility. The bile and faeces are the second most important, but we have mentioned above that other routes include expired air, nails and hair. Other less obvious routes are in milk, sweat or tears. Virtually any fluid excreted from the body can be used to lose foreign compounds.

The importance of water solubility on excretion in the urine can be seen in Table 1.2 which lists the relative proportions of a

Table 1.2. Comparative urinary and biliary excretion

Compound	Relative excretion (%) Urine	Bile
Fluorescein	30	14
4,5-Dibromofluorescein	5	29
2,4,5,7-Tetrabromofluorescein	4	63
2,4,5,7-Tetrabromo-12,13,14,15-tetrachlorofluorescein	0	66

Table 1.3. Excretion of volatile compounds in expired air

Compound	Formula	B.P. °C	Percent oral dose lost in expired air
Benzene	C_6H_6	80	*c.* 40
Monochlorobenzene	C_6H_5Cl	132	27
Monobromobenzene	C_6H_5Br	155	6
Monoiodobenzene	C_6H_5I	189	3
Nitrobenzene	$C_6H_5NO_2$	210	1
Aniline	$C_6H_5NH_2$	184	1

series of substituted fluoresceins excreted via the urine or via the bile. Water-solubility decreases with increasing substitution.

Orally ingested volatile compounds can be excreted to a significant extent via expired air, but the importance of this route decreases with the volatility of the compound at body temperature.

Table 1.3 lists the percentage of an oral dose of various substituted benzene derivatives, as compared to their boiling points.

Other factors affecting drug excretion include the reaction of the urine. Alkalinization decreases the loss of nicotine, quinine, and chloroquine, whereas acidification decreases the loss of sulphonamides. Glomerular filtration and tubal excretion and reabsorption play different roles in the urinary excretion of particular substances. Weak or moderate acids are often excreted via the tubules, e.g. penicillin, salicylic acid, carinamide, as are some alkalis, such as tetraammonium salts. Strongly acid compounds are not excreted via the tubules.

Quite a number of foreign compounds absorbed by a mother appear in her milk, and although it is unusual for them to have much effect on the baby, care is clearly needed with potent compounds. Among the compounds excreted in this manner are sulphonamides, which may actually be present in milk at concentrations higher than in plasma, ethanol, chloroform, bromides, senna and a number of others.

Sweat is also an unusual route of excretion for most compounds, but may be important for a few. Bromides are lost in this way in significant amounts, while the anti-leprosy drug, ditophal (dithioisophthalate) is excreted equally between sweat, bile and urine.

III. METABOLIC TRANSFORMATIONS OF FOREIGN COMPOUNDS

1. Microsomal Oxidations

The most important organ involved in foreign compound metabolism is the liver, but the kidneys, skin and other tissues all play some part. In all tissues involved in metabolizing foreign compounds, it is the class of sub-cellular particle known as the "microsome" which appears to possess most of the enzyme systems capable of acting on these compounds.

A number of chemical transformations of foreign compounds are possible, but oxidation within the microsomes is frequently the reaction of primary importance. Some examples of microsomal oxidations are listed in Figure 1.1.

The mechanism of microsomal oxidation has been the subject of much investigation and at least the overall details are now clear. Studies with labelled oxygen have shown that the oxygen atom incorporated into the foreign compound molecule during microsomal oxidation is derived entirely from molecular oxygen and not from water. Microsomes are known to contain cytochromes which differ from those found in mitochondria. Several different cytochromes have been described and may be involved in the oxidation of particular classes of foreign compound, however the most abundant microsomal cytochrome is known as "cytochrome P-450", from its absorption spectrum. This cytochrome is present in the microsomes of tissues actively involved in foreign compound metabolism, such as the liver, but is absent from tissues not concerned with this type of metabolism, such as the brain. Cytochrome P-450 also appears to be

(i) Aromatic oxidation

$$C_6H_5.X \rightarrow HO.C_6H_4.X$$

(ii) Acyclic oxidation

$$CH_3.X \rightarrow HO.CH_2.X$$

(iii) *O*-dealkylation

$$CH_3.O.X \rightarrow HO.CH_2.O.X \rightarrow HCHO + HO.X$$

(iv) *N*-dealkylation

$$\underset{\displaystyle H}{CH_3.\overset{}{\underset{|}{N}}.X} \rightarrow \underset{\displaystyle H}{HO.CH_2.\underset{|}{N}.X} \rightarrow HCHO + H_2N.X$$

(v) Deamination

$$\begin{matrix}X'\\X''\end{matrix}\!\!>CH.NH_2 \rightarrow \begin{matrix}X'\\X''\end{matrix}\!\!>C(OH).NH_2 \rightarrow \begin{matrix}X'\\X''\end{matrix}\!\!>C{=}O + NH_3$$

(vi) Sulphoxidation

$$\begin{matrix}X'\\X''\end{matrix}\!\!>S \rightarrow \begin{matrix}X'\\X''\end{matrix}\!\!>S^+.OH^- \rightarrow \begin{matrix}X'\\X''\end{matrix}\!\!>S \rightarrow O + H^+$$

Figure 1.1. Oxidation of foreign compounds.

involved in the metabolism of endogenous steroid hormones and so may also be found in tissues involved with steroids, such as the adrenals.

The amount of cytochrome P-450 present in the liver is significantly increased following exposure to foreign compounds. Simultaneous administration of actinomycin D, an inhibitor of messenger-RNA formation, abolishes this increase.

Like many other cytochromes, P-450 will react with molecular oxygen to form an unstable complex. The oxygen atom on this complex can then be transferred to a molecule of the foreign compound (Figure 1.2).

As only the reduced form of cytochrome P-450 will react with molecular oxygen, a mechanism for the continuous regeneration of this form is present within the microsomes. The mechanism is in many ways analogous to the electron transport pathway of the mitochondria. The process requires $NADPH_2$ specifically, and $NADH_2$ is not used. The transfer of electrons from $NADPH_2$ to cytochrome P-450 is accomplished by an enzyme known as "$NADPH_2$-cytochrome *c* oxido-reductase" and is the same enzyme present in mitochondria. This enzyme can be shown to have at least two components, a flavoprotein (FAD) and a protein containing non-haem iron. Like cytochrome P-450 the amount of this microsomal oxido-reductase increases greatly following exposure to foreign compounds.

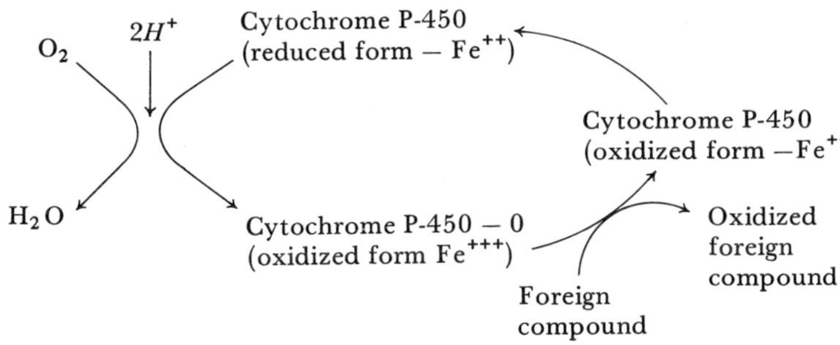

Figure 1.2. Cytochrome P_{450} in foreign compound oxidation.

The complete electron transfer pathway for foreign compound oxidation within the microsomes is shown in Figure 1.3.

The non-haem ferroprotein is shown as accepting two electrons, but is more likely to be a single electron-acceptor.

Cytochrome P-450 is present within microsomes in a protein-bound form and it is, presumably, the nature of the protein that determines the specificity of the oxygen transfer reaction to the foreign compound. In fact, these microsomal oxidations represent the class of enzymic reaction with the *lowest* degree of substrate specificity. A very large range of chemically different substances are capable of accepting oxygen from the active-form of oxidized cytochrome P-450. Thus phenobarbital, aminopyrine, steroid hormones and aromatic hydrocarbons all appear to be oxidized by the same system. As mentioned above, there appear to be several microsomal cytochromes other than

P-450, and these help to determine what specificity there is in microsomal oxidations. Hexobarbital is oxidized by a system that will not oxidize tyramine, and vice versa. Some foreign compounds may be oxidized by two different pathways by the microsomes. If the compound is given alone the oxidized products may be quite different to those formed when the

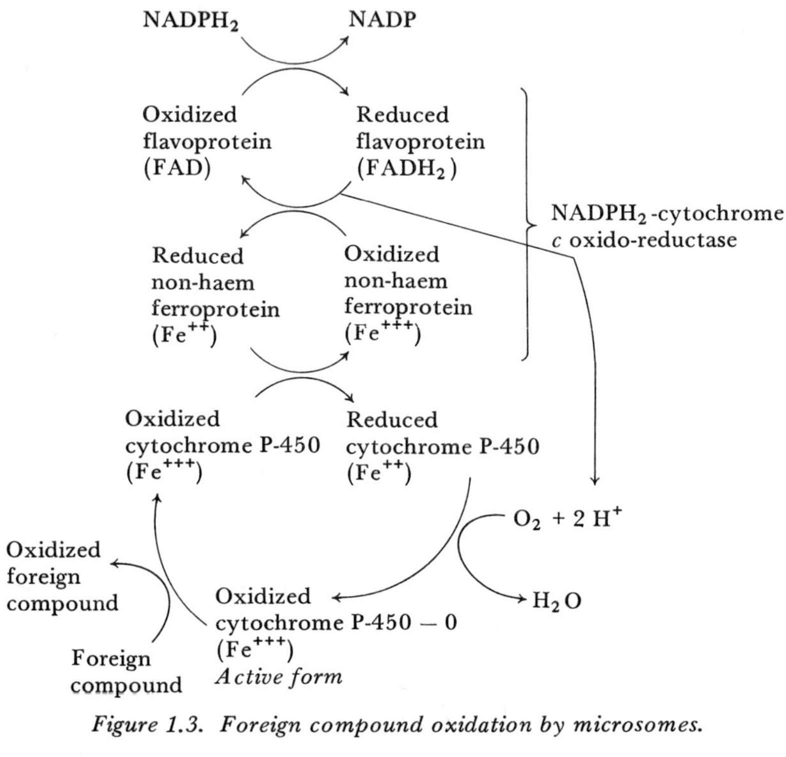

Figure 1.3. Foreign compound oxidation by microsomes.

compound is given to an animal pre-treated with a second foreign compound. A good example of this is the polycyclic aromatic hydrocarbon, 7,12-dimethylbenz(a)anthracene (DMBA). When given alone to rats the major metabolite is formed by oxidation of the 7-methyl group. It happens that this metabolite has a potent necrotic effect on the adrenal. If the rats are pre-treated by some other hydrocarbon, such as naphthalene, or even by steroids, such as betamethasone, these substances induce a microsomal oxidation system (enzyme induction) different from that induced by DMBA given alone. If DMBA is now given to the pre-treated rats, no adrenal necrosis occurs for the 7-hydroxymethyl metabolite is not formed and the major metabolites of DMBA are ring-hydroxylated compounds (Figure 1.4).

A somewhat similar effect can be seen in the interaction between certain drugs and the metabolism of endogenous steroid hormones. It was mentioned above that steroid hormones are metabolized by microsomal processes involving cytochrome P-450 and that tissues, such as the adrenal glands, which

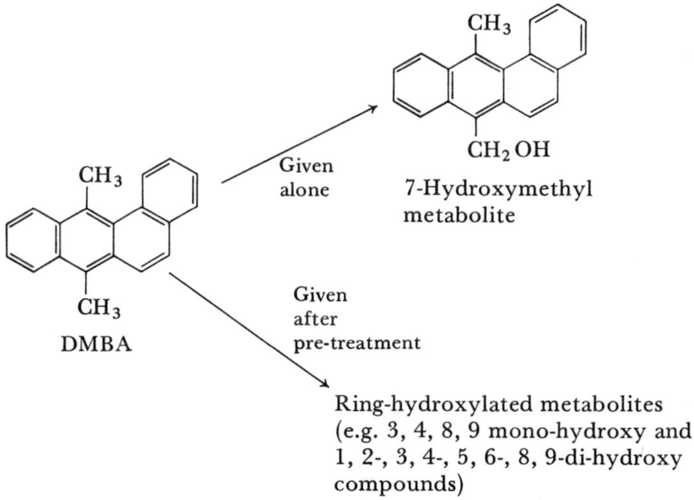

Figure 1.4. Metabolism of DMBA.

are active in steroid metabolism possess these systems. It naturally follows that steroid hormones circulating through the liver of a subject treated with foreign compounds that have induced increases in hepatic microsomal oxidases may well be acted upon in a manner unlike that in an untreated subject. Phenobarbitone, for example, has been found to increase the rate of turn-over of several steroids, including cortisol and testosterone.

All microsomal enzyme systems are closely associated with the endoplasmic reticulum and are very difficult to solubilize for study in the laboratory. A certain degree of success has been achieved by the use of snake venom.

2. Non-microsomal Metabolism

While the microsomes are the most important subcellular fraction involved in the metabolism of foreign compounds, a number of reactions occur, catalysed by soluble enzymes of the cell sap. Again the liver is usually the principal organ involved in these transformations.

An important non-microsomal enzyme is alcohol dehydrogenase, which will catalyse the reversible inter-conversion of primary alcohols and aldehydes (Figure 1.5).

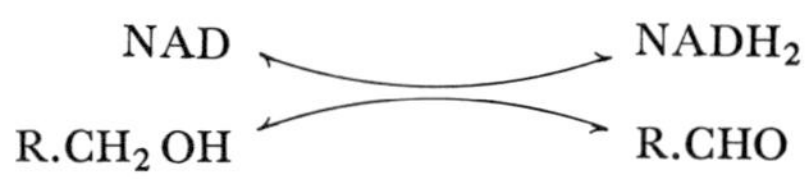

Figure 1.5. Primary alcohol oxidation.

A second important soluble enzyme is aldehyde dehydrogenase, which catalyses the conversion of primary aldehydes to the corresponding carboxylic acid (Figure 1.6).

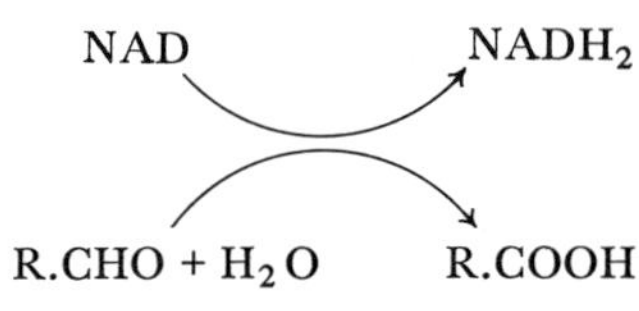

Figure 1.6. Aldehyde oxidation.

Other important non-microsomal transformations are acylation with acyl-coenzyme A, methylation with *S*-adenosyl methionine, and conjugation reactions with sulphates, glycine and glutathione. Conjugation reactions will be discussed separately below.

The non-microsomal enzymes concerned with foreign compound metabolism are enzymes involved in normal metabolic processes of the body. Like the microsomal enzymes, their activities increase when exposure to a foreign compound requiring their presence occurs.

Table 1.4 contrasts the reactions of foreign compounds requiring microsomal or non-microsomal enzymes.

Table 1.4. Types of foreign compound metabolism

Type of reaction	Examples	Intracellular enzyme
1. Oxidations	Hydroxylation	Microsomes
	Dealkylation	Microsomes
	Oxide formation	Microsomes
	Desulphuration	Microsomes
	Dehalogenation	Microsomes
	Alcohol oxidation	Soluble
	Aldehyde oxidation	Soluble
2. Reductions	Azo reduction	Microsomes
	Nitro reduction	Microsomes and soluble
	Aldehyde reduction	Soluble
3. Hydrolysis	De-esterification	Microsomes
	De-amidation	Microsomes
4. Acylations	Acetylation	Soluble
	Benzoylation	Soluble
5. Methylations	—	Soluble
6. Conjugations	Glucuronation	Microsomes
	Sulphation	Soluble
	With Glycine	Soluble
	With Glutathione	Soluble

3. Conjugation Reactions

The primary important reactions of foreign compounds, such as microsomal oxidations, involve alteration to the basic molecular structure. Conjugation, on the other hand, is the addition of a group to the foreign molecule. The commonest groupings added during conjugation are glucuronic acid or sulphate, but conjugation with glycine or glutathione is of importance for particular compounds.

It is important to realize that conjugation may occur with the unchanged foreign compound, or with its metabolites produced by some other process, e.g. microsomal oxidation. Compounds can only be conjugated if they possess a suitably reactive group, such as a hydroxyl or an amino group. For example, cyclohexane cannot form a conjugate, so it is first oxidized in the microsomes to cyclohexanol, which is then conjugated with glucuronic acid. Benzoic acid, however, can be immediately conjugated to glucuronic acid, but it is also subjected to microsomal oxidation to *m*-hydroxybenzoic acid, which is then conjugated with glucuronic acid and sulphate.

The mechanism of conjugate formation differs for each chemical class. For glucuronates, the conjugate is formed by a transfer of a glucuronic acid residue from uridine diphosphate (UDP)-glucuronate to the foreign compound, a reaction catalysed by microsomal UDP-transglucuronidase. The overall pathway is shown in Figure 1.7.

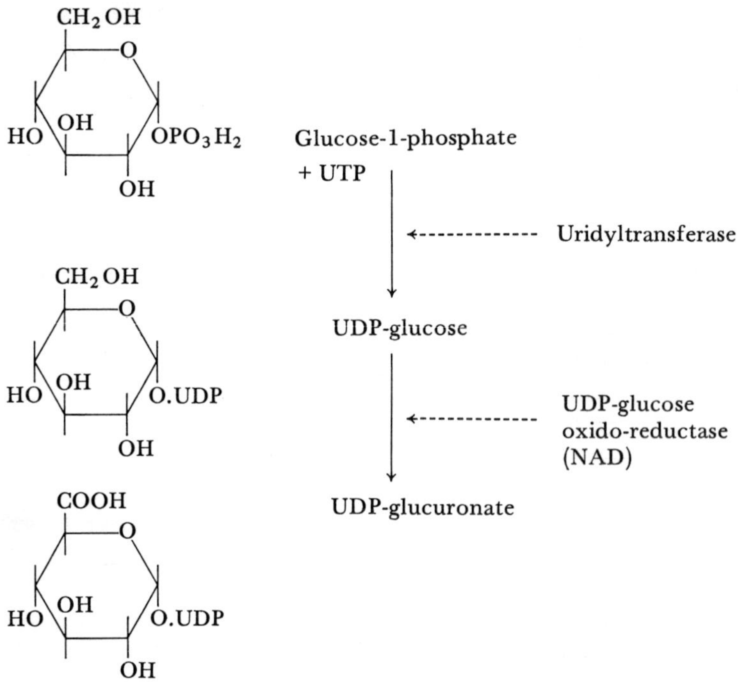

Figure 1.7. Biosynthesis of UDP-glucuronic acid.

Glucuronates of foreign compounds are of several different types. *O*-glucuronates are formed by a linkage of the β-oxygen atom on C-1 to an oxygen containing function on the foreign compound. If the latter is a phenolic or alcoholic hydroxyl, the resulting *O*-glucuronide is an ether, but if the function is a carboxylic acid group, it is an ester (Figure 1.8).

Linkage of glucuronic acid to atoms other than oxygen is also possible, but rather less common. *N*-glucuronates are formed from primary or secondary amines, while *S*-glucuronates are formed from thiols or disulphides.

O-*glucuronates*

1. UDP-glucuronate + $R.CH_2.OH$

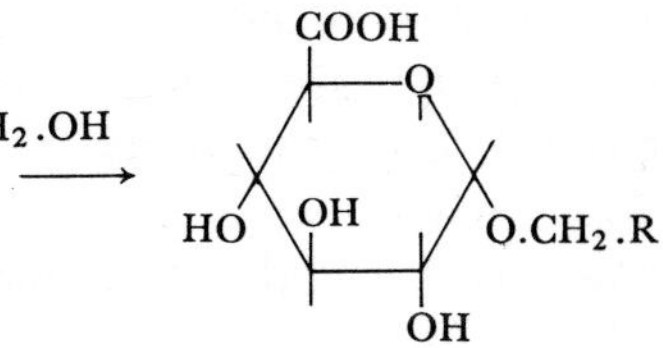

Ether type

2. UDP-glucuronate + R.COOH

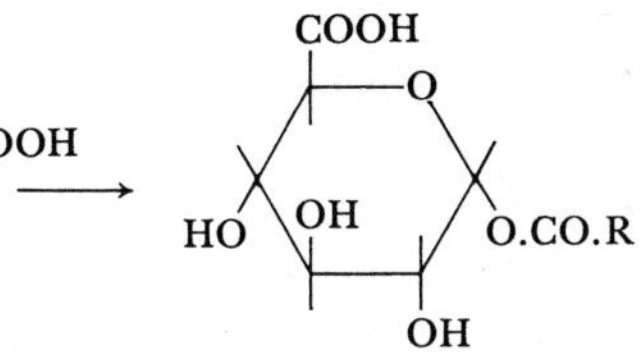

Ester type

N-*glucuronides*

1. UDP-glucuronate + $R.NH_2$

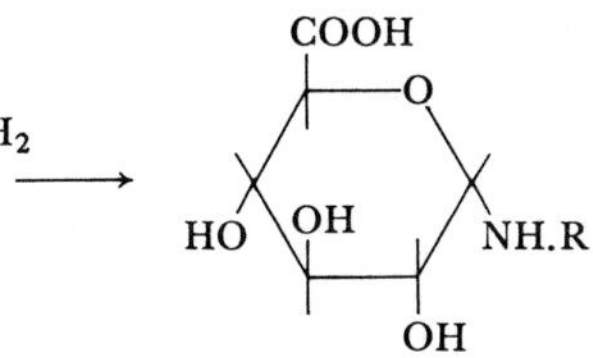

Amide type

2. UDP-glucuronate + R.NH.R′

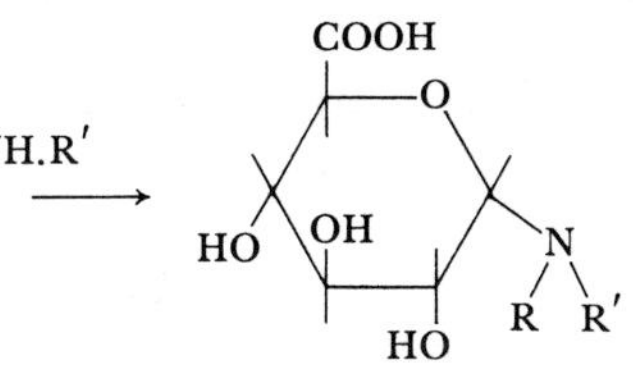

Imide type

S-*glucuronates*

1. UDP-glucuronate + R.SH

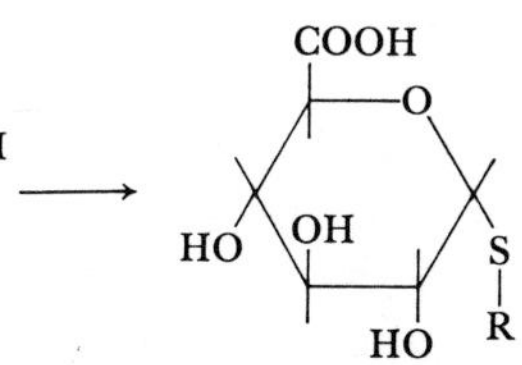

2. UDP-glucuronate + R.S.S.R.′

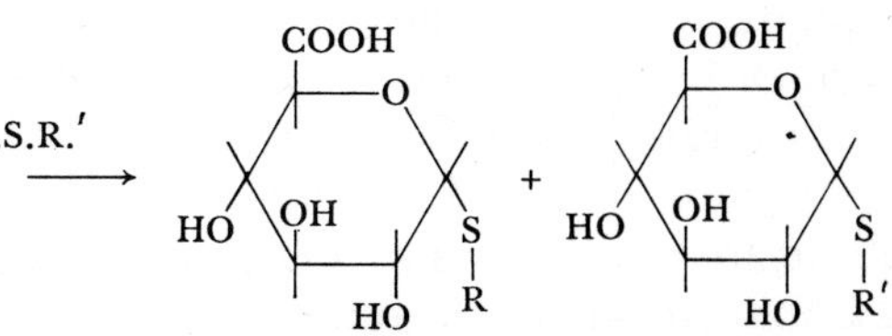

Figure 1.8. Chemical types of foreign compound glucuronates.

The other major class of conjugates are the sulphates, which are formed by the transfer of the sulphate group from 3′-phosphoadenosine-5′-phosphosulphate (PAPS) to the foreign molecule, a reaction catalyzed by the soluble enzyme sulphotransferase. PAPS is formed from inorganic sulphate by the following reactions:

$$\text{ATP} + \text{SO}_4^{2-} \rightarrow \text{Adenosine-5}'\text{-phosphosulphate (APS)}$$
$$\text{APS} + \text{ATP} \rightarrow \text{PAPS} + \text{ADP}$$

The first reaction is catalysed by ATP-sulphate adenyltransferase, and the second by ATP-APS phosphotransferase.

As with the glucuronates, several types of sulphate conjugate are possible depending on the nature of the group sulphated. *O*-sulphates are by far the commonest and are formed readily with phenols or aliphatic alcohols.

O-*sulphates*

1. $\text{PAPS} + \text{R.OH} \rightarrow \text{R.O.SO}_3\text{H}$
   (phenol) *aryl sulphate*
2. $\text{PAPS} + \text{R.CH}_2\text{OH} \rightarrow \text{R.CH}_2\text{.O.SO}_3\text{H}$
   (alcohol) *alkyl sulphate*

Linkage of sulphate with nitrogen containing groups is also possible to give *N*-sulphates, which are named "sulphamates".

N-*sulphates*

$$\text{PAPS} + \text{R.NH}_2 \rightarrow \text{R.NH.SO}_3\text{H}$$
sulphamate

Not only foreign compounds are sulphated. Connective tissues contain sulphated polysaccharides known as chondroitin sulphates, while many of the steroid hormones yield sulphates in the normal course of their metabolism.

Conjugation of a compound with either glucuronic acid or sulphate increases the water-solubility and the compound is more easily eliminated in the urine. Both types of conjugate are also excreted partly in bile, but the intestines possess both β-glucuronidases and sulphatases, so that some of the conjugates in bile are hydrolysed and may be partially reabsorbed.

Conjugation with amino acids, or their derivatives, is an important mechanism for only a small group of foreign compounds, such as aromatic carboxylic acids. The conjugation mechanism is indirect and for glycine (or glutamine) conjugates a coenzyme A derivative of the foreign compound is formed first, which then reacts with the amino acid to form a peptide-like bond, while the coenzyme A is released.

Glutathione conjugates form mainly with aromatic hydrocarbons and are known as "mercapturic acids". The glutathione may react directly with the hydrocarbon, or with an epoxide formed by microsomal oxidation.

Despite the fact that phosphate esters are abundant in all tissues, phosphate conjugates are very rarely formed with foreign compounds and only one or two isolated examples are known.

IV. HUMAN METABOLISM OF FOREIGN COMPOUNDS

1. Pharmaceuticals

The metabolism of foreign compounds is difficult to investigate in human subjects as it requires, in almost every case, administration of radioactive-compounds. Excretion of metabolites in bile can only be studied in subjects following the insertion of a T-tube. For these reasons, less is known of human metabolism of pharmaceuticals than of that in laboratory animals. What comparative data are available often reveal marked species differences.

It is clearly impossible to deal with the human metabolism of all types of pharmaceuticals in the chapter, so emphasis will be laid on knowledge of the following groups; salicylates, phenothiazine tranquillizers, oral contraceptive progestogens, barbiturates.

(i) Salicylates

These are among the most commonly used pharmaceuticals and some have been in use for well over 100 years. Salicylic acid is *o*-hydroxy benzoic acid and is itself used, sometimes as the sodium or potassium salt, as a bacteriostatic and fungicide. Its most widely used derivative is aspirin (acetyl salicylic acid), but methyl, phenyl, glycol, or salicyl esters are also used medicinally.

All commonly used esters of salicylic acid are hydrolyzed

within the body to release the free acid. This then undergoes microsomal oxidation to form gentisic acid (1,4-dihydroxybenzoic acid) and 1,2,4-trihydroxybenzoic acid, both of which appear in the urine, together with salicylic acid, mainly as glucuronates. About 20% of an oral salicylate is metabolized and eliminated in this way. The major metabolite, however, is salicyluric acid, the glycine conjugate of salicylic acid, which accounts for about 80% of the dose (Figure 1.9).

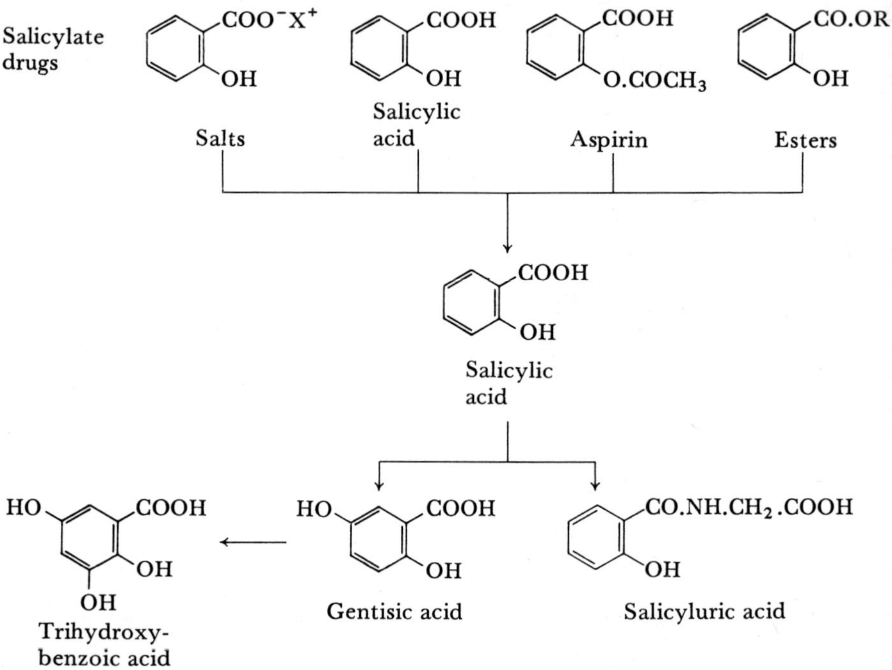

Figure 1.9. Salicylate metabolism.

(ii) Phenothiazine Tranquillizers

Phenothiazine (thiodiphenylamine) has been in use for many years as an antihelminthic and is still widely used in veterinary medicine for this purpose. It was later discovered that many of the substituted phenothiazines were tranquillizers of comparatively low toxicity and several are now in widespread human use.

There are a number of possible routes of metabolism that can be followed:

(a) Hydroxylation of the aromatic rings.
(b) Oxidation of the sulphur to sulphoxide or sulphone.
(c) Dealkylation of the N^{10} side-chain.
(d) Cleavage of the N^{10} side-chain.

Each phenothiazine tranquillizer undergoes one or more of these metabolic transformations in the human (Figure 1.10). The commonest used drugs in this series are chlorpromazine(2-methylmercapto-10-aminopropyl)-phenothiazine), thioridazine (2-methylmercapto-10-(2-(*N*-methyl-2-piperidyl)-ethyl)-phenothiazine, and imipramine (*N*-(3-dimethylaminopropyl)-iminodibenzyl).

It is difficult to show all the various metabolites as the various metabolic transformations occur at all stages. The first step in the metabolism of chloropromazine appears to be microsomal oxidation, but this can occur in three places, either at the S atom, the terminal N of the side chain, or on C-7 of the ring. Dimethylation of the terminal N of the side chain follows the *N*-oxidation and can occur in two stages. All the following compounds can therefore be formed:

1. Chlorpromazine sulphoxide
2. Chlorpromazine *N*-oxide
3. Mono-demethylchlorpromazine
4. Mono-demethylchlorpromazine sulphoxide
5. Di-demethylchlorpromazine
6. Di-demethylchlorpromazine sulphoxide
7. 7-hydroxychlorpromazine
8. 7-hydroxychlorpromazine sulphoxide
9. 7-hydroxychlorpromazine *N*-oxide
10. 7-hydroxy-mono-demethylchlorpromazine
11. 7-hydroxy-mono-demethylchlorpromazine sulphoxide
12. 7-hydroxy-di-demethylchlorpromazine
13. 7-hydroxy-di-demethylchlorpromazine sulphoxide
14. 3,7-dihydroxychlorpromazine
15. 3,7-dihydroxychlorpromazine sulphoxide,

etc.

The situation is further complicated by the fact that most of these metabolites can and do form both glucuronic acid and

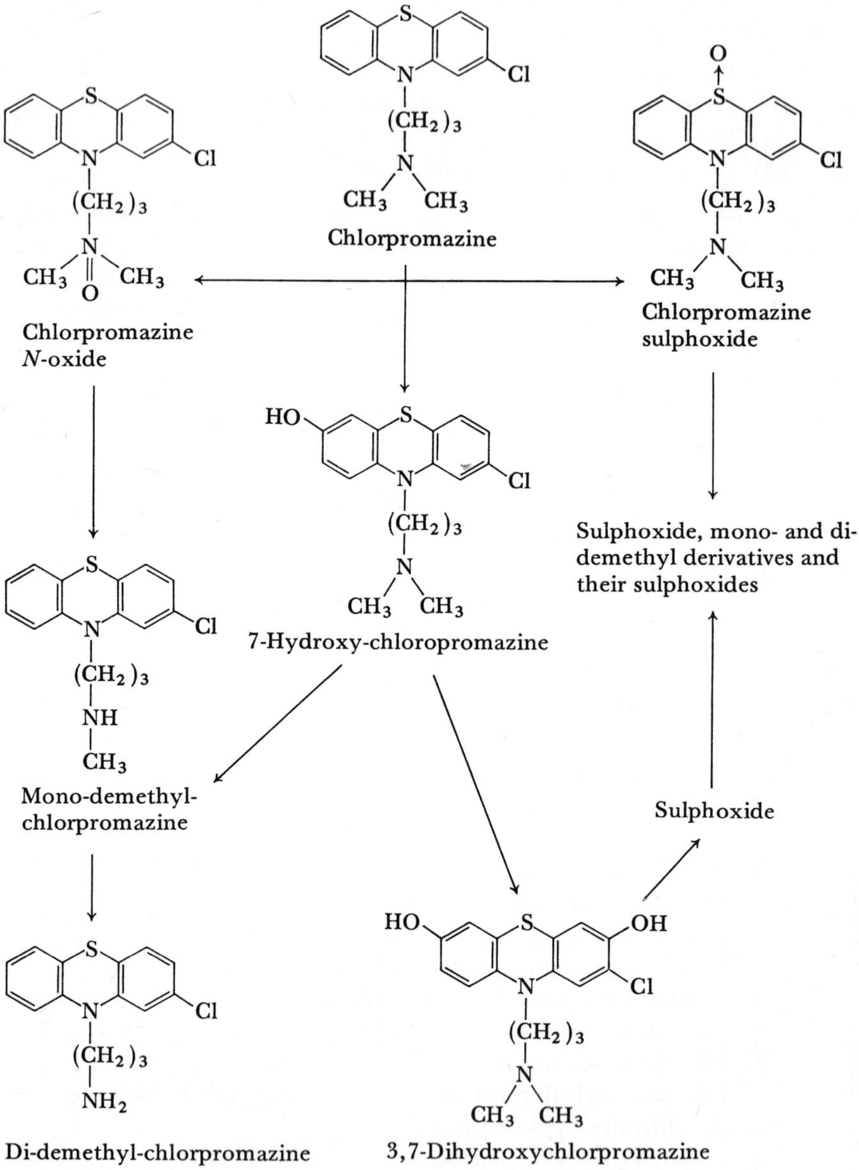

Figure 1.10. Metabolism of chlorpromazine.

sulphate conjugates, so that well over 30 individual metabolites of chlorpromazine appear in human urine. If conjugates are ignored, then the following approximate figures represent the percentages of an oral dose excreted as urinary metabolites—0.7% *N*-oxide, 5% sulphoxide, 12% unchanged, 5% mono-demethyl derivatives, 2% di-demethyl derivatives.

There is some evidence that complete elimination of the side chain of chlorpromazine is a further alternative pathway in humans.

The metabolism of thioridazine and imipramine follows lines similar to those for chlorpromazine, though the structural differences of these two compounds (Figure 1.11) somewhat

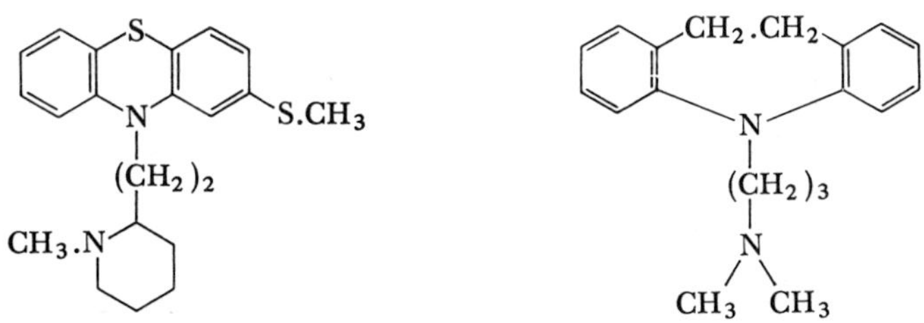

Thioridazine Imipramine

Figure 1.11. Chemical structures, tranquillizers related to chlorpromazine.

alters the pattern. It should be noted that the numbering of the atoms in imipramine differs from that in the other two compounds, for imipramine is strictly a derivative of iminodibenzyl while the others are phenothiazines.

Thioridazine has two S atoms, both of which can undergo oxidation to the corresponding sulphoxides or even the sulphone in the case of the S in the side-chain. Demethylation of the heterocyclic N of the side-chain occurs in a manner analogous to chlorpromazine. About a third of an oral dose appears in the urine in humans, of which most is a mixture of sulphoxides of thioridazine and demethyl-thioridazine. Ring hydroxylation may occur, but is probably a minor pathway.

While imipramine lacks an S atom, the ethylene bridge can be hydroxylated on either carbon. Ring hydroxylation to give the 2- or 10-monohydroxy derivatives also occurs. As with chlorpromazine, oxidation of the side-chain N results in mono- then di-demethylation, so that the range of metabolites in humans is very analogous to those for chlorpromazine. The di-demethylimipramine is formed in humans fairly rapidly, but is only slowly subjected to further metabolism, so that this is often the principal tissue metabolite.

(iii) Oral Contraceptive Progestogens

About a dozen different synthetic progestogens are used in various brands of oral contraceptive preparations, but are

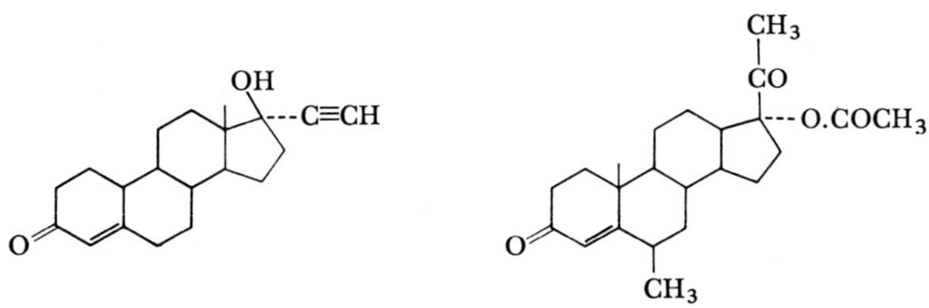

Norethisterone　　Medroxyprogesterone acetate

Figure 1.12. Structure of two synthetic progestogens.

mainly derivatives of norethisterone (17α-ethinyl-17β-hydroxy-4-estren-3-one), or substituted hydroxyprogesterones (17α-hydroxy-4-pregnene-3,20-dione), such as medroxyprogesterone (Figure 1.12).

Usually, only about one third of an oral dose of any of these progestogens appears in the urine, and it may take 3 or 4 days for this amount to be collected following a single administration. Most of the remainder appears in bile and, eventually, faeces. Many of these steroids are poorly absorbed in the intestines and only between 30 and 60% of an oral dose may enter the blood.

Aromatization of the A-ring had been reported for the estrane compounds, but not for the pregnanes. However, saturation of 4-ene and 3-one probably occurs in both series, with formation of 3α- and 3β-hydroxy compounds in both 5α- and 5β-series. Almost all steroid metabolites are excreted as glucuronate or sulphate conjugates.

(iv) Barbiturates

Many derivatives of barbituric acid are important hypnotics, sedatives and anaesthetics. They are often classified as short-, medium- and long-acting and, to a large degree, this depends upon their rate of metabolism and clearance by the body.

Chemically, three types of process can be recognized in metabolism of the barbiturates: ring scission, oxidation, addition or removal of substituents. We will consider the metabolism of two long-acting barbiturates, barbital (5,5-diethyl-barbiturate) and phenobarbital (5-ethyl-5-phenyl-barbiturate),

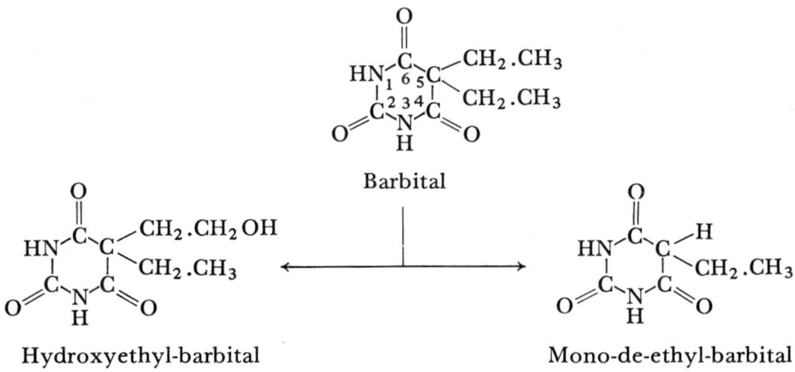

Figure 1.13. Metabolism of barbital.

and two short-acting compounds, secobarbital (5-allyl-5-(1-methylbutyl)-barbiturate) and hexobarbital (5-(1′-cyclo-hexenyl)-1,5-dimethyl-barbiturate). It will be noted that these compounds all carry two substituents on C-5 of the barbituric acid ring.

Barbital is little metabolized in the human body and up to 95% of an oral dose is excreted in the urine unchanged. The remainder is excreted in roughly equal amounts as mono-de-ethyl and hydroxy-ethyl derivatives; some of the latter being conjugated with glucuronic acid (Figure 1.13).

Phenobarbital is also long-acting, but only about 13% is excreted unchanged. However, metabolism of the compound is very slow and if radioactively-labelled phenobarbital is given, less than half the label is excreted in 24 hours. The major metabolite is the *p*-hydroxyphenyl derivative, though small amounts (about 1% of the total) appear as the *o*-hydroxyphenyl derivative; perhaps with small traces of other compounds (Figure 1.14).

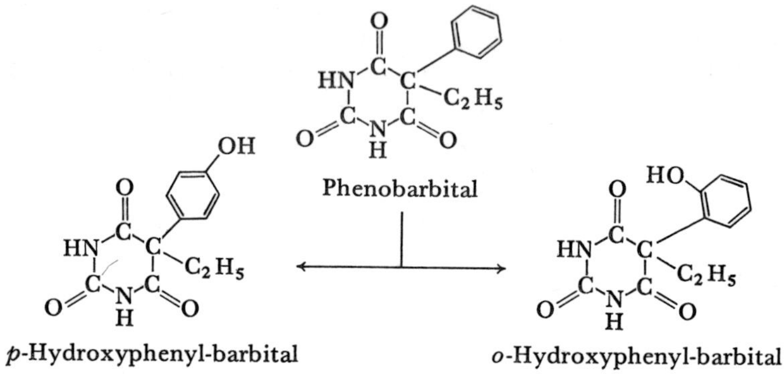

Figure 1.14. Metabolism of phenobarbital.

Short-acting barbiturates are metabolized to a far greater extent. Secobarbital is almost completely converted to other compounds by humans, so that very little unchanged substance is excreted in urine. There are two principal metabolites, formed, respectively, by hydroxylation of the two side-chains (Figure 1.15).

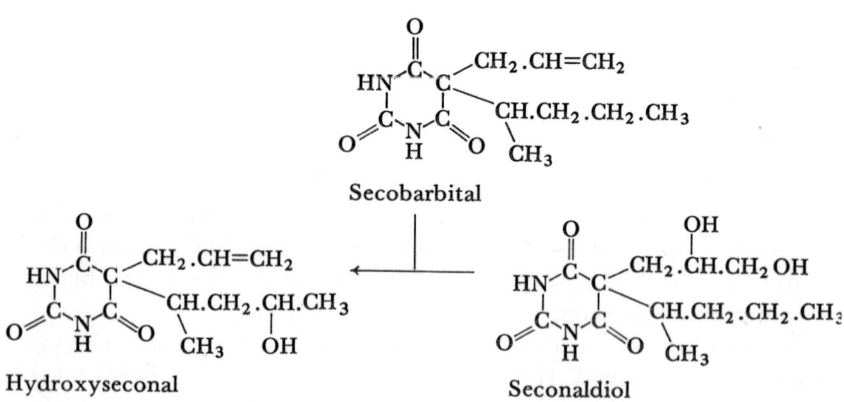

Figure 1.15. Metabolism of secobarbital.

Hexobarbital is similarly short-acting and a wide range of metabolites are excreted in urine, with little or no unchanged parent compound. Microsomal oxidation attacks the cyclic substituent on C-5, as in phenobarbital, but yields a 3-hydroxy

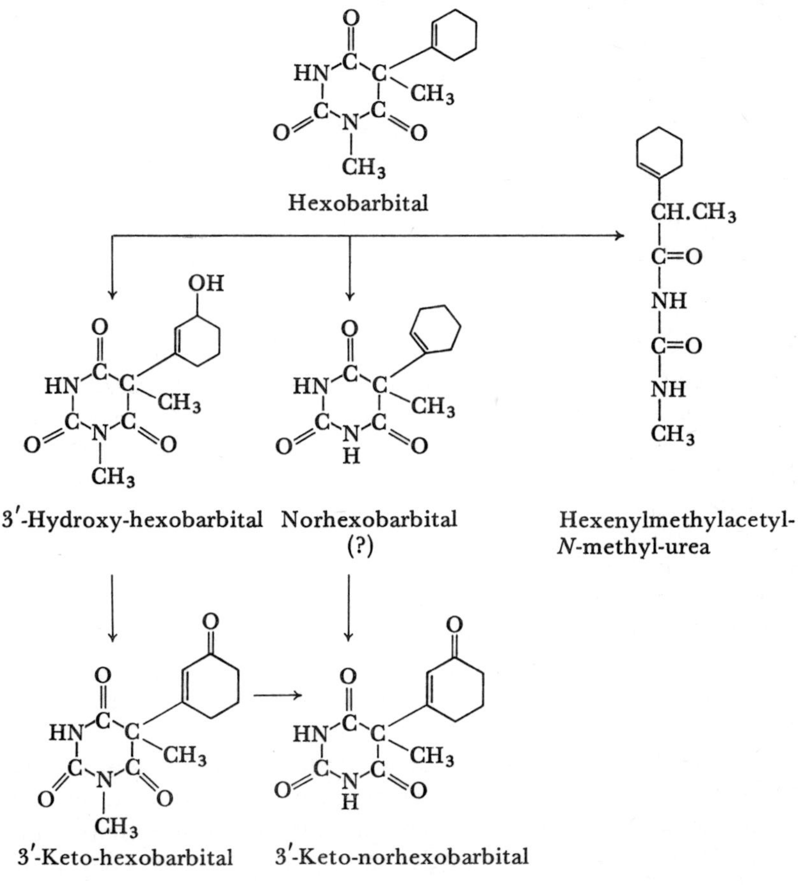

Figure 1.16. Metabolism of hexobarbital.

derivative for hexobarbital, rather than a 4-hydroxy compound as with phenobarbital. Hexobarbital is also subjected to demethylation, which can occur before or after ring hydroxylation. Finally, ring scission is a major metabolic process for this particular barbiturate (Figure 1.16).

2. Industrial Chemicals

As for pharmaceuticals, numerous chemicals used in industry come into contact with humans and it will be possible to deal with only a few widely used substances. A selection has been made from various groups.

(i) Industrial Solvents

Benzene is widely used in many industries, but is among the most difficult of foreign compounds for the human body to detoxify. It presents a uniformly hydrophobic molecule with no substituent or grouping that may be acted upon by enzymes. Fortunately, it is relatively volatile at body temperature so that about 40% of an ingested dose can be eliminated in expired air. If a large dose is given, the amount lost through the lungs can rise to as much as 70%. At high levels, a small amount of unchanged benzene appears in the urine, but its water-solubility is low. A small amount is metabolized by the body, in a variety of ways, but it is a very toxic compound, difficult to remove from tissues, and may eventually induce aplastic anaemia and liver damage.

The principal metabolite of benzene in humans is phenol, which is excreted in urine both as glucuronate and sulphate. Other di- and tri-hydroxy products may be formed in small amounts. About a quarter of an ingested dose is eliminated in this manner. A small amount of the remainder appears as *trans-trans*-muconic acid and phenylmercaputic acid (Figure 1.17).

In contrast to benzene, toluene (methylbenzene) is much less toxic and about 80% of an ingested dose is rapidly excreted in urine as hippuric acid; the remainder being eliminated unchanged via the lungs. Little or no ring hydroxylation of toluene occurs, so that phenols are not formed. Oxidation of the methyl group to yield benzoic acid presumably takes place through the intermediates benzoyl alcohol and benzaldehyde, though these are not excreted. The three isomeric dimethylbenzenes (the xylenes) are often sold together as a mixture known as "xylol". All are extensively metabolized to corresponding toluic acids (methylbenzoic acids), which are excreted in urine as conjugates with glycine and glucuronate.

Naphthalene is a widely used chemical in many industries and has been shown to be capable of inducing cateracts in eyes. This toxic action is due to its metabolite, 1,2-dihydroxy-naphthalene, which is formed via the 1,2-epoxide. Other metabolites are shown in Figure 1.18. Most are excreted as conjugates.

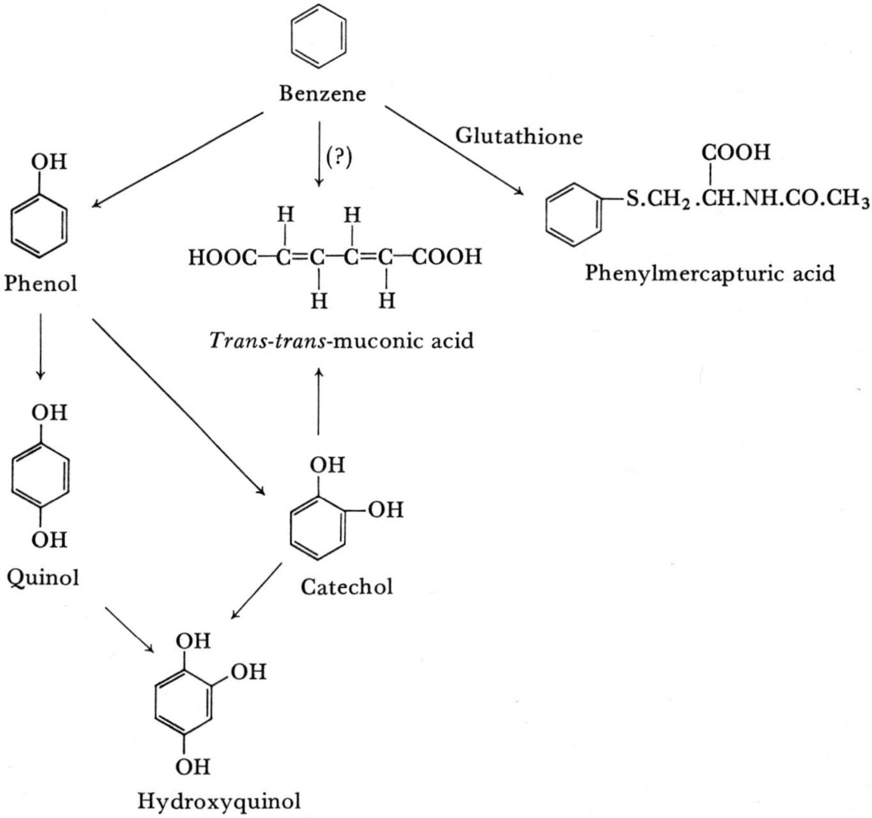

Figure 1.17. Metabolism of benzene.

In contrast, tetralin (tetrahydronaphthalene), which is widely used in polishes and paints as a turpentine substitute, is readily metabolized by hydroxylation of the saturated ring (*ac*-) to *ac*-α-tetralol and *ac*-β-tetralol, with some tetralin-1,2-diol. Hydroxylation of the aromatic ring (*ar*) occurs only to the extent of about 0.1% (Figure 1.19).

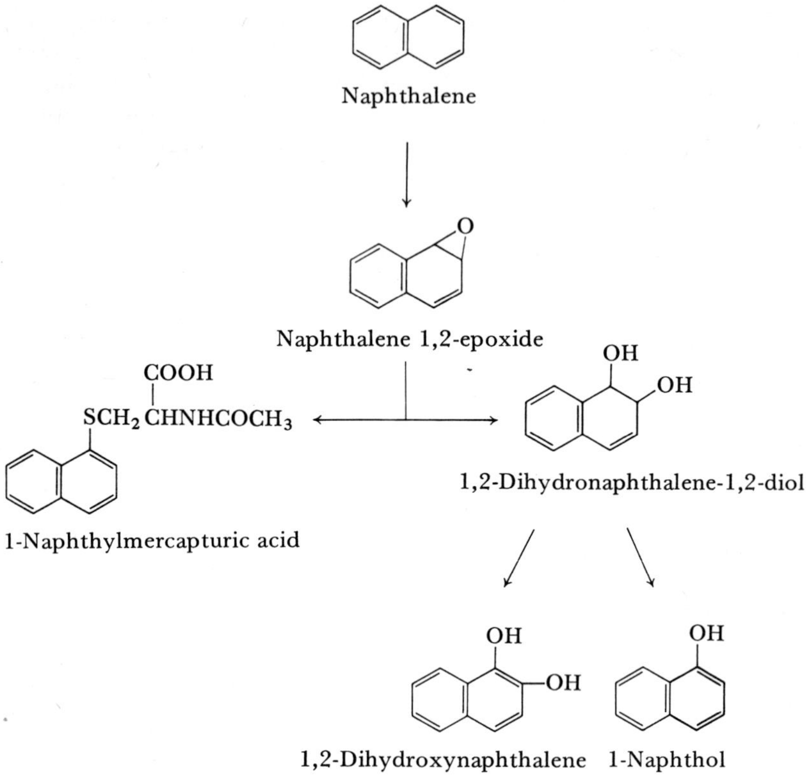

Figure 1.18. Metabolism of naphthalene.

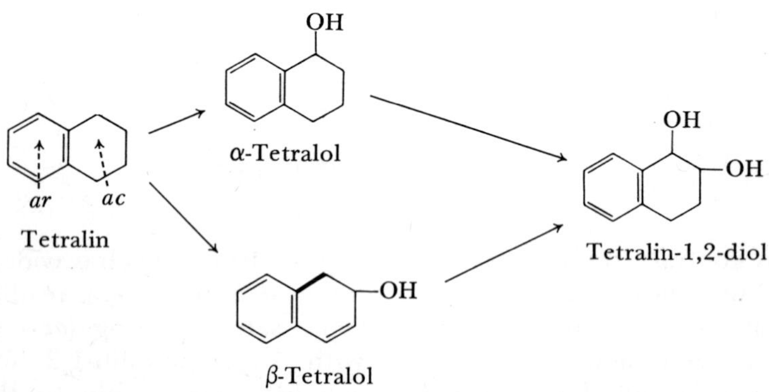

Figure 1.19. Metabolism of tetralin.

As might be expected, cyclo-hexane is metabolized to a much greater extent than benzene and is much less toxic. Due to its volatility at body temperature, about a third of an ingested dose is lost unchanged via the lungs, while about another 10% leaves the lungs as CO_2 produced by complete oxidation of the molecule. In urine, the principal metabolite is cyclohexanol (about 38%), with some *trans*-cyclohexane-1,2-diol (about 7%), both of which are present as glucuronides.

Like benzene, chlorinated aliphatic hydrocarbons present a difficult metabolic problem to body tissues. Over 80% of an ingested dose of carbon tetrachloride, and over 90% of chloroform or tetrachloroethylene, appears unchanged in expired air. All these compounds have toxic actions, particularly on liver, but carbon tetrachloride is the most dangerous. It is thought to give rise to certain reactive free-radicals *in vivo*. About 1% of CCl_4 and 4% of $CHCl_3$ are converted to CO_2, presumably through reductive dechlorination, then oxidation.

(ii) Dyestuffs

Aromatic amines used in the manufacture of dyes (and some drugs) have a particularly unfortunate reputation due to their association with bladder cancer. The simplest compound is aniline (aminobenzene), which is largely excreted in urine as conjugates of various aminophenols (Figure 1.20). The most

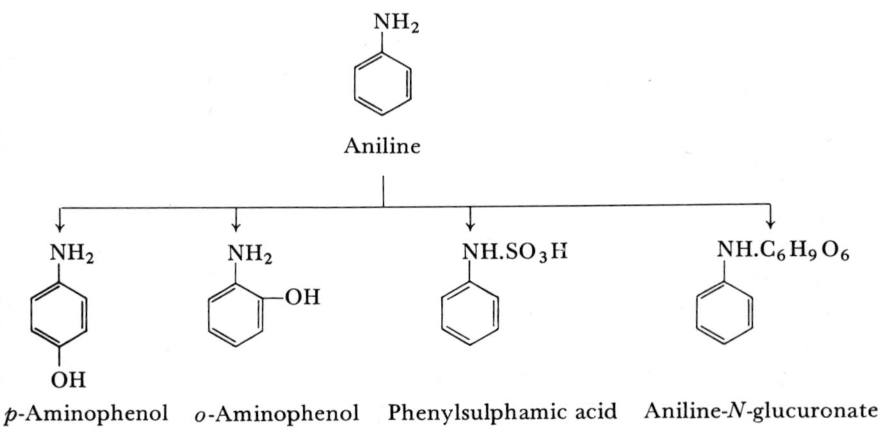

Figure 1.20. Metabolism of aniline.

abundant metabolite is *p*-aminophenol, with small amounts of *o*-aminophenol. The meta-isomer is formed in only trace quantities. About 4% of an ingested dose is excreted as free aniline, with another 6% as aniline-*N*-glucuronide. The sulphate conjugate (phenylsulphamic acid) accounts for a further 8%. Traces of several other metabolites are formed. α-Naphthylamine is metabolized similarly to aniline, with the principal metabolites in urine being conjugates of 1-amino-2-naphthol and 1-amino-4-naphthol. It is also excreted as *N*-glucuronide and *N*-sulphate (α-naphthylsulphamic acid) (Figure 1.21).

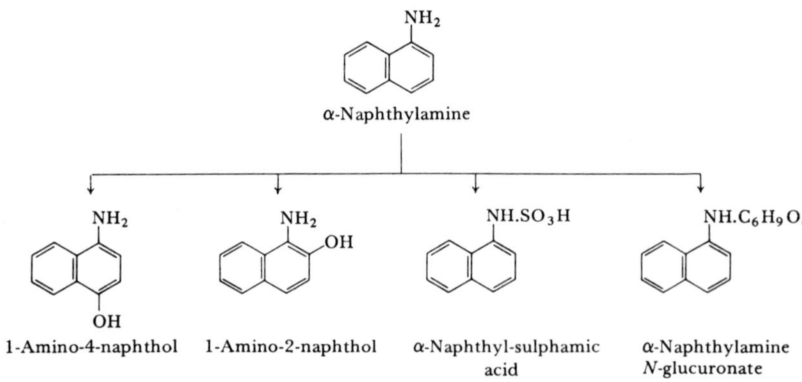

Figure 1.21. Metabolism of α-naphthylamine.

Neither aniline nor α-naphthylamine have been proved carcinogenic, in contrast to β-naphthylamine where the evidence is conclusive. Interestingly, metabolites of β-naphthylamine are unlike those of aniline or α-naphythylamine. Like these two compounds it gives rise directly to *N*-glucuronide and *N*-sulphate conjugates (β-naphthylsulphamic acid), but ring hydroxylation produces a mixture of 2-amino-6-naphthol and 2-amino-1-naphthol, both of which also appear in urine as conjugates. There is significant *N*-hydroxylation, which occurs to only a trace extent with the other two amines. 2-Amino-1-naphthylmercapturic acid is also formed in small amounts (Figure 1.22).

The bladder carcinogen appears to be either (or both) 2-amino-1-naphthol or 2-naphthylhydroxylamine. Although the

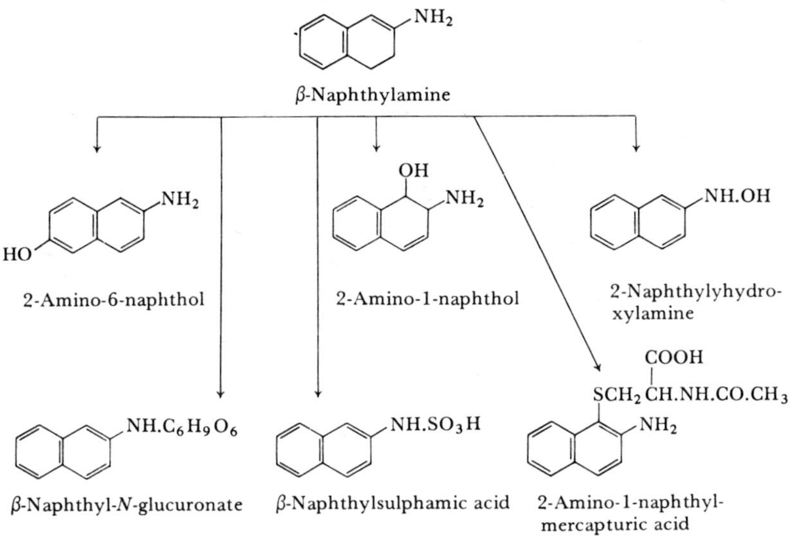

Figure 1.22. Metabolism of β-naphthylamine.

former is excreted in urine almost entirely as conjugates, these are hydrolysed to a considerable extent within the bladder to release the free compound.

Of the many dyes used in cosmetics and food colourants, we will discuss only a few widely used compounds. For azo dyes there is a liver reductase which catalyses conversion to a mixture of amines, but this enzyme is of low activity and probably most dye metabolism is conducted by gastro-intestinal micro-organisms and products are absorbed and excreted.

Chrysoidine, also known as "Orange II", is extensively metabolized and only about 2% of an oral dose is excreted unchanged. Reduction of the azo group yields one molecule of 1-amino-2-naphthol and one of sulphanilic acid. The latter compound is partly acetylated to *p*-acetamidobenzene sulphonic acid and partly excreted unchanged. The former is conjugated with glucuronate and sulphate in approximately equal proportions (Figure 1.23).

About 1% of Sudan I is excreted unchanged in urine, but the principal metabolite is *p*-aminophenol, which is formed partly from aniline released by reduction of the amine group, and

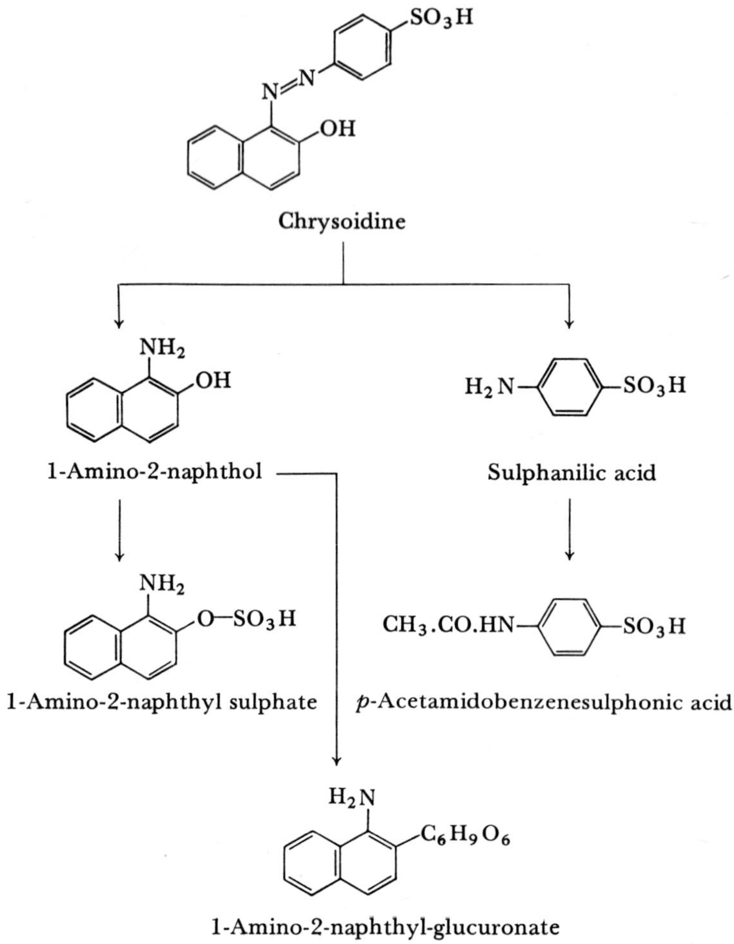

Figure 1.23. Metabolism of chrysoidine.

partly by *p*-hydroxylation of the benzene ring prior to reduction. Other metabolities are conjugates of 1-amino-2-naphthol, *o*-aminophenol, with traces of several others (Figure 1.24).

Tartrazine has no true azo-linkage, but nevertheless appears to be reduced by intestinal micro-organisms and in humans a principal metabolite is sulphanilic acid, together with its various conjugates. The pyrazoline derivative that must also be released is excreted almost entirely through bile and is present in faeces partly unchanged, but largely as an oxidized diamer (Figure 1.25).

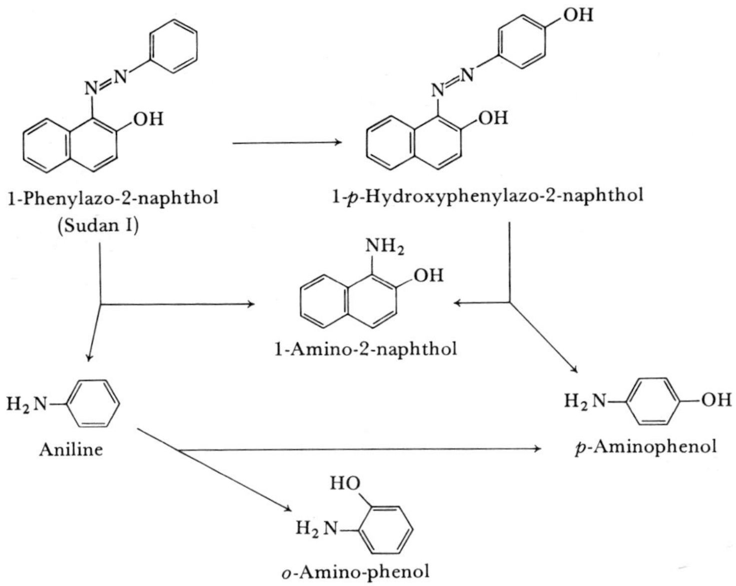

Figure 1.24. Metabolism of dyes (Sudan I).

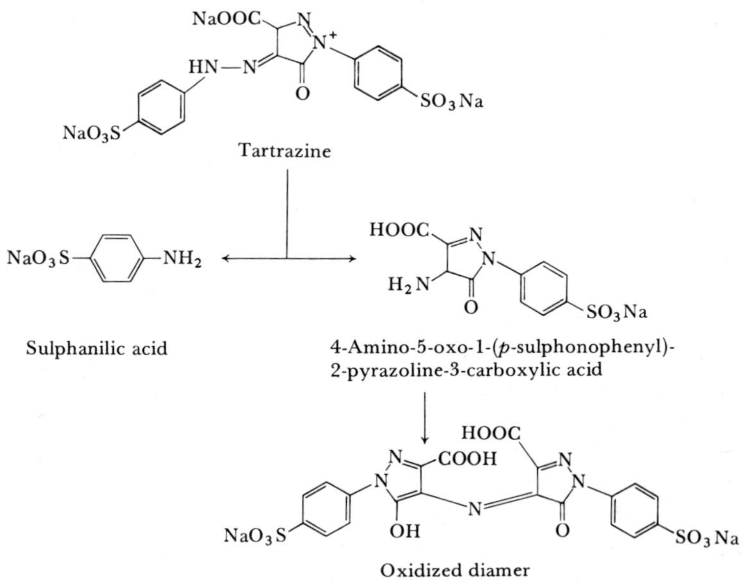

Figure 1.25. Metabolism of dyes (tartrazine).

3. Other Food Additives

(i) Sweetening Agents

Two compounds have been very widely used, cyclamates and saccharine. Almost all an oral dose of cyclamate is excreted unchanged in urine by humans, but there is some evidence that particular individuals metabolize a small amount (less than 1%) to cyclohexylamine. Saccharin is similarly rapidly excreted by humans, unchanged in urine. No metabolites have so far been detected.

(ii) Preservatives

Benzoic acid is excreted in man almost entirely as its glycine conjugate, which is called "hippuric acid". Traces of conjugates of various non-hydroxylated benzoic acids can be detected. *p*-Hydroxybenzoic acid, or its various esters, have found wide application as preservatives. The esters are rapidly hydrolysed *in vivo* and about a third of an oral dose appears in urine as the free acid, while up to another third occurs as *p*-hydroxyhippuric acid. *p*-Hydroxybenzoic acid readily forms glucuronates with either hydroxyl or carboxylic acid groups. Sulphate conjugates are also formed, though to a lesser extent.

(iii) Anti-Oxidants

Aside from naturally occurring tocopherols, the commonest food anti-oxidants are butylated hydroxyanisole (BHA) and butylated hydroxytoluene (BHT). BHA is a mixture of 3-tert-butyl-4-hydroxyanisole and 2-tert-butyl-4-hydroxyanisole; with the former in greatest abundance. The major route of excretion is via urine and in humans, the only metabolites of either compound appear to be glucuronates and sulphates. In contrast, BHT gives rise to a large series of metabolites, for the hydroxyl is sterically hindered by the tert-butyl groups and cannot form conjugates. The molecule is oxidized at the methyl or one of the tert-butyl groups to form 3,5-di-tert-butyl-4-hydroxybenzoic acid or 3-tert-butyl-5-(β,β- dimethylhydroxyethyl)-4-hydroxytoluene. These metabolites give rise to series of conjugates with glucuronate, sulphate and glycine (Figure 1.26).

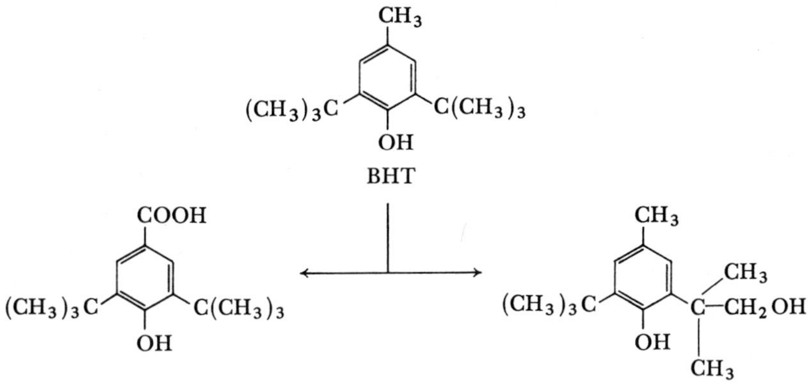

3,5-Di-tert-butyl-4-hydroxy-benzoic acid

3-Tert-butyl-5-(β,β-dimethyl-hydroxy-ethyl)-4-hydroxytoluene

Figure 1.26. Metabolism of butylated hydroxy-toluene (BHT).

4. Pesticides

(i) Herbicides

Large amounts of phenoxyacetic acids are used, but all are strong, water-soluble acids that are rapidly eliminated in urine, apparently without metabolism.

(ii) Chlorinated Hydrocarbon Insecticides

The most widely used insecticide is DDT, which is fat-soluble and only poorly metabolized by humans. In consequence, DDT ingested orally is absorbed with dietary lipids and deposited, largely unchanged, in adipose tissues. Two metabolic pathways operate in humans and other mammals. The first is a reductive dechlorination to the corresponding dichloro-ethane (DDD), which then undergoes hydrolytic dechlorination and oxidation to a carboxylic acid (DDA). The second pathway is an oxidative dechlorination to yield dichloroethylene (DDE) (Figure 1.27).

The principal urinary metabolite in humans is DDA, which, surprisingly, appears as conjugates with the amino acids serine and aspartic acid. The major metabolite in adipose tissue is DDE (together with unchanged DDT).

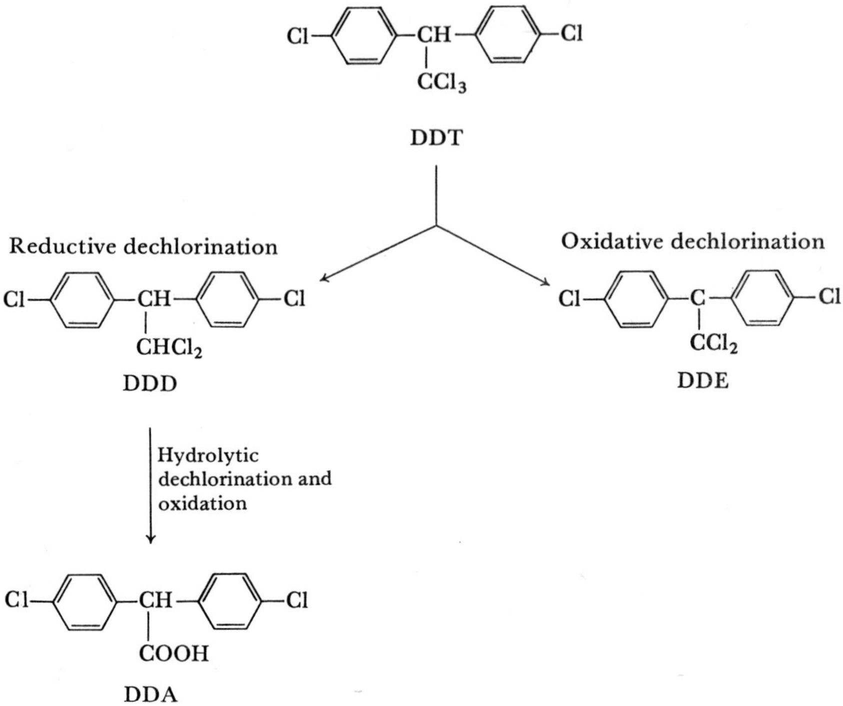

Figure 1.27. Metabolism of DDT.

Aldrin and dieldrin differ only by an epoxide group, which can be added to aldrin by metabolism in mammals. The reaction, however, is difficult and aldrin is largely unmetabolized. Dieldrin is metabolized by opening the epoxide to the corresponding dihydroxy derivative (Figure 1.28).

Several other trace metabolites have been detected.

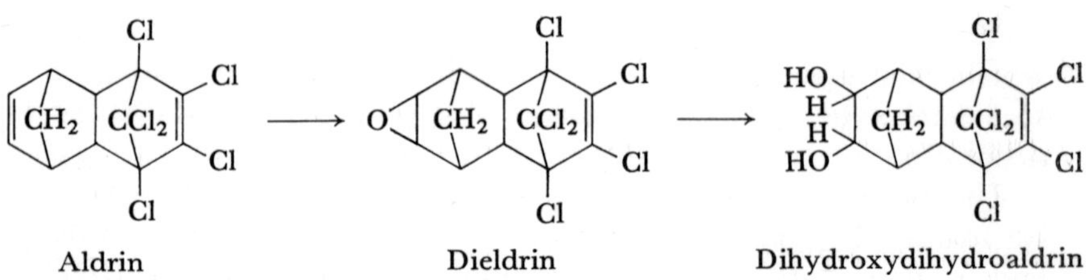

Figure 1.28. Metabolism of aldrin and dieldrin.

(iii) Organophosphorus Insecticides

Parathion is readily metabolized *in vivo* to its oxygen analog, "paraoxon", which is a potent anti-cholinesterase (an example of lethal synthesis). Further metabolism yields diethyl phosphate and *p*-nitrophenol, which is reduced to *p*-aminophenol and excreted in urine as conjugates (Figure 1.29).

The structurally related malthion is also activated *in vivo* to a potent anti-cholinesterase, "malaoxon". Both parent compound

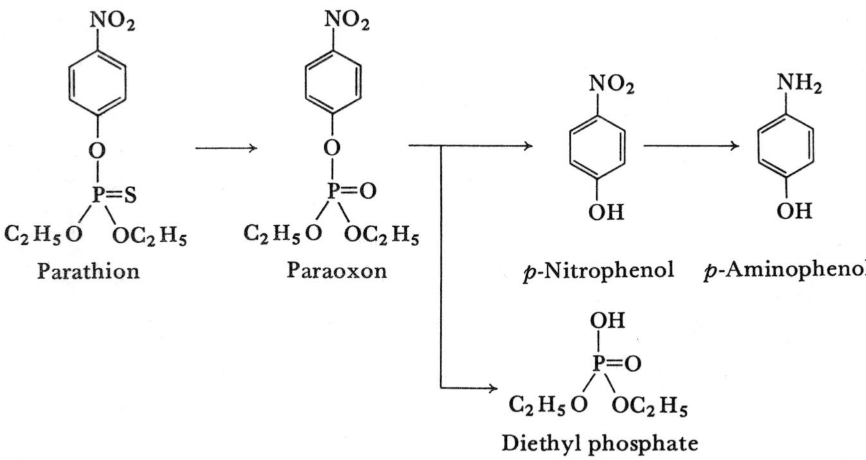

Figure 1.29. Metabolism of parathion.

and malaoxon undergo a series of successive hydrolyses to a number of metabolites (Figure 1.30).

Similar activation by conversion to oxygen analogs occurs for other phosphorothionates.

(iv) Carbamate Insecticides

Like organophosphorus compounds, these insecticides act as anti-cholinesterases, but do not require metabolic activation. All metabolites are much less active than the parent compounds. Most compounds in this series are aromatic carbamates and are subject to microsomal oxidations. Sevin yields 4-hydroxy-1-naphthylmethylcarbamate, 5-hydroxy-1-naphthylmethylcarbamate, 5,6-dihydroxy-5,6-dihydro-1-naphthylmethyl-

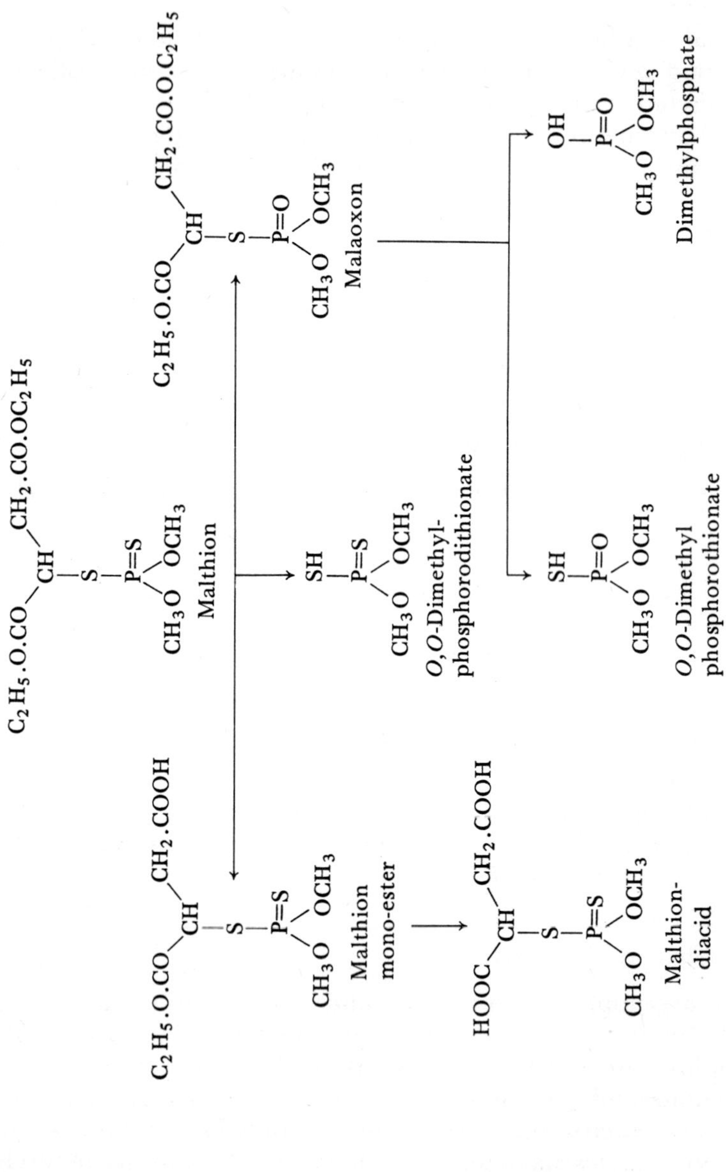

Figure 1.30. Metabolism of malthion.

carbamate, 1-naphthylhydroxymethylcarbamate and several others. Zectran has its carbamic acid residue removed by hydrolysis, then yields 2,6-dimethylhydroquinone.

5. Anutrients in Foods

Many anutrients of plant foods occur as glycosides and it is usual for extensive hydrolysis by intestinal bacteria to occur on ingestion to release aglycones. The extent of further metabolism depends on the particular substance.

Oily compounds, such as terpenes, may be metabolized by enzymes of important metabolic pathways, such as fatty acid β-oxidation. Methyl purines, such as caffeine, are subjected to demethylation, then excreted in urine as mono- or dimethyl uric acids.

Organotin compounds are metabolized in mammals by dealkylation, though much is excreted unchanged in bile. In the lower gut unchanged compounds are extensively dealkylated by bacteria, so that little unchanged compound actually leaves the body by any route.

Metabolism of food carcinogens is discussed in other chapters (page 307), as is metabolism of food toxins (page 272).

V. METABOLIC INTERACTIONS

1. General

We have already shown that the presence of a foreign compound can alter normal metabolism of the body, while the presence of two or more foreign compounds, given separately or together, can influence the metabolic products formed. Thus phenobarbital increases the rate of metabolism of endogenous steroid hormones, while naphthalene pre-treatment prevents formation of metabolites of 1,2-di-methylbenz(a)anthracene which induce hepatic necrosis.

Drugs are often given in combination to patients, or may be given to persons exposed to foreign compounds of other types. The remainder of this chapter is devoted to a brief discussion of interactions between foreign compounds that may be of clinical significance.

2. Anti-Coagulants

A widely used anti-coagulant is warfarin (4-hydroxy-3-(3-oxo-1 phenylbutyl)-2H-1-benzopyran-2-one) (Figure 1.31) which, given at the therapeutic dose of 30 to 50 mg daily (though this is normally reduced to a maintenance dose of less than 10 mg,) attains a plasma concentration of 2 to 5 mg/l. Of this, about 98% is bound to serum albumin, while only 2% is the free, metabolically active drug. It is immediately apparent that any other drug or foreign compound capable of displacing warfarin from serum albumin will increase the concentration of active

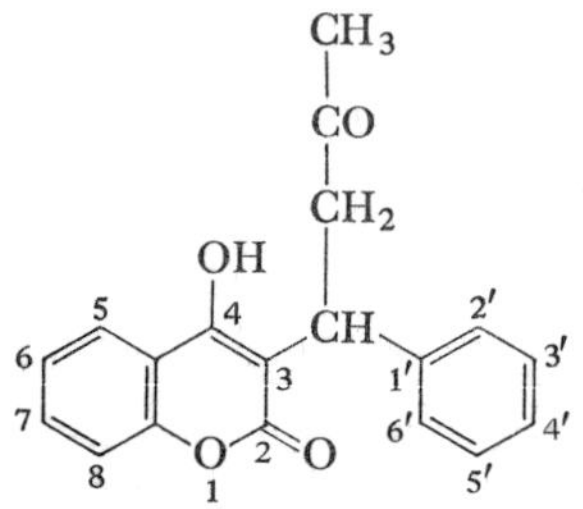

Warfarin

Figure 1.31. Chemical structure of warfarin.

drug and lead to excessive anti-coagulant effects. If only a further 2% were to be displaced, this would double the concentration of active warfarin in serum.

Drugs which displace significant amounts of warfarin from its binding to albumin include phenylbutazone, oxyphenylbutazone, mefanamic acid, chloral hydrate. Theoretically, any acidic substance could displace bound warfarin and *in vitro* tests show that simple fatty acids, such as stearic or lauric acids, and cyclamates have this property.

A second factor can also be shown to be of influence. Warfarin is metabolized in humans to a variety of monohydroxy derivatives, with substituents at 6,7,8 or 4′. These metabolites possess the ability to displace warfarin from serum albumin, though they themselves possess little or no anti-coagulant properties.. The apparently paradoxical situation arises that an increase in the rate of warfarin metabolism, while decreasing the total amount of circulating drug, could actually increase the amount of *active* free substance by metabolites displacing more unchanged warfarin from albumin. In this circumstance, the *duration* of action would be reduced, but the *degree* of action increased.

A number of clinical studies have confirmed the effects of coincident administration of acidic drugs with warfarin. The best documented effect is for chloral hydrate, for which the major human metabolite is strongly acidic trichloroacetic acid. Approximately one quarter of patients who receive warfarin and chloral hydrate together, show potentiation of anti-coagulant action.

3. Ethanol

Ethanol is normally metabolized to acetaldehyde, and then to acetic acid. Acetaldehyde is a particularly toxic compound, so than anything interfering with clearance of .acetaldehyde from the body would be expected to produce toxic actions in an individual ingesting ethanol. A number of foreign compounds are known which inhibit aldehyde dehydrogenase. The most widely used is "Antabuse" (disulphiram or diethylthiuram disulphide), which has found application in the treatment of chronic alcoholism. Other compounds which block the enzyme are cyanamides, arsenite and mercaptans. Antabuse inhibits not only aldehyde dehydrogenase, but also xanthine oxidase and, perhaps, lactate dehydrogenase.

Conversely, ethanol interacts with a number of other pharmaceuticals to prolong their action. The best known is the synergistic action with barbiturates and antihistamines (for their sedative action only). Interestingly, Antabuse has a similar action, perhaps by increasing survival time of ethanol in the body.

VI. SELECTED REFERENCES

1. Deichmann W. B. and Gerarde H. W. (1969) *Toxicology of Drugs and Chemicals.* New York: Academic Press.
2. Parke D. V. (1968) *The Biochemistry of Foreign Compounds.* Oxford: Pergamon Press.
3. Vessell E. S. (editor) (1971) *Drug Metabolism in Man.* Annals of the New York Academy of Sciences, Vol. 179.

4. Williams R. T. (1959) *Detoxication Mechanisms* London: John Wiley Ltd.
5. La Du B. N., Mandel H. G. and Way E. L. (editors) (1971) *Fundamentals of Drug Metabolism and Drug Disposition* Baltimore: Williams & Wilkins.

Chapter 2

BIOCHEMISTRY OF DRUGS

I. TYPES OF PHARMACEUTICAL*

1. Introduction

The study of drugs, in all their aspects, is known as "Pharmacology" (Greek, *Pharmakon,* meaning a "a medicine"), though the term "Pharmacognosy" is sometimes used interchangeably. In recent years the latter word has come to be used in a more restricted sense for the study of the occurrence and preparation of drugs, particularly those derived from natural sources.

Biochemical pharmacology is the study of the absorption and excretion of drugs, their metabolism by the body, and particularly their mode of action at molecular levels. We have already discussed the metabolism of many different types of pharmaceuticals in the previous chapter on foreign compounds. Here we will discuss primarily "pharmacodynamics" (the absorption, transport and excretion of drugs), together with the interaction of drugs with tissue metabolism. Drug metabolism will not be further discussed, except where it is specially relevant to these topics. First we will discuss in detail the various chemical classes of pharmaceuticals. An account of the action and effects of pharmaceuticals of natural origin will be reserved for the next chapter, where all classes of natural anti-metabolites are dealt with.

2. Body Constituents

A major class of drugs are compounds which occur naturally in the human body and are given to correct deficiencies, either in the amounts of these constituents, or to increase tissue content for other reasons. Pharmaceuticals in this class include vitamins, hormones, bile salts, anti-sera, mineral salts, urea, oxygen and other gases, etc. Only rarely are these substances

* Most compounds mentioned in the early tables are discussed in detail later in the chapter.

actually prepared for human tissues; they can usually be extracted from animal or plant sources or in some cases synthesized chemically.

Table 2.1 gives information on the major members and uses of substances found naturally in the human body.

Table 2.1. Naturally occurring substances used as drugs

Type of substance	Important examples	Therapeutic uses
A. Vitamins	1. Retinol (vitamin A)	Night-blindness, xerophthalmia, and other signs of deficiency. Acne vulgaris, and other skin conditions. Hyperthyroidism.
	2. Calciferol (vitamin D)	Rickets. Hypoparathyroidism. Lupus vulgaris.
	3. Tocopherols (vitamin E)	Intermittent claudication. Megaloblastic anaemia.
	4. Vitamin K	Hypoprothrombinaemia.
	5. Ascorbic acid (vitamin C)	Scurvy and other deficiency signs.
	6. Thiamine	Beri-beri, and other forms of neuritis.
	7. Riboflavin	Deficiency signs, dermatitis, glossitis, etc.
	8. Pyridoxine	Deficiency signs, dermatitis, anaemias, alcoholism, Raynaud's disease, etc.
	9. Pantothenic acid	Psoriasis, arthritis, some other conditions.
	10. Nicotinic acid (Niacin)	Pellagra, hypercholesterolaemia, ischaemic heart disease, schizophrenia.
	11. Cobalamins (vitamin B12)	Pernicious anaemia and other macrocytic anaemias.
	12. Folic acid	Various anaemias.
B. Hormones	1. Corticotrophin (ACTH)	Pituitary deficiency syndromes, and as for corticosteroids.

Table 2.1—continued

Type of substance	Important examples	Therapeutic uses
	2. Gonadotrophins (LH, FSH, HCG, HMG)	Pituitary deficiency syndromes, induction of ovulation, etc.
	3. Thyrotrophin (TSH)	Pituitary deficiency syndromes, etc.
	4. Calcitonin	Paget's disease.
	5. Growth hormone (somatotrophin)	Pituitary dwarfism.
	6. Thyroxine	Hypothyroidism.
	7. Corticosteroids (cortisone, hydrocortisone)	Adrenal insufficiency, as anti-inflammatory agents, and in various blood diseases.
	8. Testosterone	Male hypogonadism, breast cancer, etc.
	9. Estrogens (estradiol, estrone)	Various gynaecological conditions, malignancies, etc.
	10. Progesterone	Various gynaecological or obstetric conditions, etc.
	11. Adrenaline	Sympathetic nervous stimulant, asthma, shock, etc.
	12. Angiotensin	Hypertensive (where hypotension is due to shock).
	13. Vasopressin	Anti-diuretic.
	14. Oxytocin	Induction of labour, etc.
C. Mineral salts	1. Sodium chloride (also bicarbonate)	Saline infusions, heat cramps, salt-baths, etc.
	2. Potassium chloride	Any signs of potassium deficiency, muscular weakness, cardiac arrhythmias, etc.
	3. Sodium fluoride	Dental caries.
	4. Magnesium salts	Purgation, hypomagnesaemia.
	5. Calcium salts	Some bone disorders, hypocalcaemia, etc.

Table 2.1—continued

Type of substance	Important examples	Therapeutic uses
	6. Phosphates	Purgation, disorders of mineral metabolism, etc.
D. Anti-sera	1. Gas-gangrene anti-toxin	Gas-gangrene.
	2. Leptospira anti-serum	Weil's disease.
	3. Rabies anti-serum	Rabies.
	4. Tetanus anti-toxin	Tetanus.
	5. Venom anti-sera	Snake or scorpion bites.
	6. γ-globulin	Virus infections.
E. Gases	1. Oxygen	Anoxia.
	2. Carbon dioxide	Respiratory stimulant.
	3. Nitrogen	Warts (as liquid nitrogen).
F. Intermediary metabolites	1. Lactic acid (and lactates)	Gastro-enteritis, acidosis, leucorrhoea, diabetic coma, etc.
	2. Acetic acid (and acetates)	Expectorant, linaments, etc.
	3. Citric acid (and citrates)	Diuretic, to alter reaction of urine, etc.
	4. Malic acid	Dental cleaning.
	5. Glucose	Intra-venous feeding, hypoglycaemia, etc.
	6. Urea	Diuretic, hypotensive, sickle-cell crisis, etc.
	7. Cholesterol	Anti-irritant in ointments, etc.
	8. Bile salts	Various gastro-intestinal disorders.

3. Modifications and Analogues

Once a therapeutic use has been found for a particular substance, and the chemical composition is known, it becomes of interest to know whether the therapeutic value can be enhanced by modifying the structure. This is usually done initially by relatively simple procedures, such as the formation of different salts or esters, then by the introduction or deletion of various organic groupings, such as methyl or ethyl. Finally a whole series of analogues will be prepared and tested. Occasionally it is found that one of these analogues has properties quite unlike those of the parent substance, so that a new drug has been discovered, rather than merely a modification of an existing preparation.

Listed in Table 2.2 are some examples of modifications and analogues to drugs based on compounds present in the human body.

4. Natural Plant Products

The early history of pharmacology is concerned largely with medicinally active compounds derived from plants. In general, metabolism of plant tissues is very similar to that of animals but with some major differences. First, plant cells are phosynthetic and can make use of the energy of light, via the green chlorophyll pigments, to split water into oxygen and hydrogen for the reduction of nicotinamide coenzymes. The latter are then used in a complex metabolic pathway (known as the Calvin cycle) to convert carbon dioxide to sugars. Part of this pathway resembles the pentose pathway of animal cells. Hence the major difference in metabolism between animal and plant tissues is the much greater synthetic abilities of the latter. While *catabolism* is similar, *anabolism* is much better developed in plants. No preformed organic molecules are required. Inorganic nitrate is reduced to ammonia, which is incorporated into organic nitrogenous compounds, so that all the amino acids and vitamins are made by plants.

Perhaps the most curious feature of plant metabolism, however, is the way in which these powerful biosynthetic abilities appear to have run wild, for plants make numerous compounds for which no function has yet been found within their biochemistry and physiology. Many of the medicinally active constitutents of plants fall into this category of substances with unknown botanical function.

In the next chapter we will consider in detail the way in which various compounds of plants interfere with human metabolism. For the present, we will list in Table 2.3 some

Table 2.2. Chemical modifications of some naturally occurring substances

Type of substance	Naturally occurring compound	Modification or analogue	Uses
A. Vitamins	1. Retinol (vitamin A) 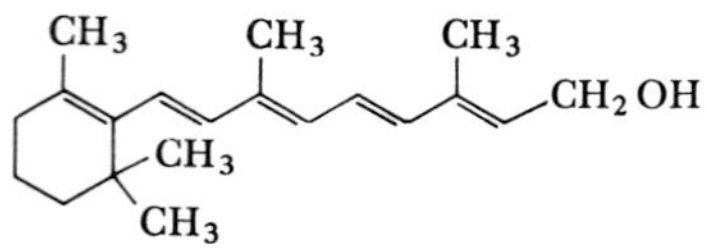	Retinoic acid (vitamin A acid) 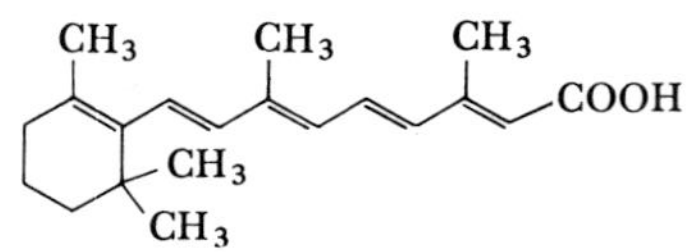	Skin conditions unresponsive to retinol.
	2. Vitamin K (phytomenadione) 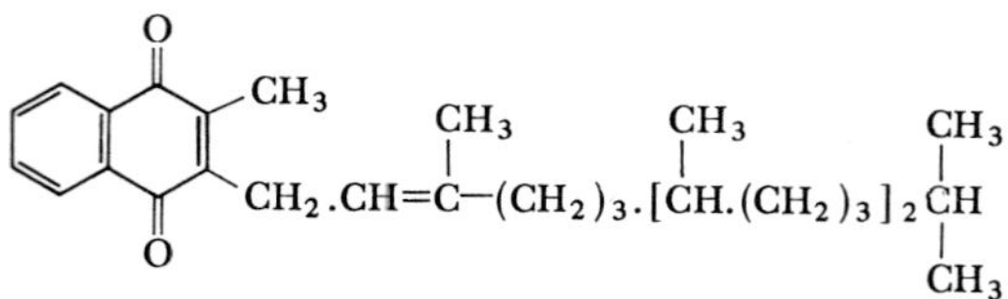	Menadiol sodium diphosphate 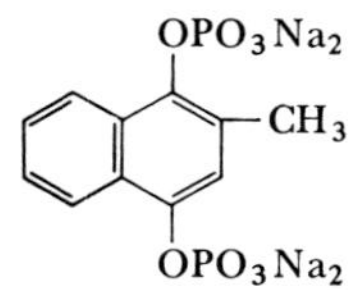	Analogue is water-soluble, whereas natural vitamin K is not.
	3. Folic acid 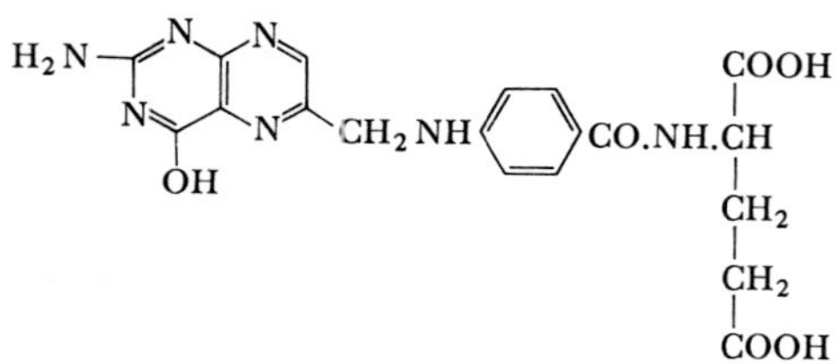	Aminopterin 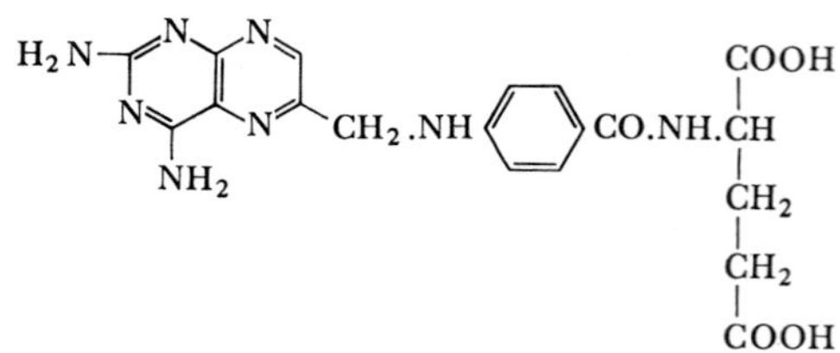	Cytostatic

		Methotrexate (amethopterin) 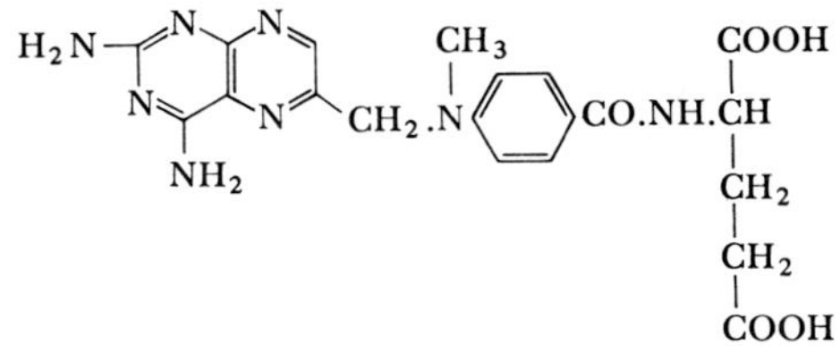	Cytostatic
B. Hormones	1. Angiotensin II Asp.Arg.Val.Tyr.Ile.His.Pro.Phe	Angiotensin amide NH$_2$.Asp.Arg.Val.Tyr.Val.His.Pro.Phe	More potent hypertensive action.
	2. Adrenaline 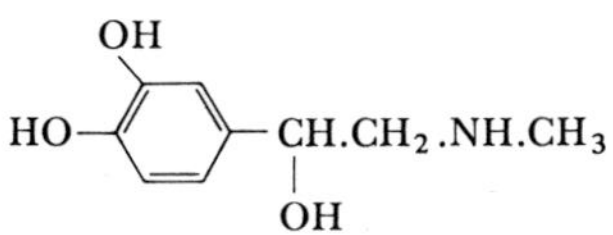	Clorprenaline 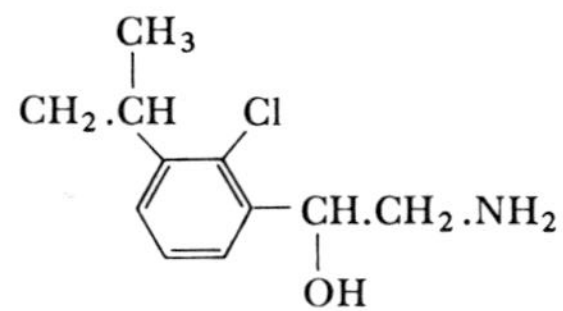	Orally active
	3. Estradiol 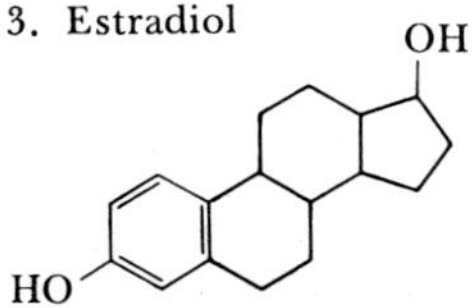	17α-Ethinyl-estradiol 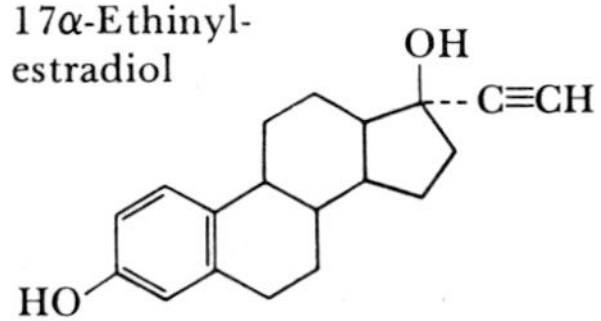	Orally active and more potent estrogen.

Table 2.2—continued

Type of substance	Naturally occurring compound	Modification or analogue	Uses
	4. Testosterone 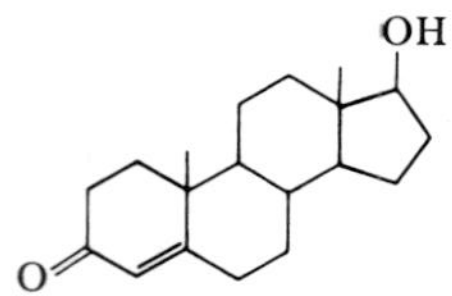	Fluoxymesterone 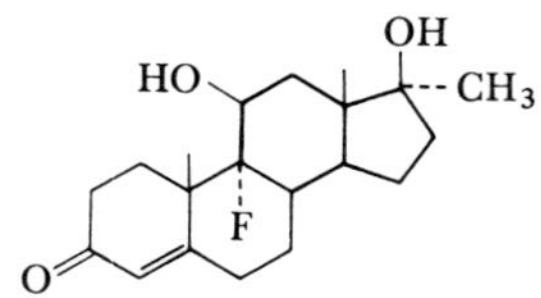	Orally active and more potent androgen.
	5. Progesterone 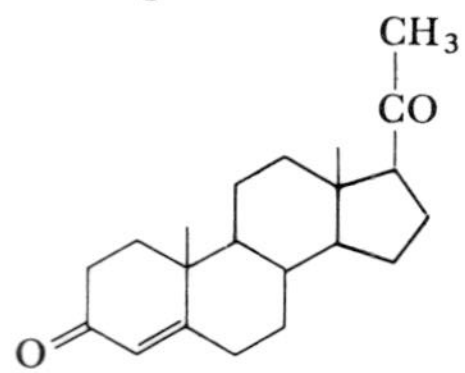	Medroxyprogesterone acetate 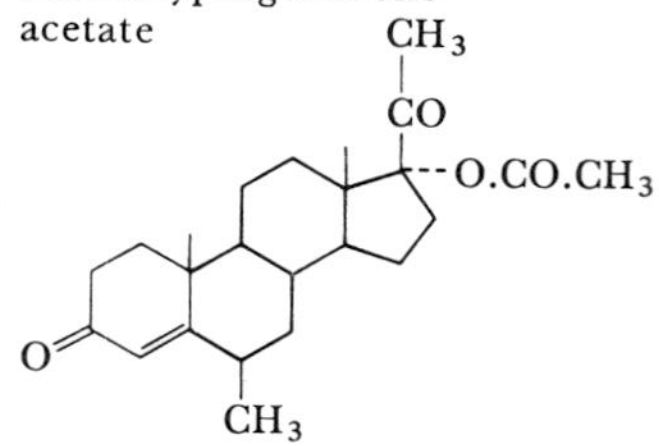	Orally active and more potent progestogen.
	6. Cortisol (hydrocortisone) 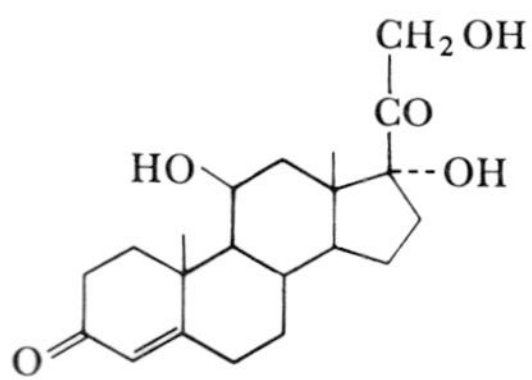	Triamcinolone 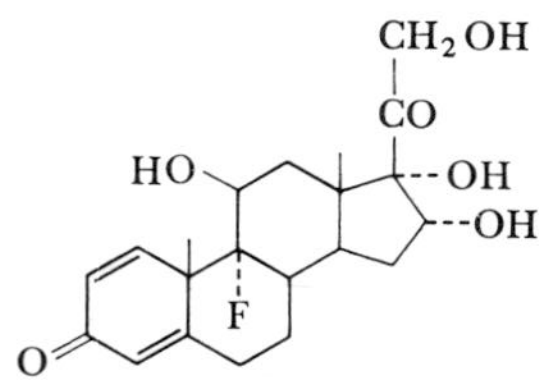	More potent anti-inflammatory properties.

Prednisolone

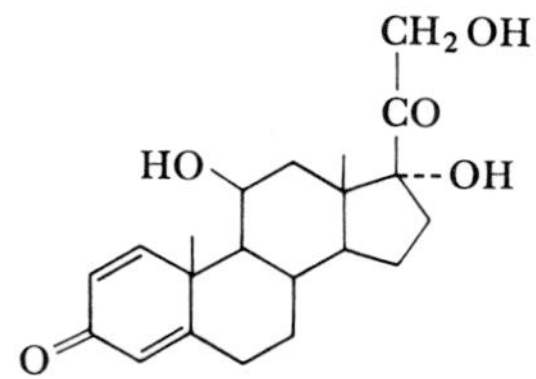

Methylprednisolone

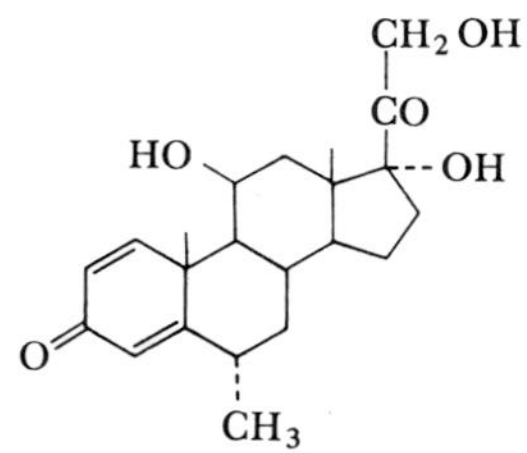

Fluoroprednisolone

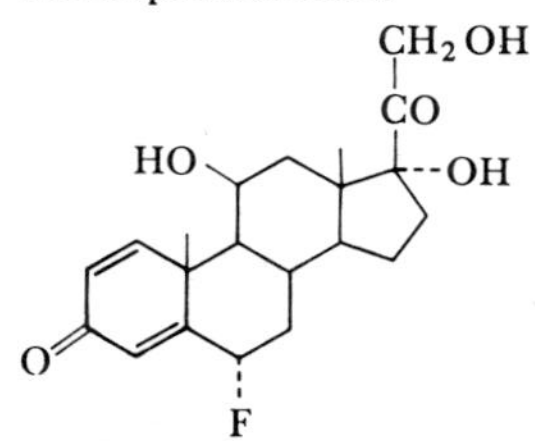

Table 2.2—continued

Type of substance	Naturally occurring compound	Modification or analogue	Uses
		Paramethasone 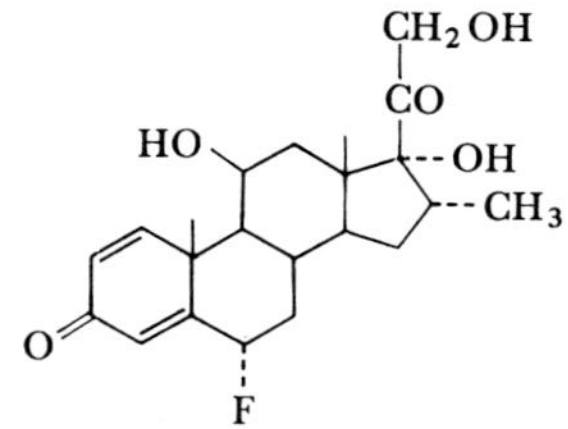	
		Dexamethasone 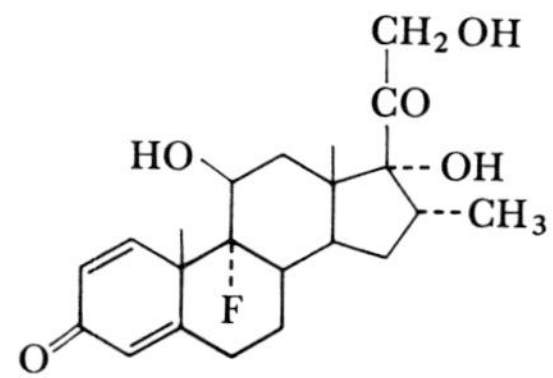	
		Betamethasone 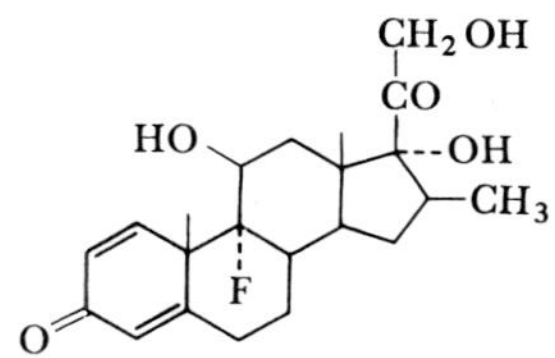	

		Flucinolone 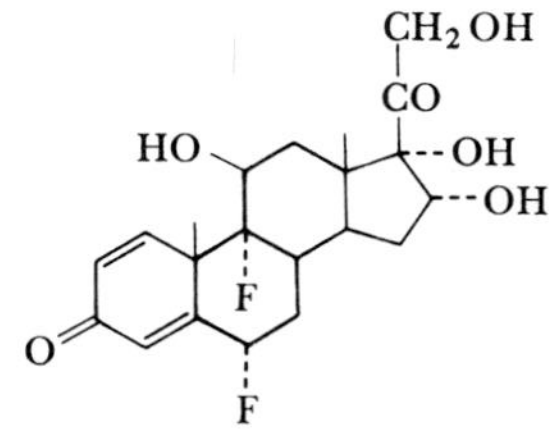	
		Flucinolone acetonide 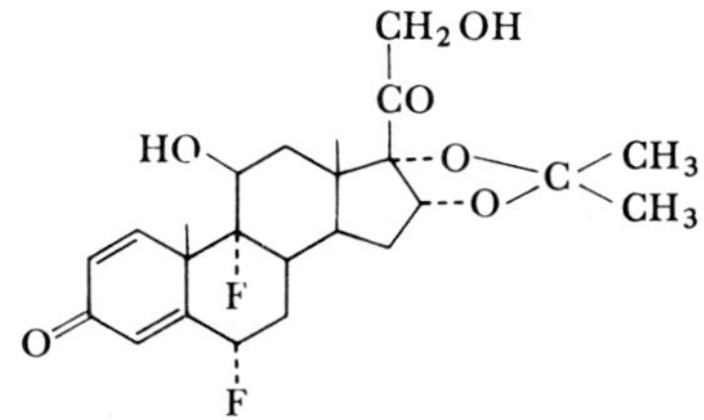	
C. Mineral salts	1. Calcium phosphate $Ca_3(PO_4)_2$	Calcium gluconate $Ca(C_6H_{11}O_7)_2$	Tasteless and non-irritant, also safe by i.m. injection.
	2. Sodium chloride NaCl	Lithium carbonate Li_2CO_3	Penetrates body water like natural alkali metal ions, but has value for manic-depression.

Table 2.2—continued

Type of substance	Naturally occurring compound	Modification or analogue	Uses
D. Intermediary metabolites	1. Urea $NH_2CO.NH_2$	Hydroxy-urea $NH_2.CO.NH.OH$	Of value in some malignancies.
	2. Glucose $C_6H_{12}O_6$	Gold thioglucose $C_6H_{12}O_5S.Au$	Used in arthritis.
E. Purine and pyrimidines	1. Hypoxanthine 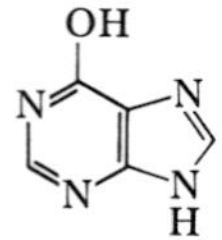	6-Mercaptopurine 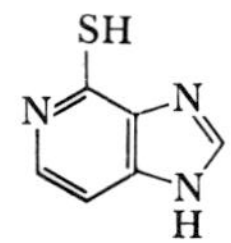	Anti-cancer drugs
	2. Uracil 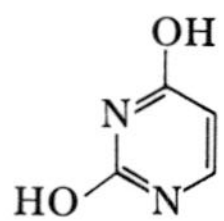	5-Fluorouracil 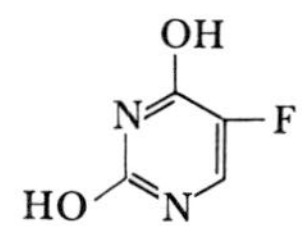	
		5-Iodouracil 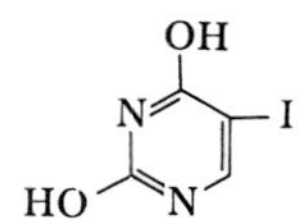	

3. Deoxyuridine

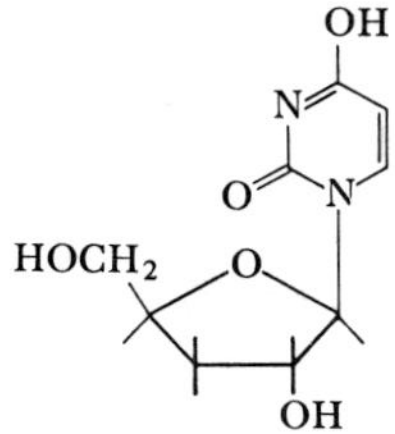

5-Fluorodeoxyuridine

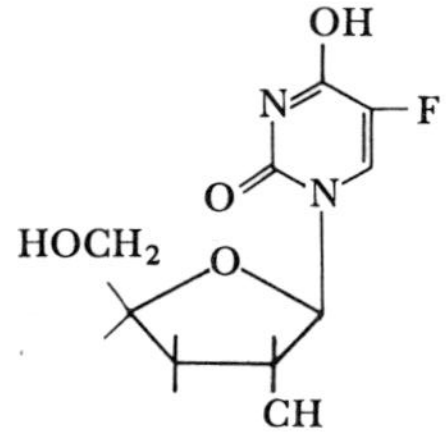

4. Orotic acid

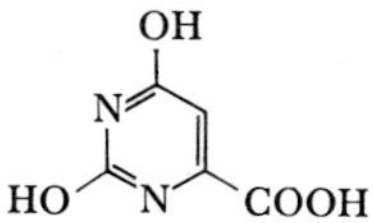

5-Azaprotic acid

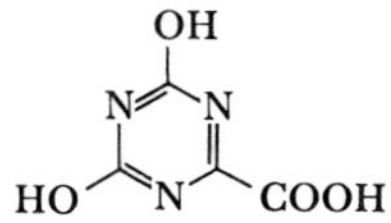

5. Guanine

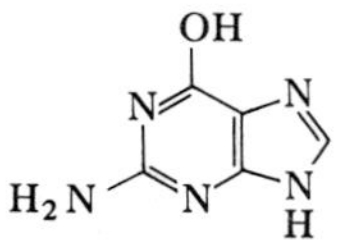

8-Azaguanine

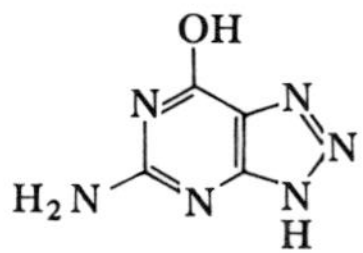

Table 2.3. Naturally occurring, pharmacologically-active plant compounds

Site of principal actions	Examples	Plant sources	Principal action
A. CNS	1. Morphine 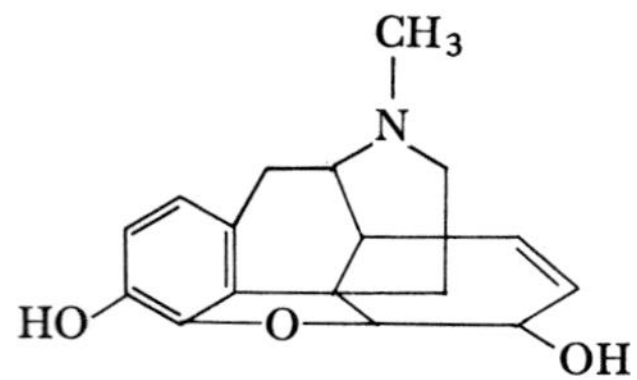	Opium poppy	Depresses brain functions, especially of the higher centres.
	2. Caffeine 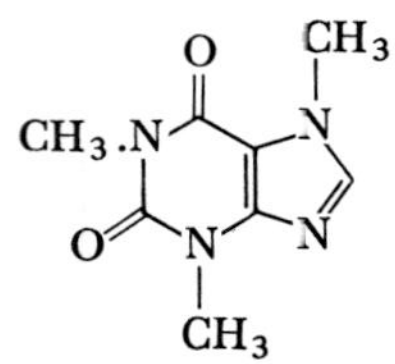	Tea, coffee, etc.	Increased irritability (also acts on muscles, vascular system, and kidneys).
	3. Cannabinol 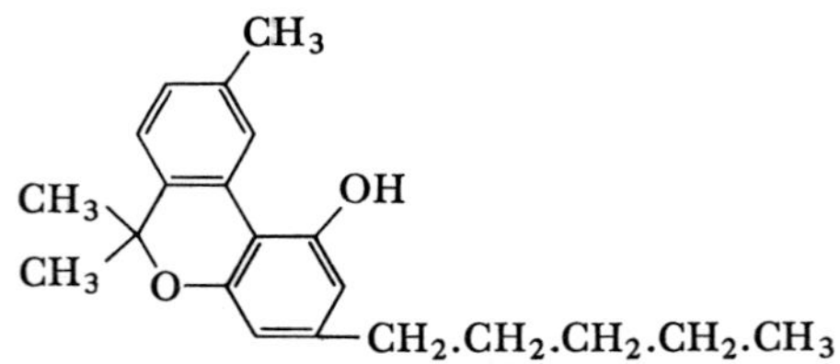	*Cannabis*	Inebriation (also acts on vascular system and muscles).

4. Mescaline 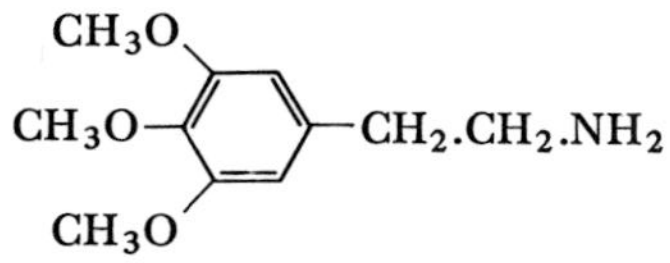	*Lophophora* cactus	Hallucinogen.
5. Lysergic acid (and esters) 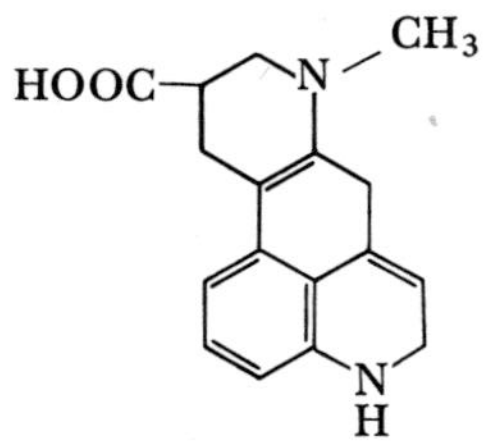	Ergot	Hallucinogen.
6. Reserpine 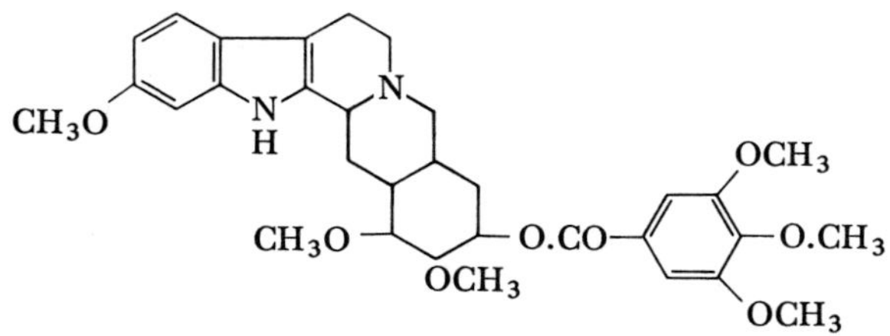	*Rauwolfia*	Tranquilizer. Hypotensive

Table 2.3—continued

Site of principal actions	Examples	Plant sources	Principal action
	7. Hyoscine 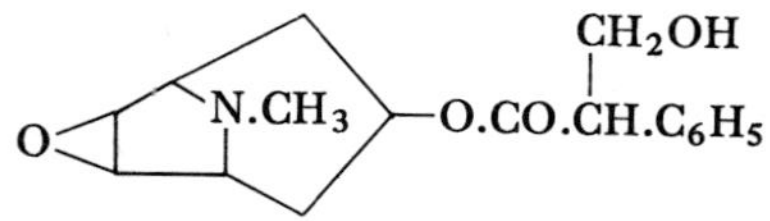	*Hysoscamus* and *Scopola*	Hypnotic with atropine-like actions.
B. Autonomic nervous system	1. Atropine 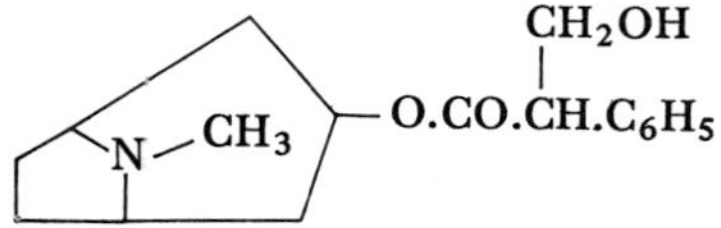	*Solanaceae*	Parasympathetic blocking
	2. Pilocarpine 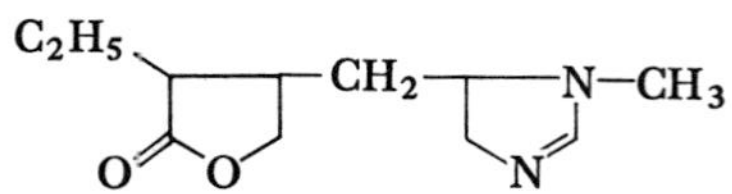	*Pilocarpus*	Parasympathetic stimulant.
	3. Physostigmine 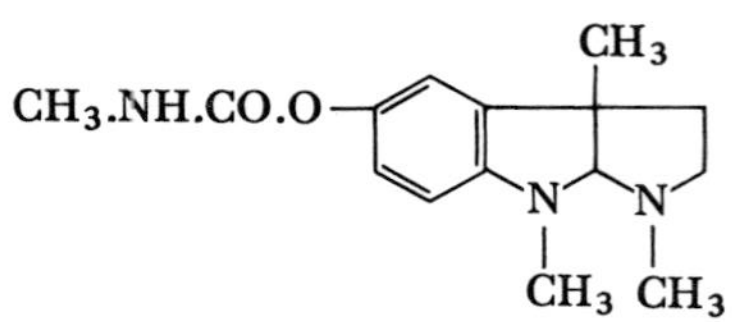	"Calabar beans"	Cholinesterase inhibitor.

	4. d-Tubocurarine 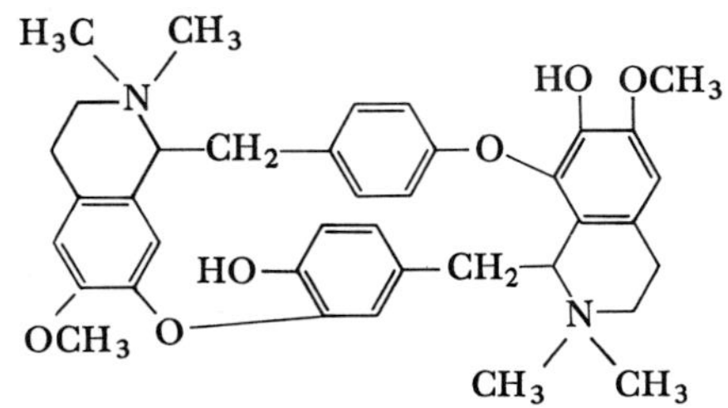	Curare	Neuromuscular transmission blocking.
	5. Digitoxigenin 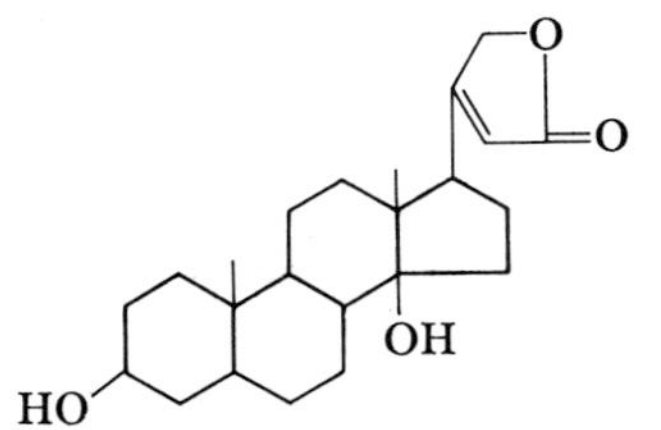	Digitalis	Cardiac innervation.
C. Hormonal actions	1. Ephedrine 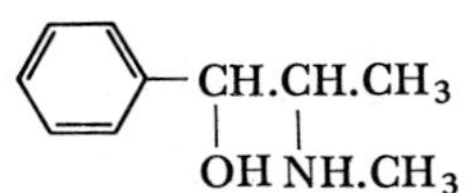	*Ephedra equistina*	Adrenaline-like effects.

Table 2.3—continued

Site of principal actions	Examples	Plant sources	Principal action
	2. Ergometrine, ergonovine 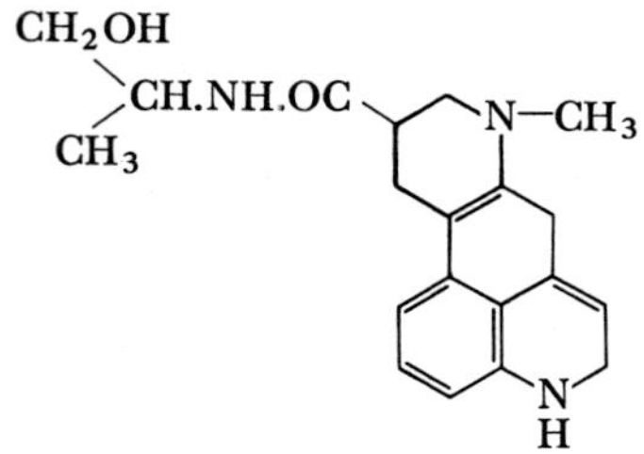	Ergot	Oxytocin-like effects.
	3. Glycyrrhetic acid 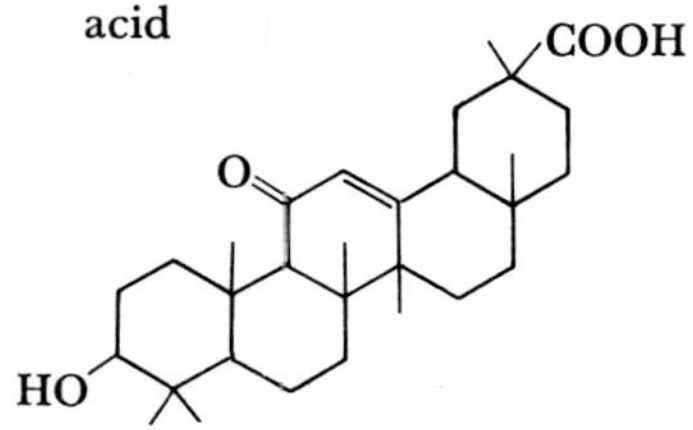	Liquorice	Cortisone-like effects.
	4. Coumestrol 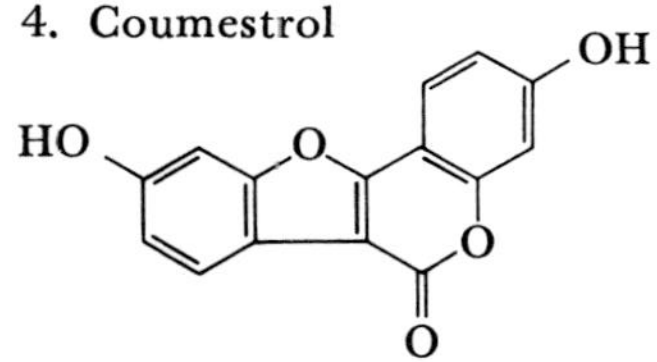	White clover	Estrogen-like effects.

D. Anti-microbial actions	1. Quinine 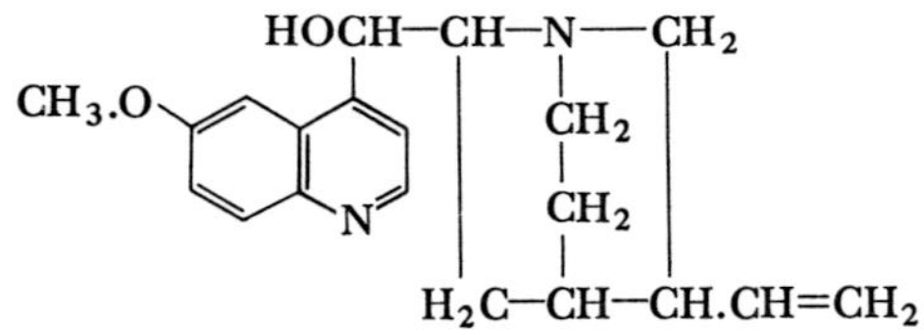	Chinchona bark	Toxic to malaria parasites.
	2. Allicin 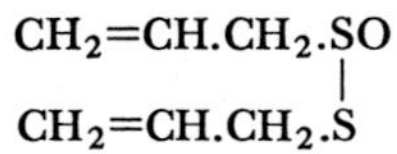	Garlic	Anti-bacterial agent.
	3. Penicillin 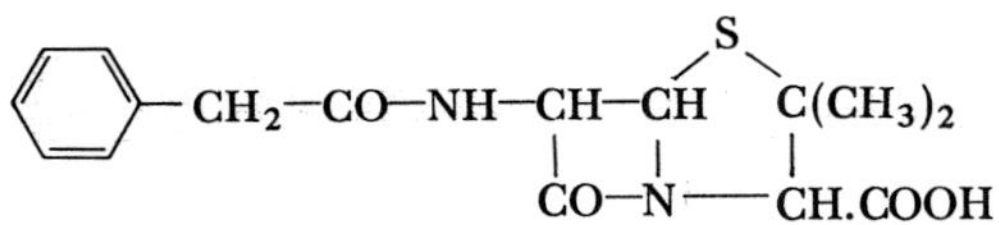	*Penicillium* fungi	Anti-bacterial agent.

Table 2.3—continued

Site of principal actions	Examples	Plant sources	Principal action
	4. Streptomycin 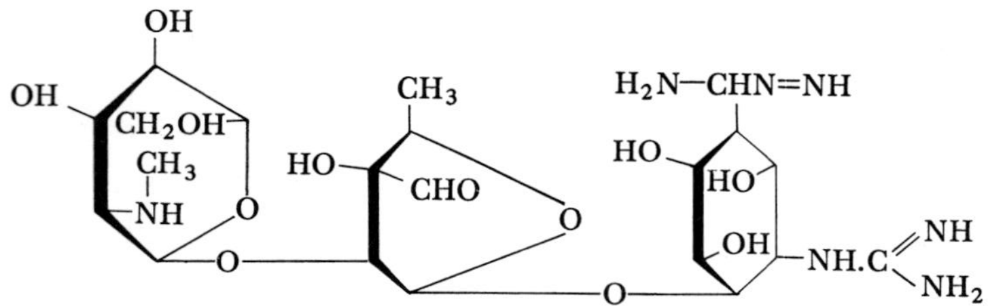	*Streptomyces* fungi	Anti-bacterial agent.
	5. Tetracycline 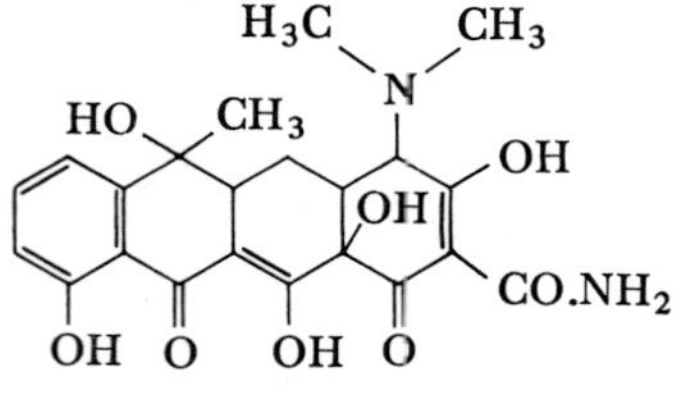	*Streptomyces* fungi	Anti-bacterial agent.

6. Griseofulvin 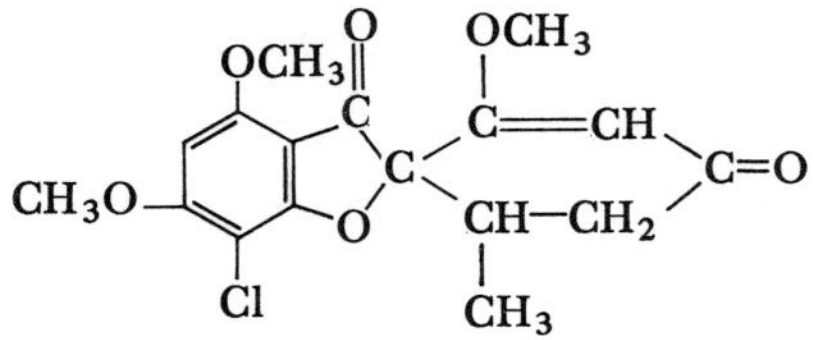	*Penicillium* fungi	Anti-fungal agent.
7. Ascaridol 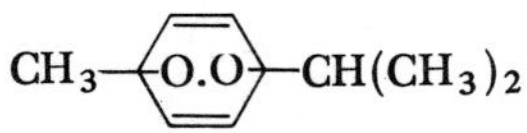	Chenopodium oil	Toxic to protozoa and nemotodes.
8. Filicic acid 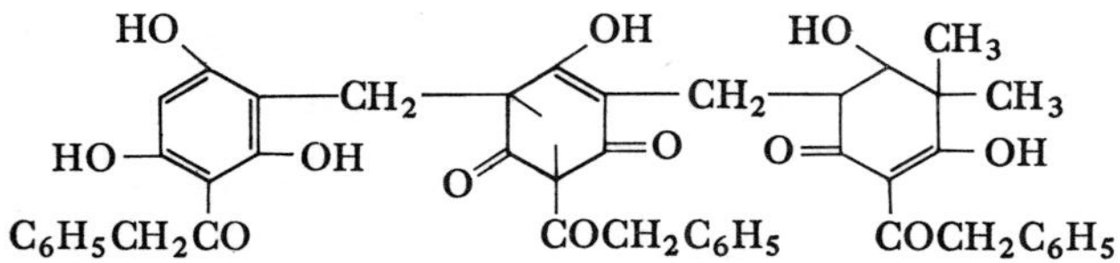	*Aspidium* (male fern)	Toxic to helminths and other parasites.

Table 2.3—continued

Site of principal actions	Examples	Plant sources	Principal action
	9. Actinomycin D 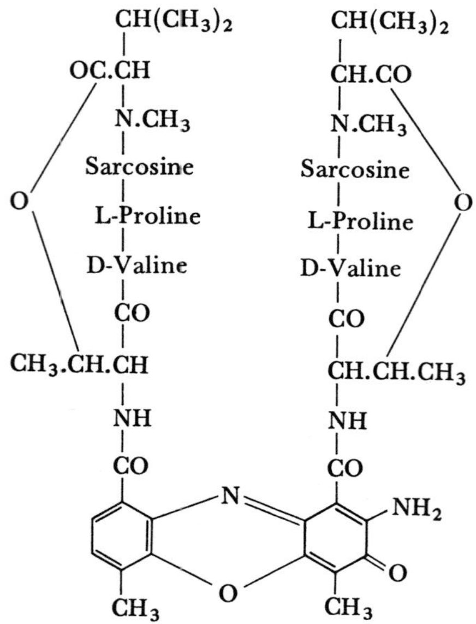	*Streptomyces chrysomallus*	Antineoplastic.

10. Mitomycin A	*Streptomyces caespitosus*	Antibacterial; cyctostatic.

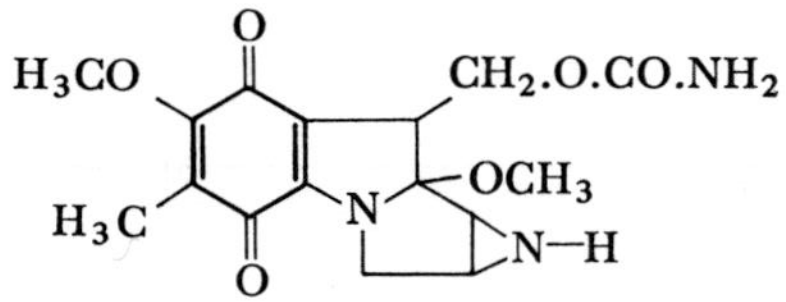

11. Novobiocin	*Streptomyces niveus*	Antibacterial.

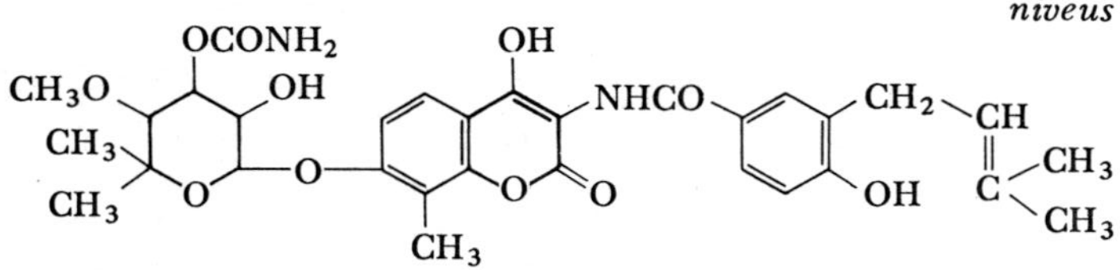

12. Chloroamphenicol	*Streptomyces venezuelae*	Antibacterial.

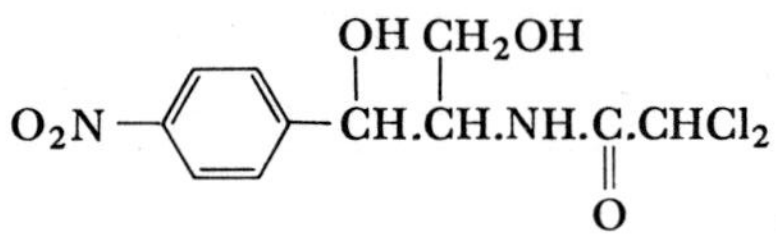

Table 2.3—continued

Site of principal actions	Examples	Plant sources	Principal action
	13. Erythromycin 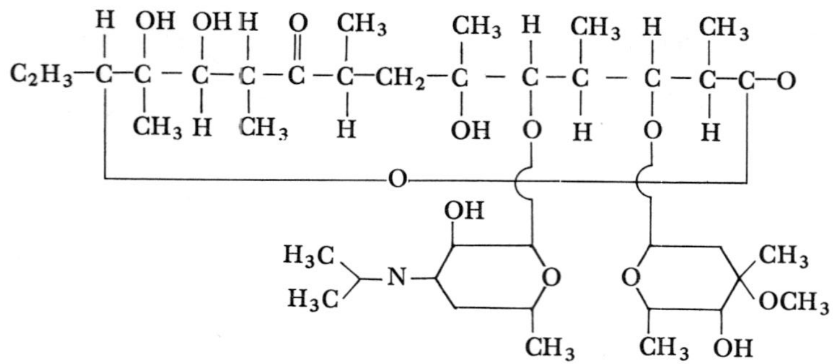	*Streptomyces erythreus*	Antibacterial.

pharmaceuticals derived from plant tissues, which will be grouped for convenience depending upon their major actions in humans. These compounds are merely a small selection from the thousands of medicinally active substances isolated from plants. They illustrate the high degree of chemical complexity that is often associated with drugs of plant origin. Like the drugs which are substances occurring naturally in the human body, these botanical compounds have all been subjected to chemical modification in the laboratory. A few structural analogues of some of these compounds are listed in Table 2.4.

Table 2.4. Structural analogues of plant pharmaceuticals

Natural plant compound	Synthetic analogues
Morphine	1. "Dionine" (ethyl-morphine) 2. Heroin (diacetyl-morphine)
Caffeine	1. Caffeine citrate 2. Chloroethyl-dimethylxanthine
Atropine	1. Methylatropine nitrate 2. "Genatropine" (atropine-*N*-oxide hydrochloride)
Physostigmine	1. Physostigmine salicylate 2. Physostigmine sulphate
Ephedrine	1. Clorprenaline 2. Isoprenaline
Glycyrrhetic acid	1. Carbenoxolone sodium (disodium glycyrrhetic succinate)
Quinine	1. Quinine bisulphate 2. Quinine ethyl carbonate
Penicillin	1. Nafcillin (6-(2-ethoxy-1-naphthamido)-penicillin) 2. Oxacillin (5-methy-3-phenyl-4-isoxazolyl-penicillin) 3. Ampicillin (α-amino-benzyl-penicillin)
Chlortetracycline	1. Tetracycline (dechloro-) 2. Terramycin (oxytetracycline)

5. Synthetic Compounds

The final group of drugs (and the largest) are compounds not occurring naturally, nor derived from natural products. Many drugs resembling compounds found in animal or plant tissues

Table 2.5. *Synthetic compounds used as drugs*

Therapeutic use	Type of compound	Examples
A. Anti-microbial	1. Sulphonamides (anti-bacterial)	Sulphanilamide 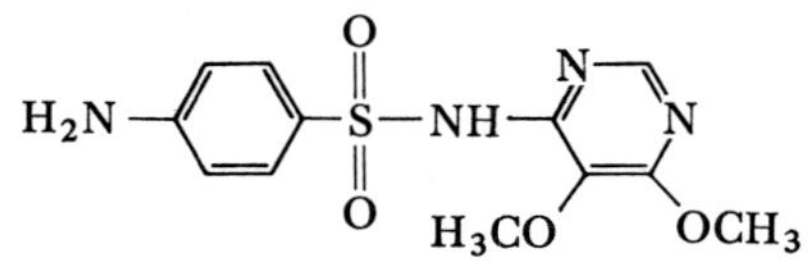
		Sulphathiazole 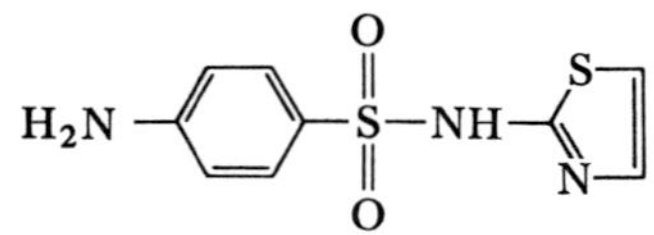
		Sulphadimethoxine 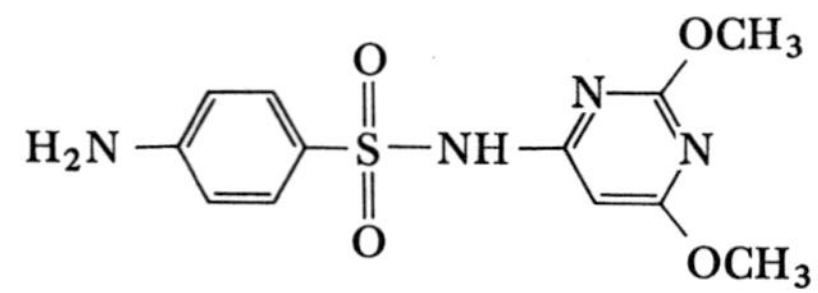
	2. Sulphones (anti-leprosy)	"Dapsone" 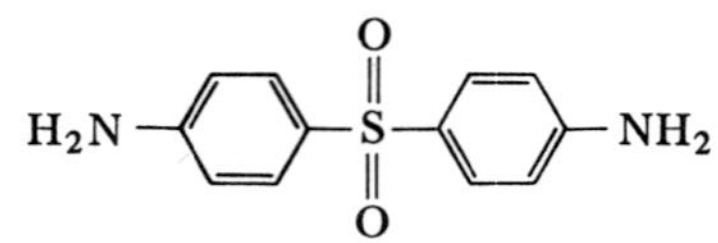

Glucosulphone ("Promin")

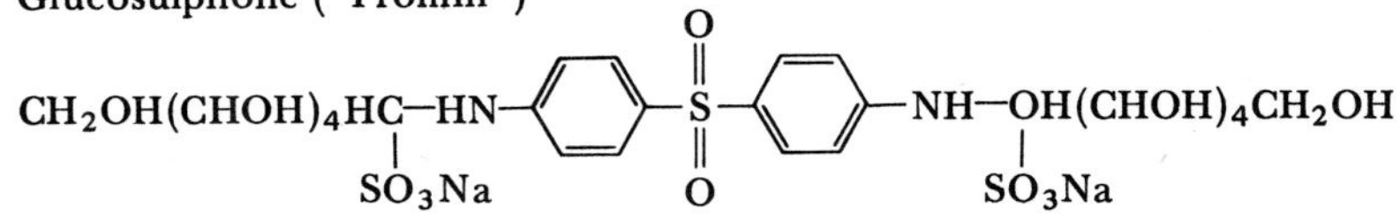

Sulphoxone ("Diasone")

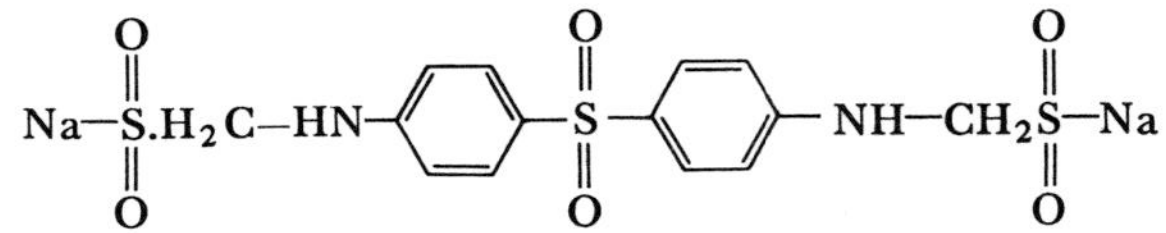

B. Anti-protozoal

1. Arsenicals (intestinal amebiasis)

Carbarsone

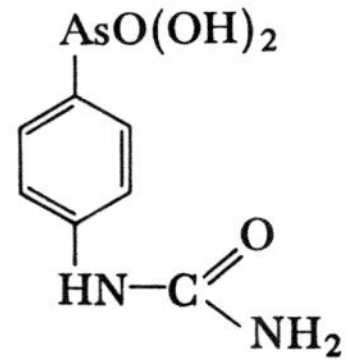

Glycobiarsol

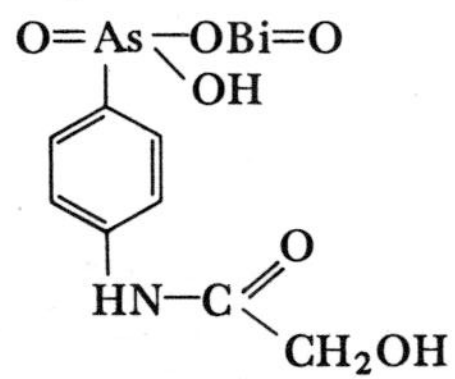

Table 2.5—continued

Therapeutic use	Type of compound	Examples
		Arsthinol 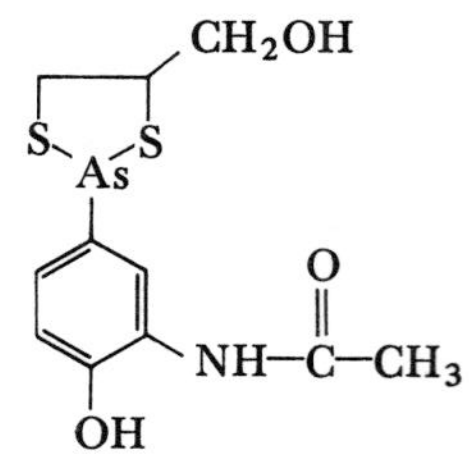
	2. Quinoloines (intestinal amebiasis, except for chloroquine)	Chiniofon 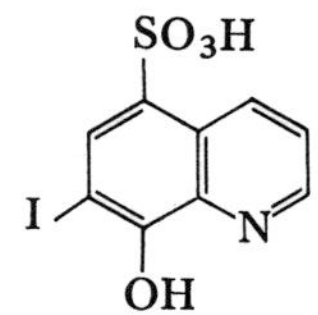
		Iodochlorohydroxyquin 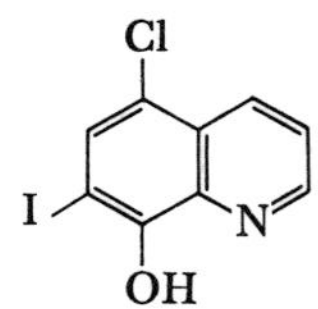

3. Other anti-malarials	Chloroquine (anti-malarial) 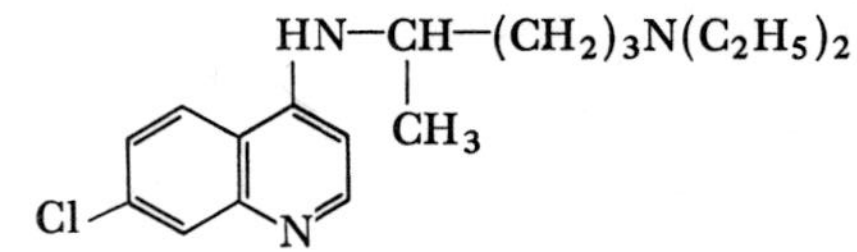 Primaquine 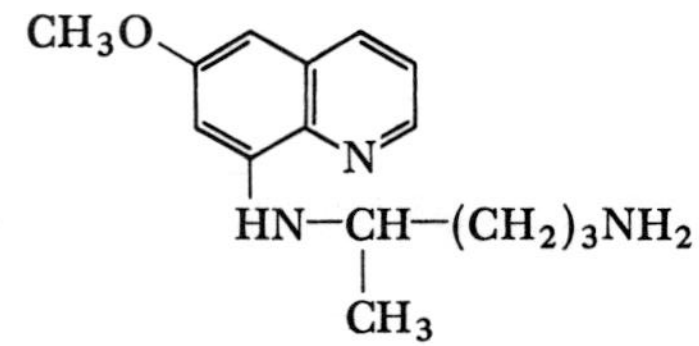 Pyrimethamine 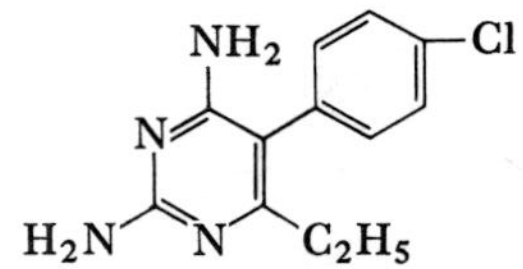 Chloroguanide 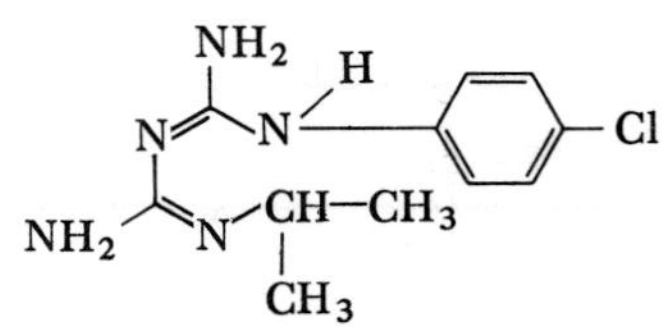

Table 2.5—continued

Therapeutic use	Type of compound	Examples
	4. Anti-trypanosomes	Melarsen 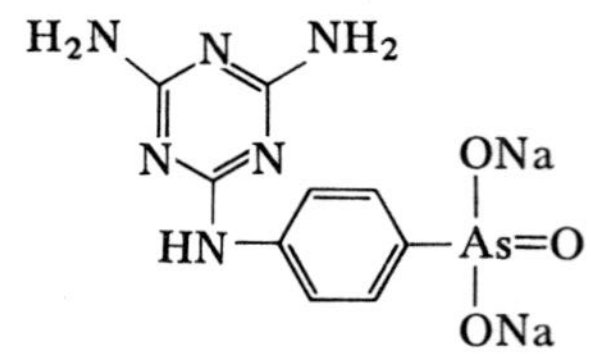 Pentamidine 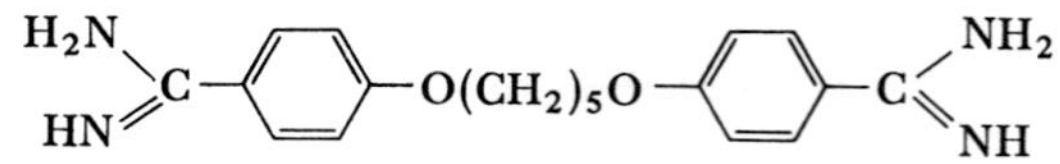 Tryparasamide 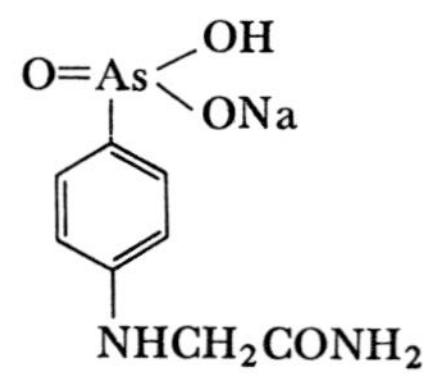

C. Anti-fungal	1. Various	Tolnaftate 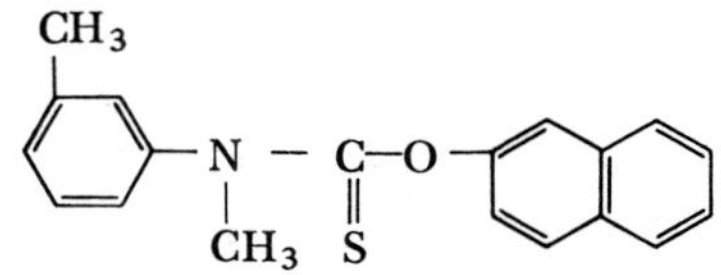
		Undecylenic acid $H_2C{=}CH[CH_2]_8COOH$
		Hydroxystilbamidine 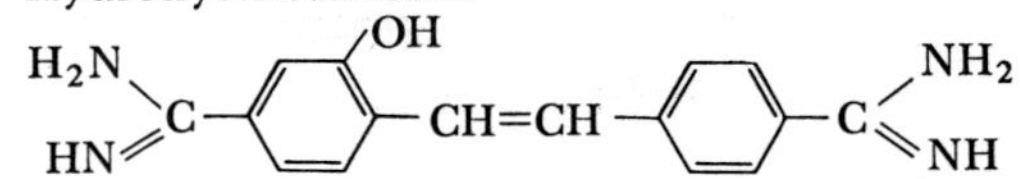
D. Anti-helminthics	1. Piperazines	Piperazine 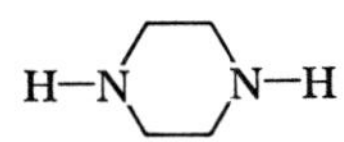
		Diethylcarbamazine 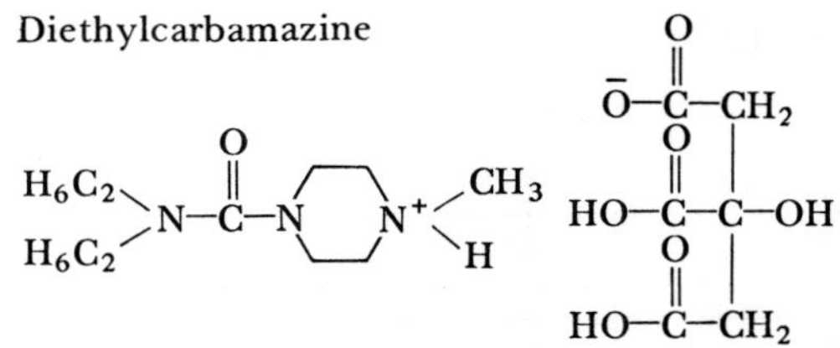

Table 2.5—continued

Therapeutic use	Type of compound	Examples
	2. Cyanines	Pyrvinium pamoate 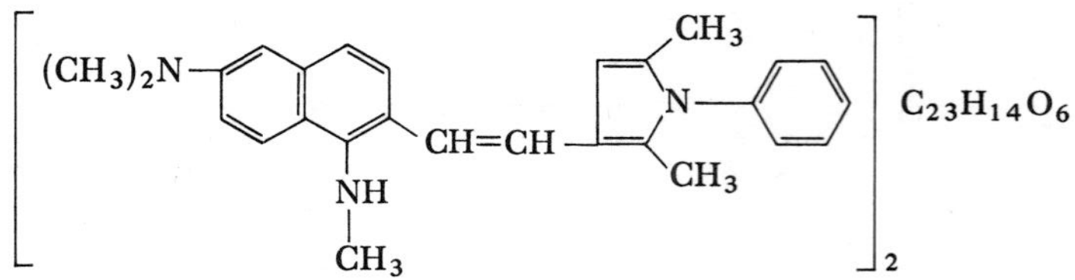
		Dithiazanine iodide 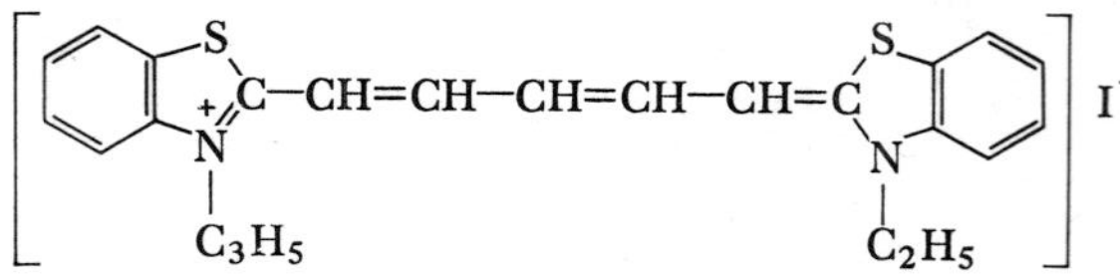
	3. Antimonals	Tartar emetic 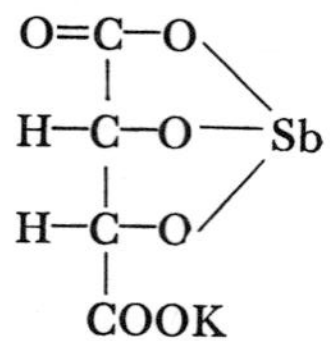

Stibophen

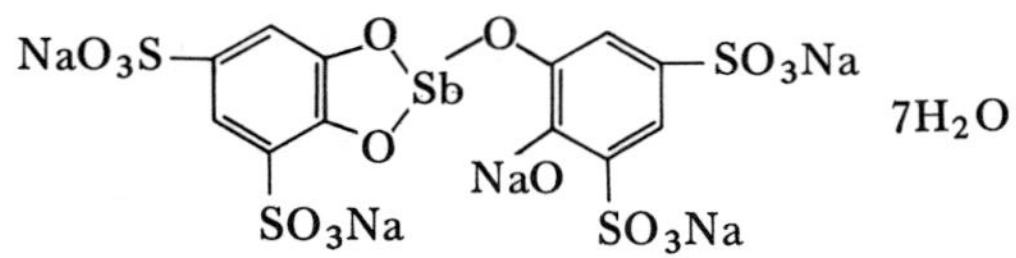

4. Various

Bephenium

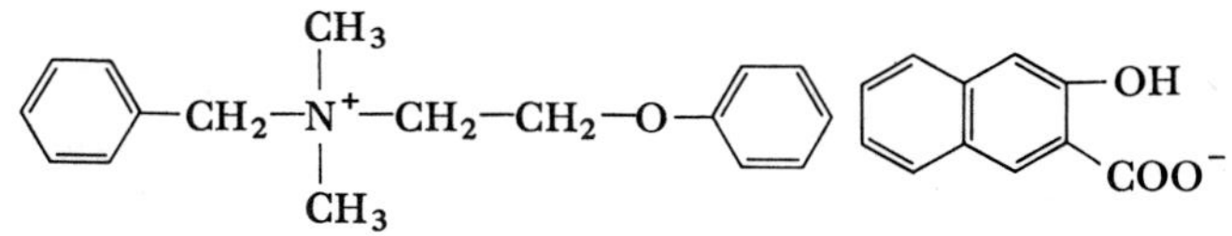

Lucanthone

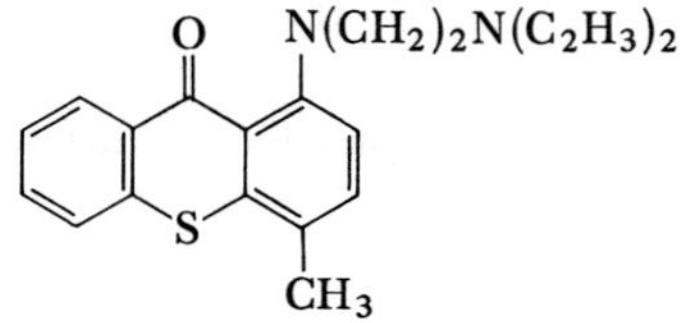

Dichlorophen

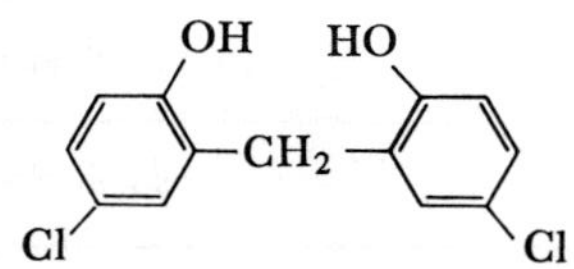

Table 2.5—continued

Therapeutic use	Type of compound	Examples
		Thiabendazole 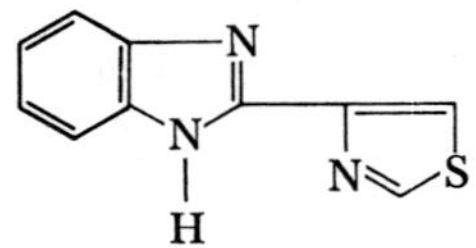
E. Anti-virus	1. Various	β-Isatin thiosemicarbazide 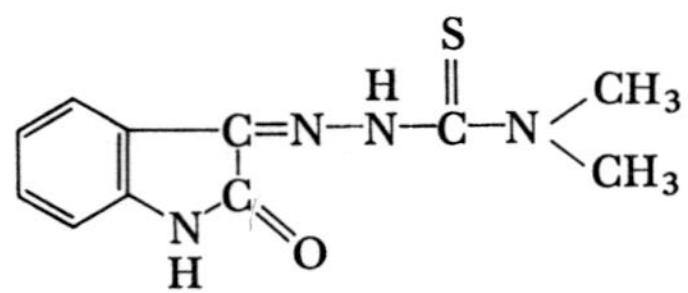
		Amantidine 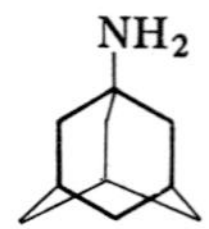
F. Anti-septics	1. Phenols	Meta-cresylacetate 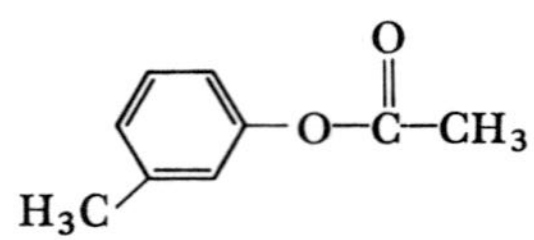

	Hexachlorophene 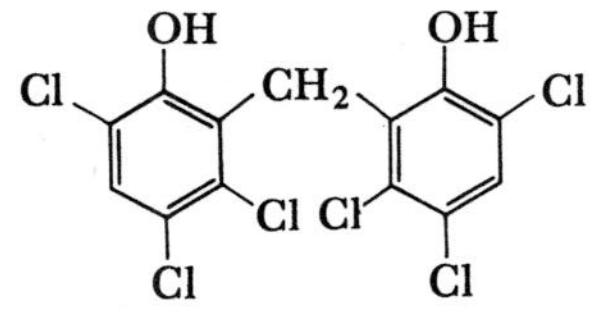
2. Quaternary ammonium salts	Benzalkonium chloride 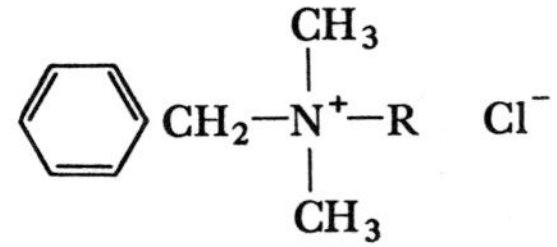 Hydrophobic groups represented by R range in size from C_8H_{17} to $C_{18}H_{37}$
	Benezethonium chloride 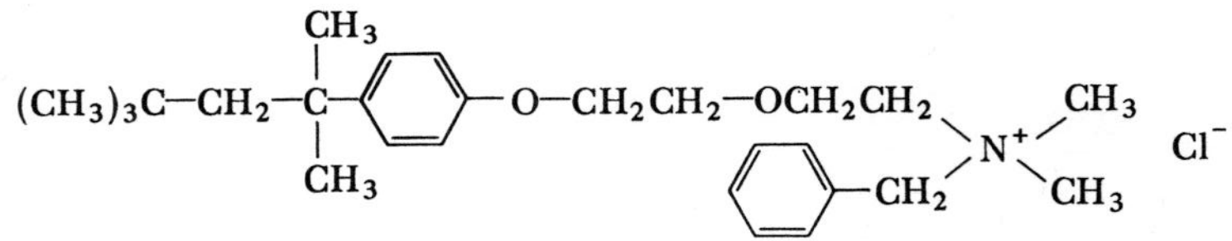
3. Organo-mercury compounds	Thimerosal 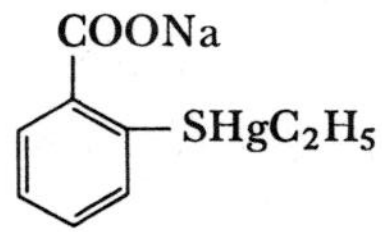

Table 2.5—continued

Therapeutic use	Type of compound	Examples
		Mercurochrome 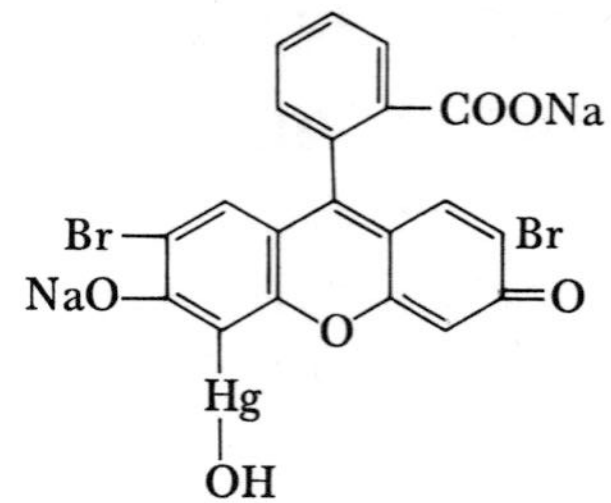
	4. Urinary anti-septics (systemic)	Nalidixic acid 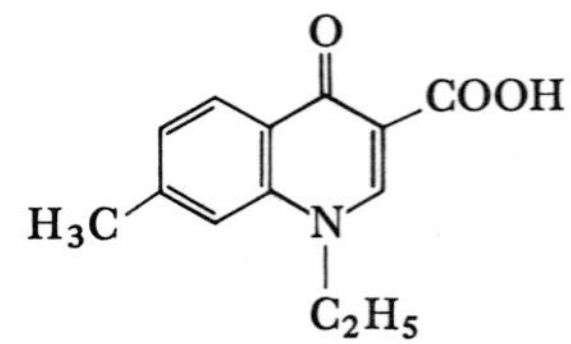
		Nitrofurantoin 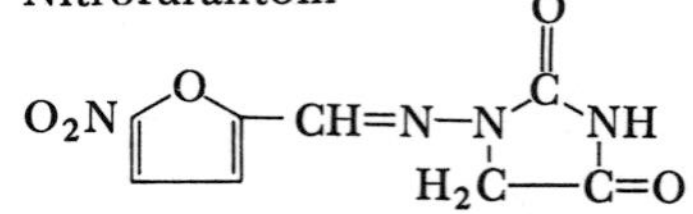

	Mandelic acid	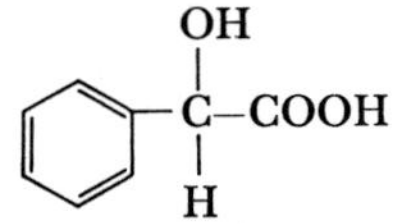
5. Acridines	Acriflavin	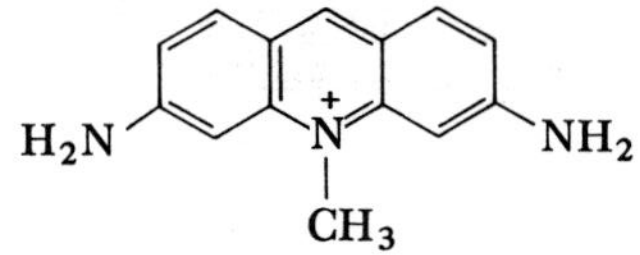
	Proflavin	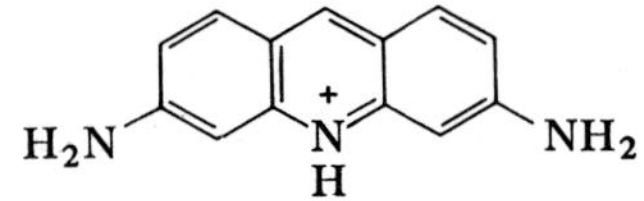
	Aminacrine	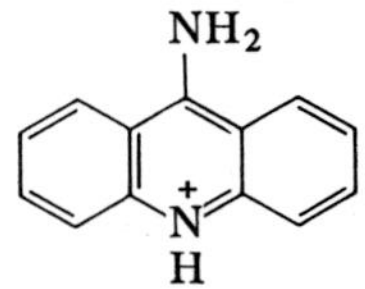

Table 2.5—continued

Therapeutic use	Type of compound	Examples
G. Cancer chemotherapy	1. Alkylating agents	Chlorambucil 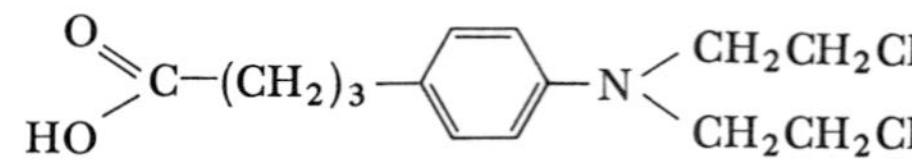 Melphalan 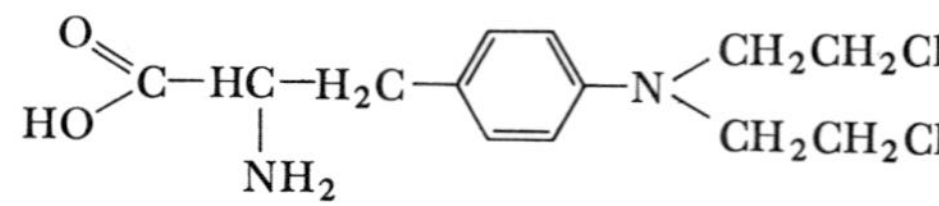 Busulphan 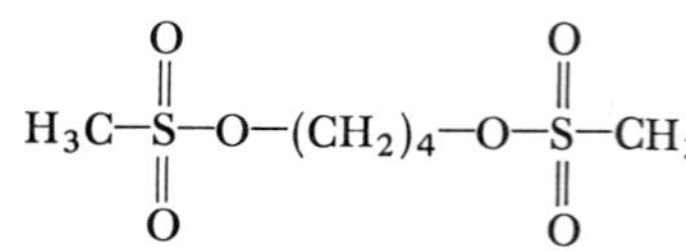 Thio-TEPA 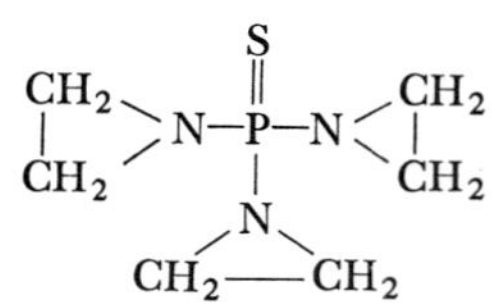

		Nitrogen mustard $CH_3.N(CH_2.CH_2.Cl)_2$
		Cyclophosphamide 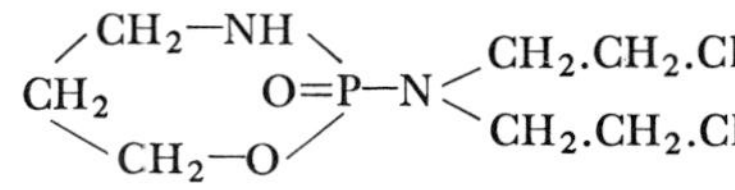
H. Hormone-like actions	1. Estrogen	Stilbestrol 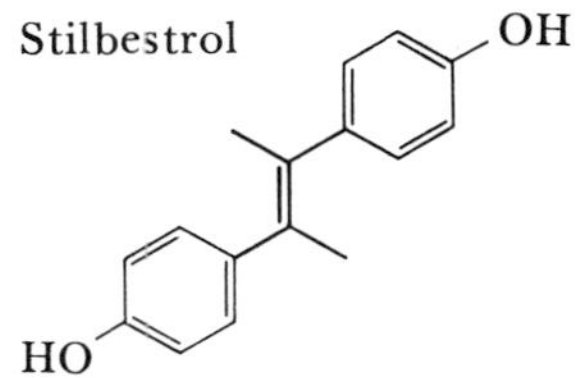
	2. Anti-diabetic	Phenformin
	3. Anti-thyroid	Propylthiouracil
		Methimazole
I. CNS drugs (analgesics, hypnotics, sedatives, tranquillizers, etc.)	1. Salicylates and related compounds (analgesics)	Acetylsalicylic acid 

Table 2.5—continued

Therapeutic use	Type of compound	Examples
		Acetanilid 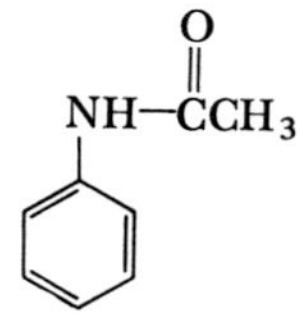
		Paracetamol 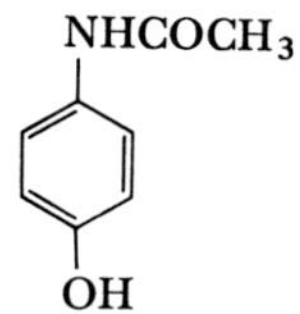
		Phenacetin 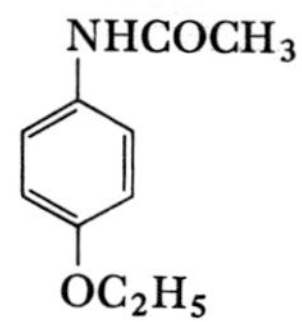
		Salicylamide 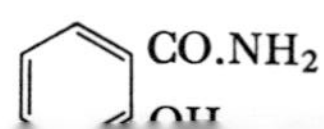

2. Pyrazolines (analgesics)	Phenylbutazone 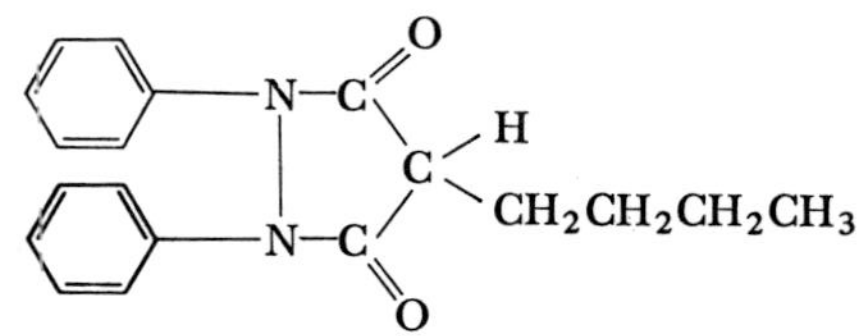	
3. Other analgesics	Amidopyrine (aminopyrine) 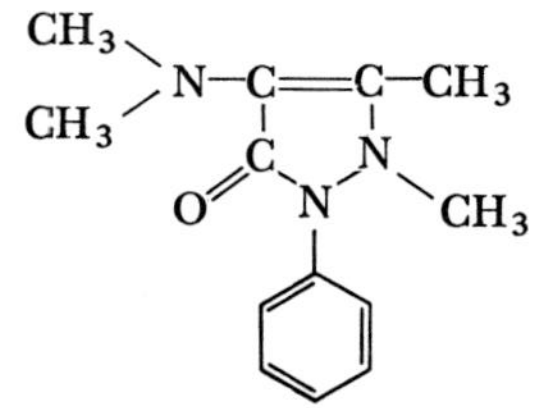	
	Indomethacin 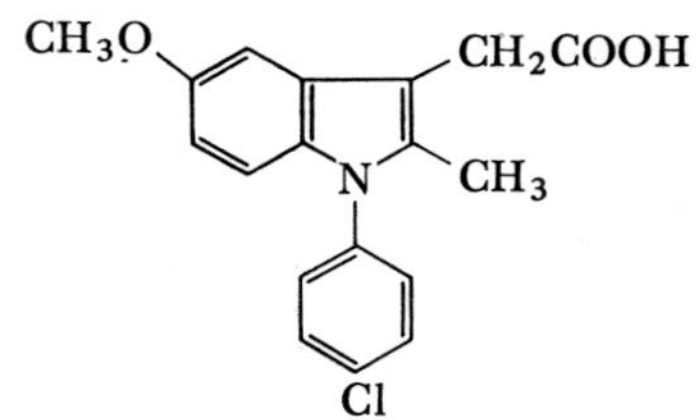	

Table 2.5—*continued*

Therapeutic use	Type of compound	Examples
		Mefenamic acid 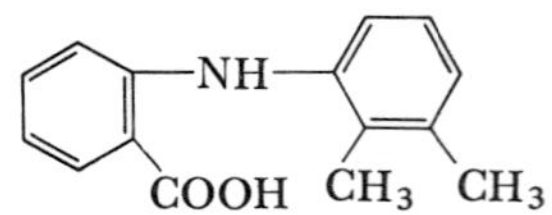
	4. Amphetamines (CNS stimulants)	Amphetamine $C_6H_5CH_2CH(NH_2)CH_3$
	5. Other stimulants	Phenelzine $C_6H_5CH_2CH_2NHNH_2$
		Phenmetrazine 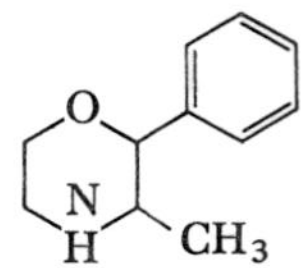
	6. Barbiturates (hypnotics)	Barbitone 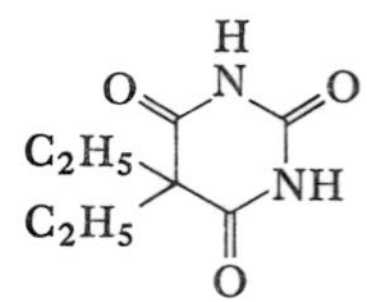

Phenobarbitone

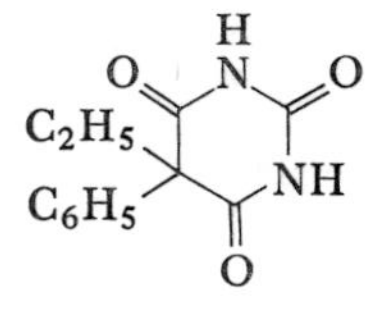

Amylobarbitone

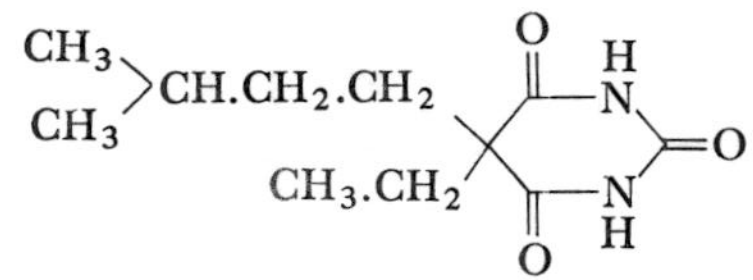

Pentobarbitone

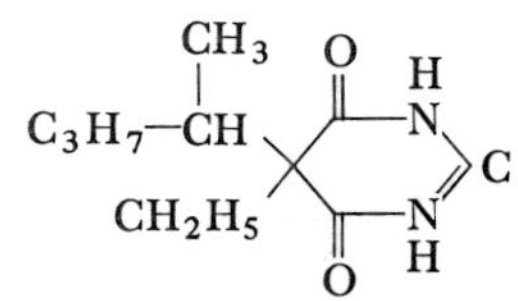

Thiopentone

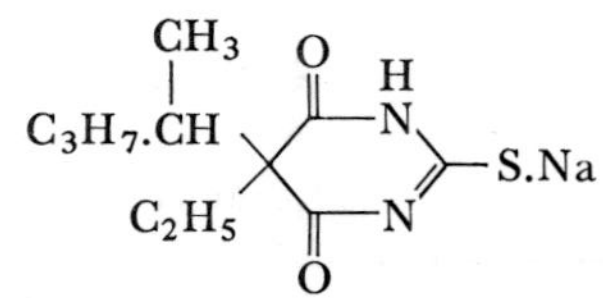

Table 2.5—continued

Therapeutic use	Type of compound	Examples
		Diallylbarbitone 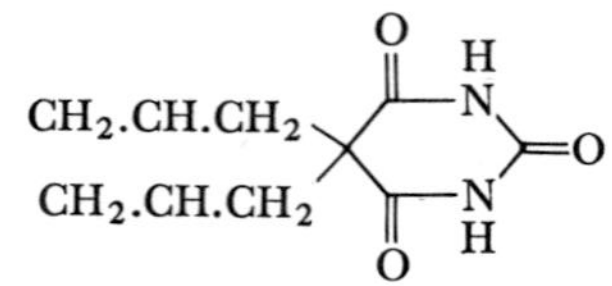
	7. Other hypnotics	Chloral hydrate $C.Cl_3.CH(CH)_2$
		Glutethimide 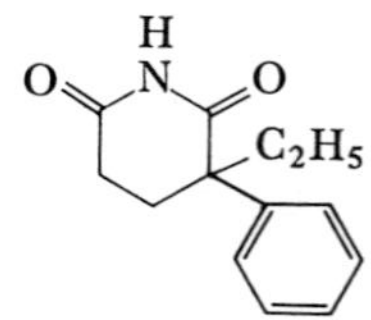
	8. Anti-depressants (hydrazines)	Iproniazid 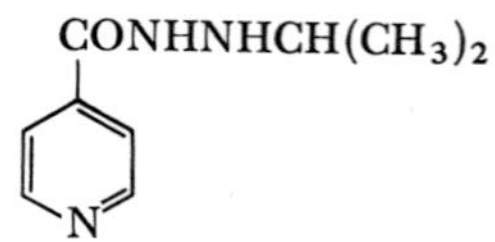

9. Anti-depressants (dibenzazepines)

Nortriptyline

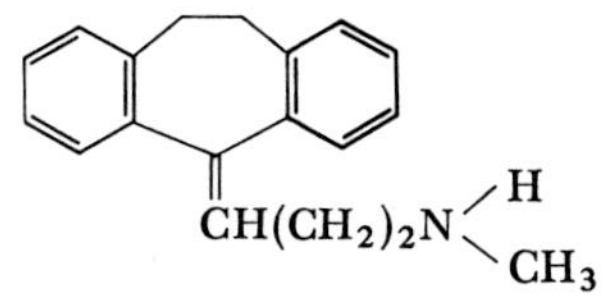

Imipramine

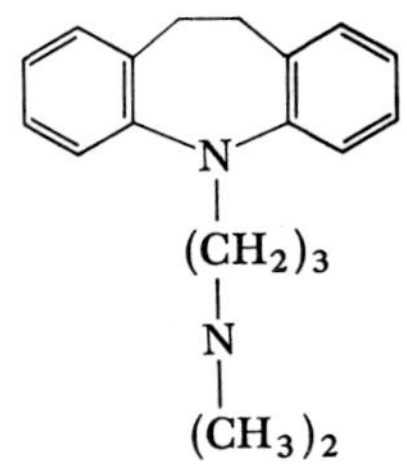

10. Tranquillizers

Chlorpromazine

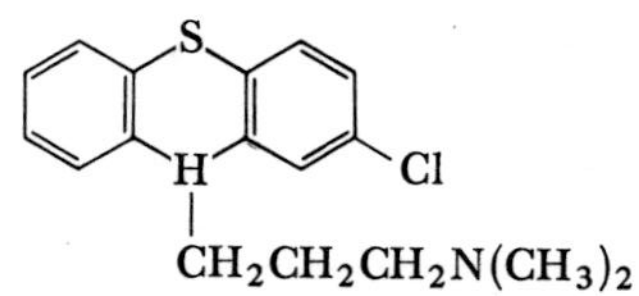

Table 2.5—continued

Therapeutic use	Type of compound	Examples
		Chlodiazepoxide 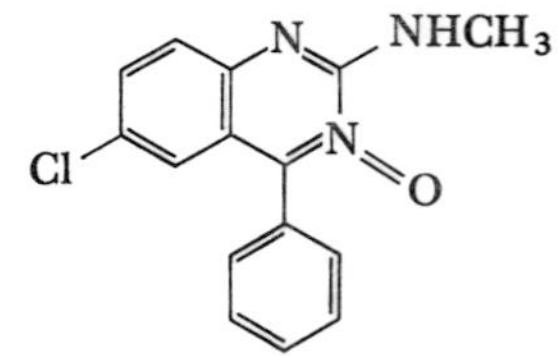
		Diazepam 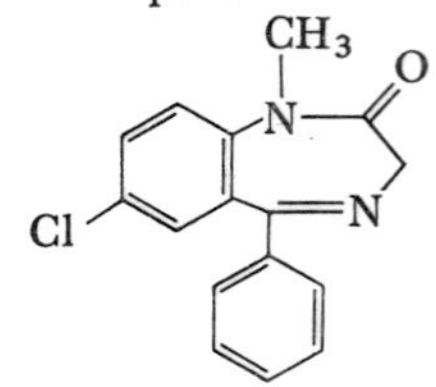
		Meprobamate $NH_2COOCH_3C(CH_3)(CH_2CH_2CH_3)CH_2OCONH_2$
		Trifluoperazine 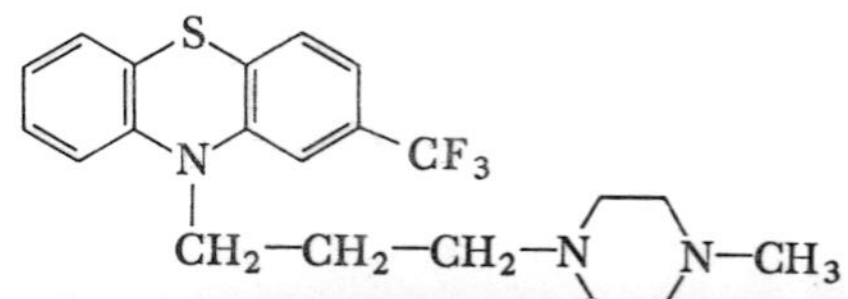

11. Anti-convulsants

Diphenylhydantoin

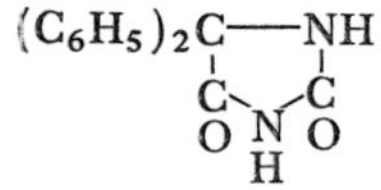

Primidone

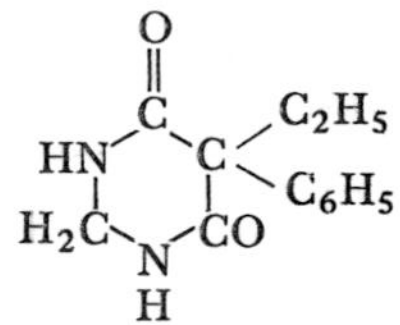

12. Anti-cholinesterases

Neostigmine

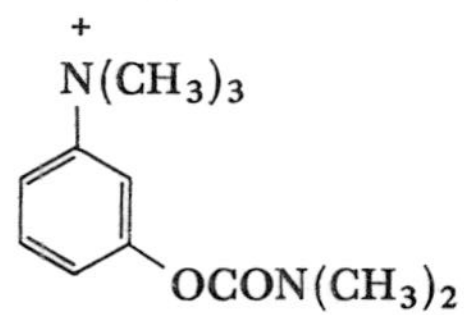

Physostigmine

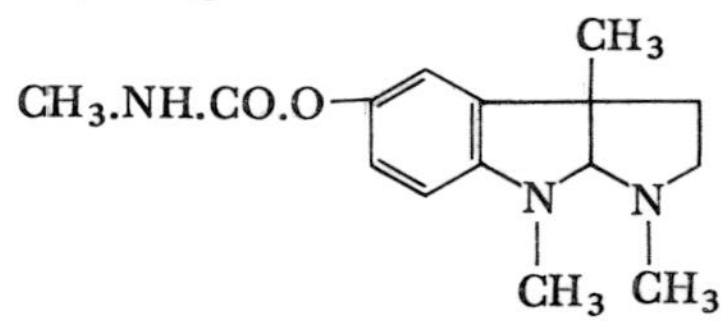

Therapeutic use	Type of compound	Examples
	13. Acetyl-choline blockers	Succinylcholine 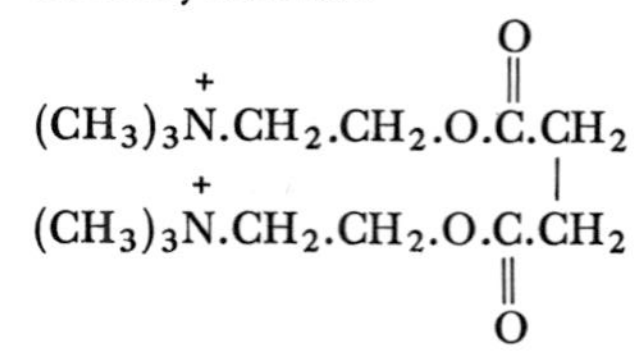
		Pentamethonium $(CH_3)_3-\overset{+}{N}-[CH_2]_5-\overset{+}{N}(CH_3)_3$
		Decamethonium $(CH_3)_3\overset{+}{N}-[CH_2]_{10}-\overset{+}{N}(CH_3)_3$
J. Anesthetics (general)	1. Various	Cyclopropane 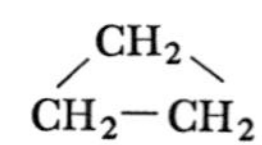

		Halothane	$CF_3.CHBrCl$	
		Chloroform	$CHCl_3$	
		Nitrous oxide	N_2O	
		Carbondisulphide	CS_2	
		Ethylene	C_2H_4	
		Diethyl ether	$C_2H_5.O.C_2H_5$	
K. Diagnostic agents	1. Radiological contrast media	Barium sulphate	$BaSO_4$	
		Thorium dioxide	ThO_2	

Table 2.5—continued

Therapeutic use	Type of compound	Examples
		Acetrizoic acid 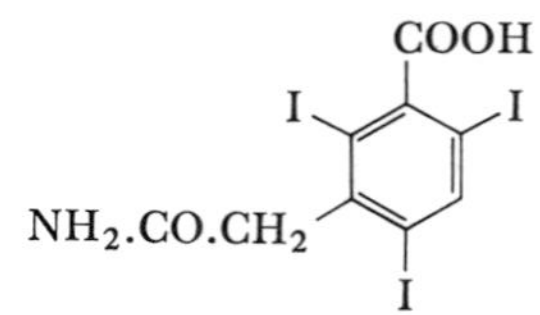
		Diatrizoate 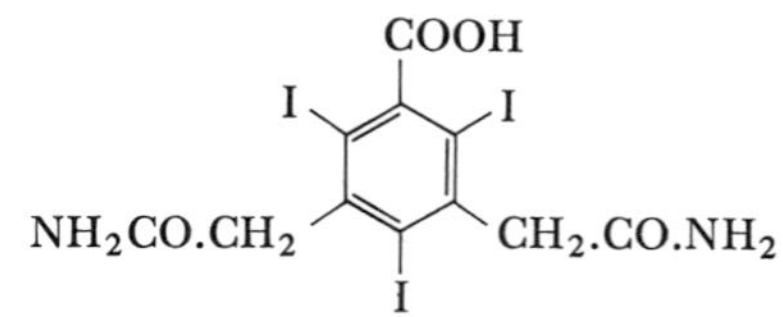
		Iopanoic acid 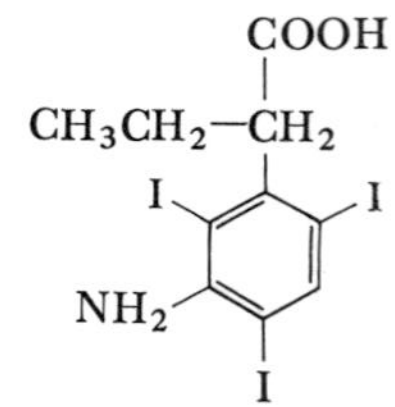

2. Dyes

Bromosulphophthalein

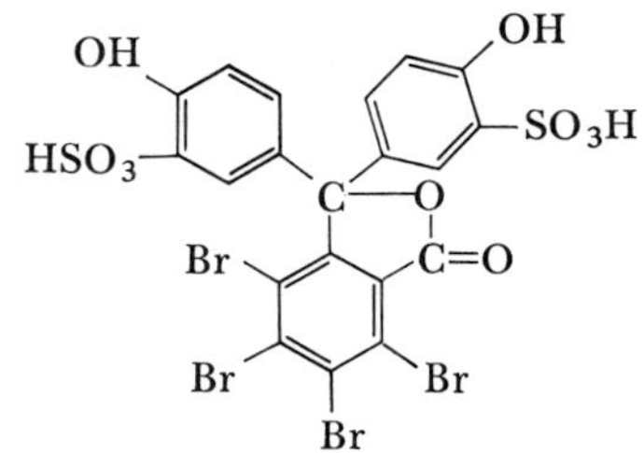

Phenol red

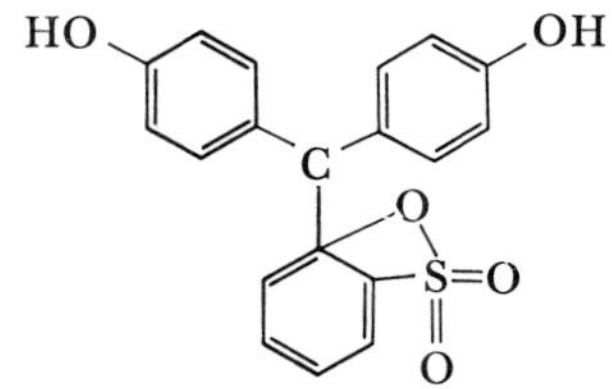

L. Anticoagulants

1. Coumarins

Warfarin

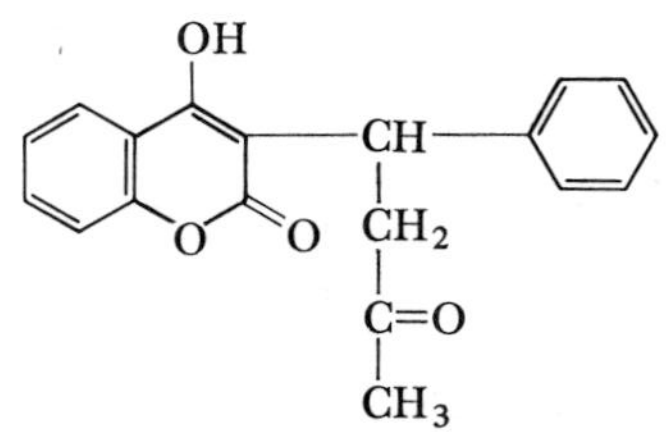

Table 2.5—continued

Therapeutic use	Type of compound	Examples
		Bishydroxycoumarin 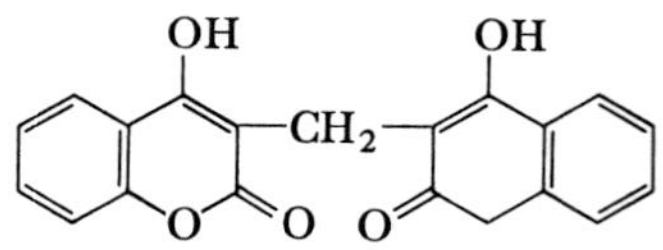
		Ethylbicoumacetate 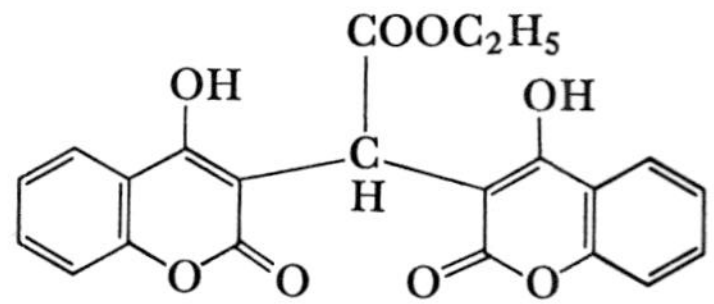
		Dicoumarol 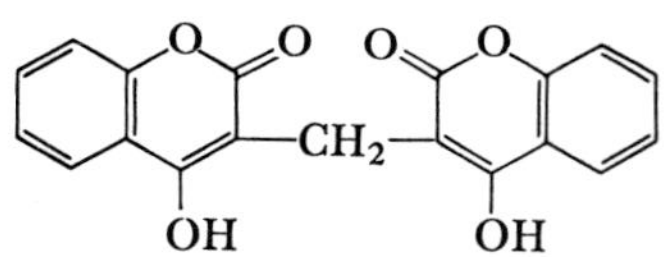

M. Miscellaneous drugs	1. Antihistamines	Meclozine 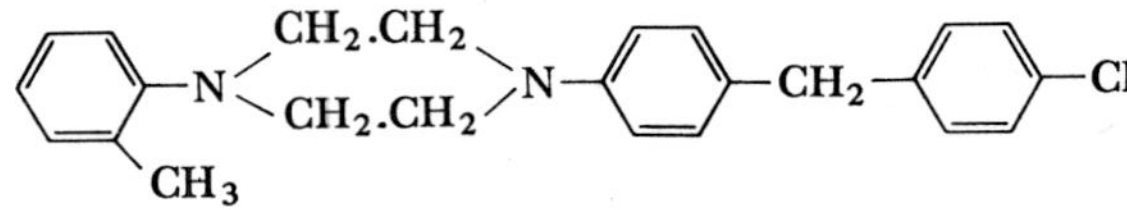
		Cyclizine 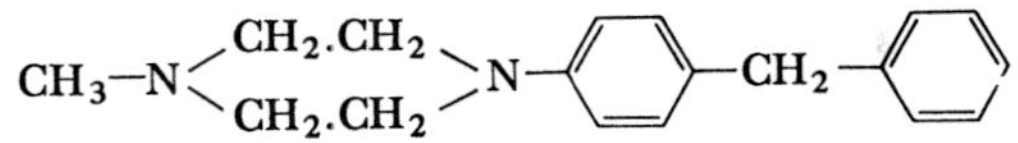
	2. Diuretics	Chlorthiazide 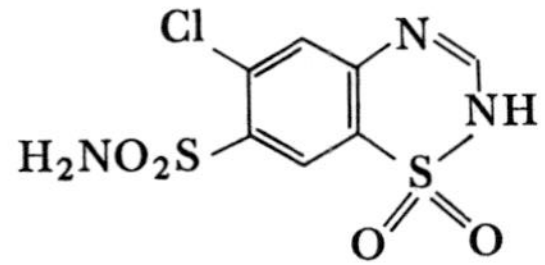
	3. Uricolytics	Probenecid 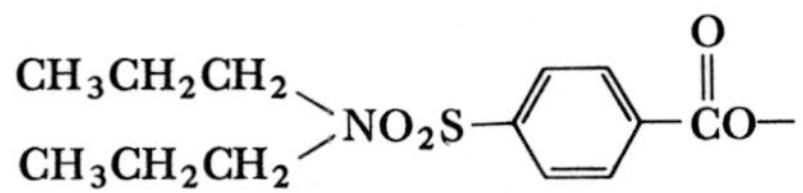

Table 2.5—continued

Therapeutic use	Type of compound	Examples
		Allopurinol 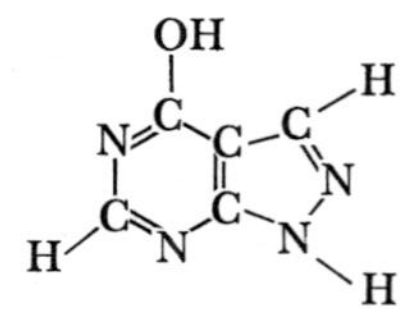
	4. Anti-alcohol	Disulfiram 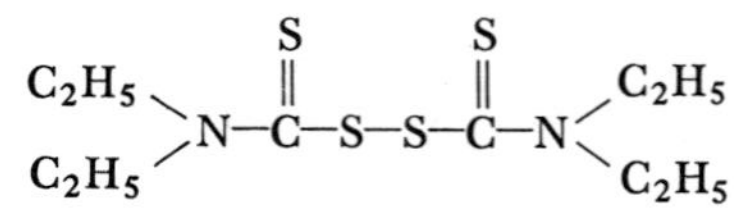
	5. Cardiovascular	Trinitrotoluene (TNT) 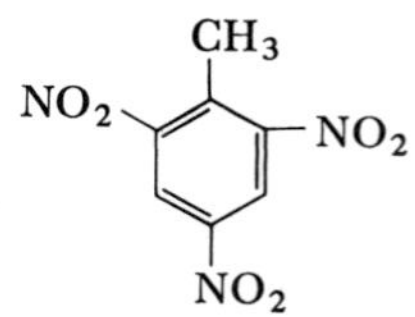

are manufactured synthetically, rather than by extraction from tissues, but compounds mentioned in this section do not bear any chemical resemblance to natural products.

Table 2.5 includes some of the more important groups.

II. ABSORPTION TRANSPORT AND ELIMINATION

1. Routes of Administration

In the early years of the 20th century, Paul Ehrlich and J. N. Langley independently arrived at the concept of drug *receptors*. These were hypothetical sites within body tissues where drugs act to produce their effects. These sites may be within the cells of parasites. Different drugs were assumed to have different receptors and some type of specific chemical interaction between the drug molecule and substances at the receptor site was assumed. It is now known that in many cases drug-receptor substances are biological macro-molecules, usually either specific proteins or nucleic acids, or both. For example anticholinesterases, such as physostigmine, combine chemically with a serine-residue of the cholinesterase enzyme protein; the antibiotic mitomycin combines specifically with DNA molecules to prevent their replication, while the alkylating agent mechlorethamine combines with a wide variety of proteins, RNAs and DNAs. One or two drugs may be an exception and act in a more general manner without requiring tissue receptors and chemical combination with macro-molecules. The radiological contrast media for the digestive tract (barium salts and organo-iodine derivatives) are inert and their function is merely to fill the lumen of the gut with a radio-opaque substance. Similarly, osmotic diuretics (such as urea and mannitol) act by increasing the osmolarity of tubular urine. Certain drugs act by reacting with *small* molecules within the body, rather than with macro-molecules. Antacids neutralize HCl in the stomach, chelating agents (such as EDTA and penicillamine) combine with inorganic ions, like Pb^{++} and Cu^{++}, and so neutralize their toxic effects; while inorganic purgatives (such as $MgSO_4$) and urinary antiseptics (such as mandelic acid) are drugs acting by altering the ionic composition of body fluids. The gaseous anaesthetics (such as ether

Table 2.6. Routes of drug administration

Type of route	Methods	Comments
1. Enteral	Sublingual	Only applicable to a few drugs (e.g. TNT, methyltestosterone), but can be useful as an alternative to injection for compounds likely to be destroyed in lower gut.
	Swallowing (oral administration, *per os*, or *p.o.*)	Most usual method, but drug may need formulation to reach site of absorption (usually small intestine) without being decomposed by gastric acid (e.g. penicillins); obvious method for drugs acting on gastro-intestine tract (e.g. digestive enzymes, anti-ulcer products, etc.).
	Rectal (*per rectum* or *p.r.*)	Route of choice for most (products) acting locally (e.g. haemorrhoidal preparations), but most drugs active by mouth may also be given by p.r.
2. Parenteral	Intravascular (usually *intravenous* (*i.v.*) but occasionally intra-arterial (*i.a.*)	Commonest route is into the antecubital vein; has the advantage that the drug enters the circulation without delay, but drug cannot be recalled if reactions occur (whereas it often can after enteral administration); other possible complications are embolisms, infections at injection site, pyrogen contaminants, etc. intra-arterial route used only for special drugs (e.g. contrast media, anti-cancer products).
	Intramuscular (*i.m.*) injection	Usually into the deltoid or gluteal muscles; small molecules are absorbed into capillaries, while large molecules (and most lipid-soluble compounds) go via lymphatics; drugs in oily vehicles (or those insoluble at tissue pH) remain in the muscle as a long-acting depot, but most aqueous solutions are absorbed in 10 to 30 minutes.

Table 2.6—continued

Type of route	Methods	Comments
	Subcutaneous (*s.c.*) injection	Absorption is usually slower than by i.m. injection but vasodilators (e.g. methacholine) are taken up rapidly; conversely, the inclusion of small amounts of a vasoconstrictor (e.g. adrenaline) will delay absorption; *s.c.* administration can be very painful and necrosis or sterile abcesses can result.
	Subcutaneous implants	Relative insoluble drugs in small pellets or sialastic capsules can be implanted under the skin to slowly release their constituents (e.g. sex hormones, insulins, etc.).
	Intradermal (*i.d.*) injection	Occasionally used for particular products (e.g. TAB vaccine) and gives less reactions than *s.c.*; absorption is usually very slow.
	Transdermal application	Application of drugs to the skin in special vehicles (e.g. DMSO) to allow rapid systematic absorption; mainly experimental.
	Inhalation	Gaseous drugs (e.g. many anaesthetics), or aerosols containing drugs in particles less than 1 μ (e.g. adrenaline, aminophylline, etc.), give rapid systemic effects via the lungs.
3. Local applications	Topical application	Many drugs for skin conditions (e.g. corticosteroids) are formulated in ointments and creams to give skin penetration, but very low systemic absorption.
	Intra-lumenal administration	Drugs (e.g. contrast media, antibiotics, etc.) can be introduced into the lumen of most body orifices in a form which prevents any significant absorption (e.g. ears, eyes, vagina, bladder, nasal sinuses, etc.).

and cyclopropane) may also be exceptions to the general rule, although there is evidence that they act on particular lipoprotein complexes in the brain.

Once the concept of specific macro-molecular receptors for most drugs is accepted, it becomes clear that a drug will act while in combination with its receptors and for some drugs the duration of action is proportional to the half-life and concentration of the drug-macromolecular receptor complex. We have surveyed in the previous section the wide range of pharmaceutical products used in medicine, and it was seen that most drugs are used to bring about a specific change in some part of the body. With the exception of a few drugs used for strictly local effect (such as skin ointments, dental antiseptics, etc.), most pharmaceuticals enter the circulation and are thus transported to their receptors. We must, therefore, consider the possible routes of administration of a drug and the way in which they may alter its action.

A drug may enter the body, generally speaking, in two main ways, *enteral* (via the gastro-intestinal tract) or *parenteral* (by-passing the gastro-intestinal tract; which in effect means injection). Table 2.6 lists the various alternatives.

As the principal routes of drug administration are oral, intravenous and topical, the factors affecting the absorption and uptake by particular receptors of important drugs given by these three routes are discussed in detail.

(i) Oral

We have already mentioned that a drug given by the oral route is usually absorbed through the wall of the small intestine and so has to pass through the acid conditions of the stomach. Generally speaking, the pH of the stomach is about 1.0, of the small intestine about 6.9 and of blood plasma about 7.4. Many drugs are capable of ionization, but are usually only weakly so. At the pH of blood plasma, most will occur as unionized molecules, with only a small number of ionized forms. However, at the pH of the stomach drugs that are weak acids will be almost entirely *un*ionized, whereas drugs that are weak alkalis will be predominantly *ionized*. Generally speaking, only unionized forms of drugs can be considered to be in equilibrium across the gastro-intestinal wall with the drug in plasma. By applying the Henderson–Hasselbalch equation to the situation in the stomach and the small intestine, the effect of the different pH of these regions on the absorption of drugs of different chemical types can be clearly seen (see Figure 2.1).

Lumen of gastro-intestinal tract	Blood plasma
$H^+ + A^- \rightleftharpoons HA \rightleftharpoons$	$HA \rightleftharpoons H^+ + A^-$
pH·Stomach = 1.0	7.4
Small Intestine = 6.9	7.4

Figure 2.1. pH of G-I tract and blood plasma.

Assuming an equilibrium situation, the Henderson–Hasselbalch equation:

$$\mathrm{pH} = \mathrm{p}K_a + \log \frac{(\text{base})}{(\text{acid})}$$

may be re-written:

$$\frac{(\text{base})}{(\text{acid})} = \text{antilog}\ (\mathrm{pH} - \mathrm{p}K_a)$$

We can now calculate the relative proportions of the unionized drugs across both the gastric and intestinal walls provided that the $\mathrm{p}K_a$ is known. Let us consider salicylic acid (a weak acid with a $\mathrm{p}K_a$ of 3.0) and quinine (a weak base with a $\mathrm{p}K_a$ of 8.4). Substituting into the re-written Henderson–Hasselbalch equation for the two sites we obtain the following results:

(i) Salicylic acid (acid moiety nonionized)

(a) Stomach

$$\frac{(\text{base})}{(\text{acid})} = \text{antilog}\ (1.0 - 3.0)$$

$$= \text{antilog} - 2.0$$

$$= c.\ 0.01$$

(b) Intestine

$$\frac{(\text{base})}{(\text{acid})} = \text{antilog}\ (6.9 - 3.0)$$

$$= \text{antilog}\ 3.9$$

$$= c.\ 8000$$

(c) Plasma

$$\frac{(\text{base})}{(\text{acid})} = \text{antilog}\ (7.4 - 3.0)$$

$$= \text{antilog}\ 4.4$$

$$= c.\ 25\ 000$$

(ii) Quinine (base moiety nonionized)

(a) Stomach

$$\frac{(\text{acid})}{(\text{base})} = \frac{1}{\text{antilog}\ (1.0 - 8.4)}$$

$$= \frac{1}{\text{antilog} - 7\text{–}4}$$

$$= \frac{1}{0.000\ 000\ 25}$$

$$= c.\ 4\ 000\ 000$$

(b) Intestine

$$\frac{(\text{acid})}{(\text{base})} = \frac{1}{\text{antilog}\ (6.9 - 8.4)}$$

$$= \frac{1}{\text{antilog} - 1.5}$$

$$= \frac{1}{0.32}$$

$$= c.\ 3$$

(c) Plasma

$$\frac{(\text{acid})}{(\text{base})} = \frac{1}{\text{antilog}\ (7.4 - 8.4)}$$

$$= \frac{1}{\text{antilog} - 1.0}$$

$$= \frac{1}{0.1}$$

$$= c.\ 10$$

These calculations indicate that the relative concentrations of drugs in plasma and gastro-intestinal tract for these two examples are as follows:

(i) Salicylic acid (relative total drug concentrations)

Plasma 25 000 + 1 = 25 001 (unionized + ionized)

Stomach 0.01 + 1 = 1.01

Intestine 8000 + 1 = 8001

Ratio:

$$\text{Plasma to stomach} = \frac{25\ 001}{1.01} = 24\ 800$$

$$\text{Plasma to intestine} = \frac{25\ 001}{8001} = 3.1$$

(ii) Quinine (relative total drug concentrations)

Plasma 10 + 1 = 11

Stomach 4 000 000 + 1 = 4 000 001

Intestine 3 + 1 = 4

Ratio:

$$\text{Plasma to stomach} = \frac{11}{4\ 000\ 001} = c.\ 4 \times 10^{-5}$$

$$\text{Plasma to intestine} = \frac{11}{4} = 2.7$$

Hence, for salicylic acid the equilibrium strongly favours plasma over the stomach, implying *good* absorption, but only moderately favours plasma over intestines, indicating *poor* absorption. For quinine, the situation is reversed. The equilibrium between plasma and stomach strongly favours the latter, so that not only would no quinine pass into the blood from the stomach, but quinine might *enter the stomach from the plasma.* Only in the intestine does the equilibrium favour absorption of quinine into the blood, and at the normal pH of the small intestine, only poor absorption would be expected. This is borne out in practice, where less than half of an oral dose of quinine is absorbed.

From the previous discussion it might be thought that the degree of absorption from, or secretion into, the stomach and intestines can be predicted simply from a knowledge of whether an orally administered drug is an acid or an alkali, together with its pK_a value.

That this is not the only determinant of absorption is shown by the values given in Table 2.7. In considering these data, a number of experimental problems must be borne in mind. First, experiments of this type are difficult to perform and can, in any case, be done only with experimental animals. Hence, the gastric absorption values are for rats, while gastric secretion was measured in dogs. For absorption studies, compounds were

Table 2.7. Absorption and secretion of drugs by stomach and intestines

Type of drug	Name	pK_a	Per cent absorbed: Stomach	Per cent absorbed: Intestine	Gastric secretion—gastric to plasma concn. ratio
Acids	Salicylic acid	3.0	61	30	0
	Acetylsalicylic acid	3.5	35	5	0
	Benzoic acid	4.2	55	35	0
	Thiopental	7.6	46		0.1
	p-Hydroxypropiophenone	7.8	55	67	0.1
	Barbital	7.8	4		0.6
	Secobarbital	7.9	30		
Bases	Acetanilid	0.3	36		1.0
	Caffeine	0.8	24		
	Antipyrine	1.4	14		4.1
	Aniline	4.6	6	58	42
	Aminopyrine	5.0	2	48	42
	p-Toluidine	5.3	0	65	
	Quinine	8.4	0	41	40
	Levorphanol	9.2	0		42
	Tolazoline	10.3	7		10

added by intubation to ligated stomachs or intestines and the residues determined after 1 hour, but for gastric secretion, compounds were given *i.v.* and stomach contents sampled by an indwelling catheter after 30 to 60 minutes.

It would seem from these results that our previous conclusions, that weak acids are significantly absorbed from the stomach but weak bases are not, are generally correct. Similarly, it is clear that significant gastric secretion of bases occurs, while that of acid drugs is negligible. There are, however, some apparent anomalies. Gastric absorption of barbital is very small, even though it has the same pK_a as *p*-hydroxypropiophenone; also gastric secretion of tolazoline is low, despite its very high pK_a, while it shows a small, but significant, gastric absorption.

One factor which explains some of these problems is the extent of binding of a drug to plasma proteins. In our calculations based on the Henderson–Hasselbalch equation it was assumed that *all* the unionized drug in the plasma was in equilibrium with the unionized drug in the gastric or intestinal lumen. In reality, only a part of the drug in blood plasma can enter into this simple equilibrium; drug molecules combined with proteins cannot. The situation is written in Figure 2.2.

Drug in lumen of gut		Drug in blood plasma
$H^+ + A^- = HA$	=	HA (free) = HA (protein-bound) ‖ $H^+ + A^-$

Figure 2.2. Drug absorption from gut to blood plasma.

The extent to which drugs are bound to plasma proteins is very variable and ranges from 0 to over 90%. Table 2.8 lists the degree of protein-binding for several drugs present in plasma at therapeutic concentrations.

Thus, for barbital, which is not bound to protein at all, the Henderson–Hasselbalch equation can be used to estimate gastric absorption without modification:

$$\text{Barbital in (a) Stomach } \frac{(\text{base})}{(\text{acid})} = \text{antilog}\,(1.0 - 7.8)$$

$$= \text{antilog} - 6.8$$

$$= c.\ 6.3 \times 10^{-6}$$

$$\text{(b) Plasma} \quad = \text{antilog } (7.4 - 7.8)$$
$$= \text{antilog} - 0.4$$
$$= c.\ 2.5$$

hence relative ratio of drug concentration for

$$\text{plasma to stomach} = \frac{2.5 + 1}{6.3 \times 10^{-6} + 1}$$
$$= \frac{3.5}{1}$$

Table 2.8. Binding of drugs to plasma proteins

Type of drug	Name	Per cent bound to plasma proteins
Acids	Barbital	0
	p-Hydroxypropiophenone	75
	Phenylbutazone	90
	Probenecid	75
	Salicylic Acid	75
	Thiopental	75
Bases	Acetanilide	0
	Aminopyrine	15
	Aniline	25
	Antipyrine	0
	Levorphanol	50
	Quinine	75
	Theophylline	15
	Tolazoline	23

p-Hydroxypropiophenone has the same pK_a, but 75% of the drug in plasma is protein-bound.

Hence, while the same calculations apply, i.e.

$$\text{(a) Stomach } \frac{\text{(base)}}{\text{(acid)}} = c.\ 6.3 \times 10^{-6}$$

$$\text{(b) Plasma} \quad = c.\ 2.5$$

the latter value represents the amount of unionized drug in equilibrium with unionized drug in the stomach, but as 75% of the drug in plasma is protein-bound, this represents only 25% of the *total* plasma unionized drug, so that the true total is *c.* 10.0.

Using this value to calculate the relative ratio of drug concentration for plasma to stomach we obtain:

$$\frac{10.0 + 1}{6.3 \times 10^{-6} + 1} = \frac{11.0}{1}$$

Taking into account the binding of *p*-hydroxypropiophenone to plasma proteins shows that the plasma to stomach relative-concentration ratio is over 300% that of barbital, which has the same pK_a but is not protein-bound in plasma. This appears to be a fairly general effect, so that drugs showing a high degree of plasma protein-binding are absorbed more readily into the blood from the stomach if they are weak acids. For bases the exact opposite is the case, so that the lower the percentage protein-binding in plasma, the lower the chance of secretion into the stomach.

Other factors influencing drug uptake from the intestines include the relative lipid–water partition coefficient and drug stability to gut pH and enzymes. The first of these two factors is rather complex, for a drug with high lipid solubility also can be absorbed into the intestinal lymphatics, particularly if ingested with a fatty meal, although the *rate* of absorption may be low. High lipid solubility may also lead to association with plasma lipoproteins. Stability within the gastro-intestinal tract is clearly of importance. Many drugs cannot survive the acid conditions of the stomach. Penicillin G and methicillin are both hydrolyzed by acid, though some of the former (about 25%) may escape. In contrast, ampicillin is acid stable. Further hydrolysis of penicillins can occur in the large intestine due to the secretion of penicillinases by micro-organisms. Enteric-coated products have been designed to protect the acid-labile drug during its passage through the stomach, but which are decomposed in the small intestine.

(ii) Intravenous

We will discuss the pharmacodynamics of drug administration later in this chapter, but a few calculations are offered here on some implications of intravenous injections.

If the required therapeutic dose of a drug for an average 70 kg man is 1.0 g, and this is administered intravenously, the speed at which the injection is being given can be very important. If the total body water is about 40 l, the 1.0 g of drug will (in many cases) distribute evenly through this volume to give the desired therapeutic concentration (in this case about 0.25 mg/ml) in a relatively short space of time. However, if the drug is rapidly injected into the vein in, say, a second, the entire dose passes with the venous blood to the heart, through which it will require several passages to equilibrate with the blood, before being distributed in the total body water. In 1 second the drug cannot mix with more than the venous return to the heart, which is about 100 ml/sec, so that 1.0 g drug is presented to the heart in about 100 ml blood, i.e. a concentration of 10 mg/ml, which is about 400 times the optimum therapeutic concentration. If such a high concentration of drug were generally present in the body, the outcome would probably be fatal. As it is, the high concentration passes through some of the most important body organs (heart, lungs, aorta) before being diluted.

The rapid entry of drugs into venous blood can occasionally lead to undesirable effects. If the osmolarity of the drug solution differs from that of blood, haemolysis or agglutination of erythrocytes may occur within the vein. Embolism can result if the drug precipitates out of solution in the vein, or if a drug containing particulate matter (intended for sub-cutaneous or intra-muscular administration) is given accidentally by the intravenous route. Drugs may also damage the venous walls, particularly if given over a considerable length of time by i.v. infusion.

Finally, some individuals develop allergic and anaphylactic reactions to certain drugs. Occasionally, these are life-threatening and the situation is most critical should they develop after a rapid *i.v.* injection. Single *i.v.* doses must always be given *slowly*; ideally over a period of time approximately equivalent to a complete circulation of the blood (i.e. about one minute). Intravenous infusions must always be given with equal care so that *both* the dose of the drug *and* the volume of fluid administered per minute do not exceed the body's capacities. Excessive fluid infusion can lead to increased blood pressure and cardiac failure, particularly if renal function is impaired. A safe *maximum* fluid infusion rate for adults is 2 ml/min, but the nature of any drug present in the infusion solution must be taken into account in deciding the optimum rate. If high-molecular weight substances are present, the maximum rate should rarely exceed 0.5 ml/min.

Because of all the attendant dangers, the intravenous route of administration for a drug should be selected only after all alternatives have been considered. It should then be carried out slowly and with great care. The most frequent reason for selecting this route is emergency. Advances in pharmaceutical formulation and synthesis of analogues have now provided alternatives to most drugs that once had to be given intravenously.

(iii) Topical

A very large number of drugs are formulated as ointments (fatty bases), creams (water base), lotions, sprays, plasters, poultices, or tinctures for application to the skin. Among the more important classes of dermatological drugs are antipruritics and sedatives, antiseptics and antibiotics, fungicides, corticosteroids, and anaesthetics. These are all intended to act locally (i.e. their receptors are in the skin cells) and not to be absorbed into the circulation. In fact, many dermatological drugs can be shown to enter the blood, although the extent to which they do so varies considerably with a number of factors. If they are applied to broken skin, or to skin with a structure altered by disease (increased vascularity, decreased thickness, etc.), systemic absorption is more likely. Secondly, the rate of application can be important, for excessive use can saturate the skin receptors.

However, many dermatological products, used at the correct dose on healthy skin, have been shown to enter the blood. Topical corticosteroids can be shown to give some degree of adrenal suppression, while sensitive persons may develop a generalized allergic response to any topical drug.

Histological examination of normal skin reveals below the outer keratinized tissue a clear, dense layer (called the "barrier layer" or "barrier area") overlying the epithelial cells of the lower epidermis. Provided that the barrier layer is intact, the skin is penetrated by compounds at a rate determined largely by their lipid–water partition coefficient. Water-soluble substances tend to be excluded and even highly lipid-soluble drugs penetrate at a rate slower than for other body membranes.

2. Transport and Distribution

(i) In Blood

Body water consists of a number of inter-related, yet separate, compartments. The volumes of these compartments can be estimated by administering compounds which are excluded from some types of body water, yet freely enter others. For example, antipyrine appears to mix freely and equally with all body water, in a manner similar to isotopically-labelled water. On the other hand, compounds such as inulin or mannitol freely permeate all forms of extra-cellular water, but cannot enter cells. Finally, dyes such as Evan's Blue, after i.v. injection cannot leave the vascular system to other compartments and so may be used to estimate the volume of blood plasma.

Drugs used to treat human diseases behave in many different ways. Some distribute throughout all body water, others cannot enter cells, while others cannot leave the blood. A knowledge of the distribution of a particular drug within the body is clearly very important, for, depending on the route of administration, it may not be able to reach its receptors. For example, several anti-microbial drugs (e.g. sulphaguanidine, some penicillins, etc.) penetrate the CSF only with great difficulty and so are not the best drugs to use for CNS infections.

In the discussion in the previous section on drug absorption from the gastro-intestinal tract it was pointed out that many drugs are bound by serum proteins and that the degree of protein-binding determines the concentration of active free drug in the plasma. Elsewhere (p. 46) we gave the example of warfarin, (an anti-coagulant) which is normally 98% bound to serum proteins, but can be displaced by various other acidic drugs, so that the amount of active warfarin in the plasma is increased. Binding to serum proteins is also important for drug distribution, for only the drug in free solution is able to cross from blood plasma into other compartments of body water. If a drug exists in blood largely as the protein-bound form, the amount available for equilibration with extravascular water is small, so that, even if the drug is capable of crossing the dividing membranes, little will actually do so.

The proteins present in blood plasma are a complex mixture that can be separated into several components by various methods. Here we will discuss drug binding by the various electrophoretic fractions of plasma proteins.

Serum protein fractions, which are mainly formed and released by the liver, contain a number of chemically reactive components essential for the transport and distribution of essential body compounds. Some examples are given in Table 2.9.

Table 2.9. Binding of metabolites to plasma proteins

Electrophoretic fraction	Substances bound
Albumin	Bilirubin Uric acid Ascorbic acid Fatty acids Histamine Thyroid hormones Steroid hormones Calcium ions Copper ions (some) Zinc ions (some)
α-1 globulins	Steroid hormones Cobalamins Sialic acids Thyroid hormones
α-2 globulins	Copper ions Haeme Fatty acids Cholesterol Vitamins A, D, E, and K
β-globulins	Iron ions Fatty acids Cholesterol Vitamins A, D, E and K
γ-globulins	Antigens

In some cases, the binding of these substances to a particular protein is highly specific; for example, ceruloplasmin (an α-2 globulin) reacts strongly and specifically with copper ions, while transferrin (a β-globulin) reacts similarly with ferrous ions. On the other hand, all types of serum protein are capable of numerous non-specific interactions with small molecules. To take the albumins as a typical example, the average molecular weight is 69 000 and at the isolectric point (pH 5) each molecule has about 100 negatively charged groups (mainly $-COO^-$ from aspartate and glutamate residues, but also a few $-O^-$ and $-S^-$ from tyrosine and cysteine respectively), together with about the same number of positively charged groups (mainly $-NH_3^+$ from lysine and *N*-terminal residues, but also $-N^+-$ from histidine and $=NH_2^+$ from arginine). At pH 7.4 in blood plasma some of the positive charges are lost, so that the albumins carry a net negative charge, though the molecules still possess both negative and positive groupings. Each of these groups is capable of weakly binding an ion of the opposite charge.

Any substance ionized at the pH of blood plasma, be it a normal body metabolite or a drug, is able to associate loosely with the various serum proteins by means of coulombic interactions. This mechanism clearly applies only to ionized drugs and is, moreover, very easily reversed. In most cases this type of protein-binding of drugs is of little physiological significance and can be disregarded. A far more important type involves a *specific* interaction between the drug molecule and a small number of sites on a particular plasma protein. These interactions are less easily reversed than the simple ionic associations, but nevertheless an equilibrium exists between the drug in solution and its protein-bound form. As the interaction is specific, a particular drug is usually bound by only one fraction of serum proteins.

The nature of specific protein-binding of many drugs has not been investigated in detail, but a good deal is known of the general principles by a study of dye-binding by serum proteins, which appears to be a completely analogous process. One of the best investigated dyes is methyl orange (see Figure 2.3). This azo-dye will bind reversibly with serum albumin, much the same as many drugs, e.g. tetracyclines, chloramphenicol, penicillins, sulphonamides, salicylates, barbiturates, etc. The interaction between methyl orange and albumin occurs in aqueous solution at a wide range of pH and at most is accompanied by a shift in the absorption spectrum of the dye. A similar shift can be seen by comparing the absorption spectrum of the dye in water and in non-aqueous solvents and is caused by suppression of ionization of the sulphonate group. It seems likely, therefore, that it is this group which interacts primarily with the albumin molecule.

Further information about the binding has come from a study of various structural analogues of methyl orange. Replacing the sulphonate group by carboxylic acid (methyl red) had little effect on protein-binding, but replacement by phosphonate or arsenate reduced the extent of binding. Replacement by a non-ionic group prevented binding completely. In methyl orange the sulphonate group is *para* to the azo linkage. Synthesis of the compound with sulphonate in the *meta* position showed that this was also bound by albumin, but the *ortho* compound was not. However, if the sulphonate was left *para* and a methyl introduced *ortho* to the azo link, this compound interacted strongly with albumin.

These studies show that the sulphonate is concerned in the reaction with albumin, but that some other part of methyl orange is also involved as the *ortho*-sulphonate analogue will not bind. Evidence that this other part of the molecule is the substituted amine group came from studies of the binding of diethyl, dipropyl and dibutyl analogues of methyl orange. Binding to albumin decreased progressively with increase in size of the alkyl substituents on the amine group.

Studies of molecular models suggested that separation of the amine-N from the sulphonate-S is about 12 Å in both methyl orange and its *meta*-sulphonate analogue, but is significantly less in the *ortho*-sulphonate analogue. The most likely explanation of these facts is that one molecule of methyl orange is linked to *two* sites on the albumin molecule, through sulphonate *and* substituted amine groups; and that the two sites on the protein are about 12 Å apart. The linkage of sulphonate is presumably to a strongly cationic group of the protein, such as the epsilon-NH_2^+ of a lysine residue, while the amine-N is probably hydrogen-bonded to another suitable grouping. Most likely to form a hydrogen bond with a substituted amine is a phenolic —OH from tyrosine and, to test this possibility, studies were made on the binding of methyl orange to albumin in which an average of 13 out of total 18 tryosyl residues had been iodinated. Iodination of tyrosine reduces the pK_a of the phenolic hydroxyl from 10.1 to 8.7 for a mono-iodo substitution and to 6.5 for a di-iodo substitution; so that the hydroxyl is more likely to be ionized in the iodinated deriva-

	Analogue	Binding to serum albumin
1.	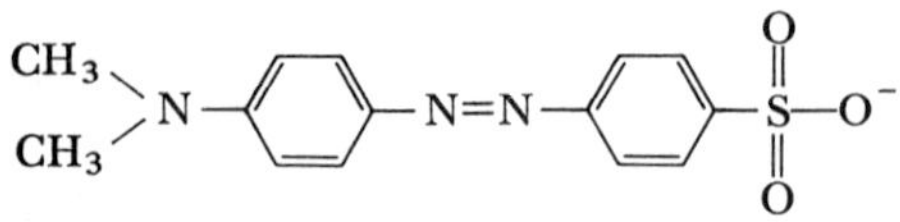 Methyl orange	Strong
2.	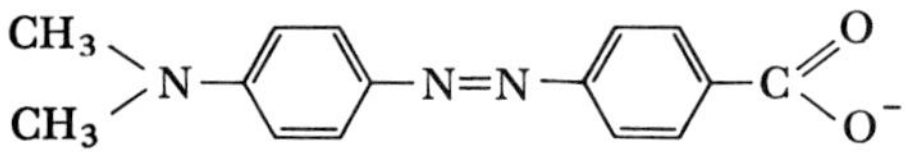 Methyl red	Strong
3.	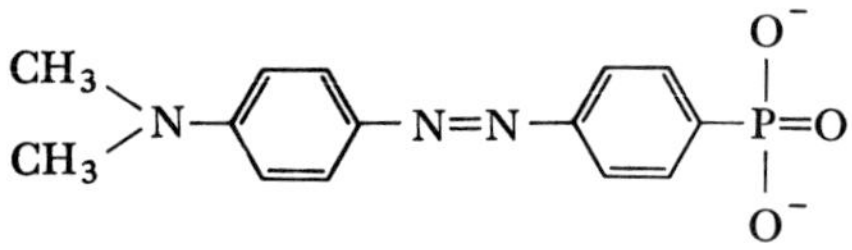 Phosphonate analogue	Moderate
4.	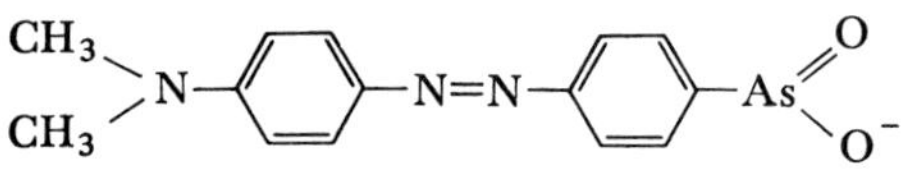 Arsenate analogue	Moderate
5.	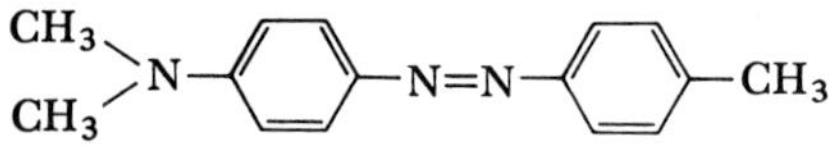 Non-ionic analogue	Nil
6.	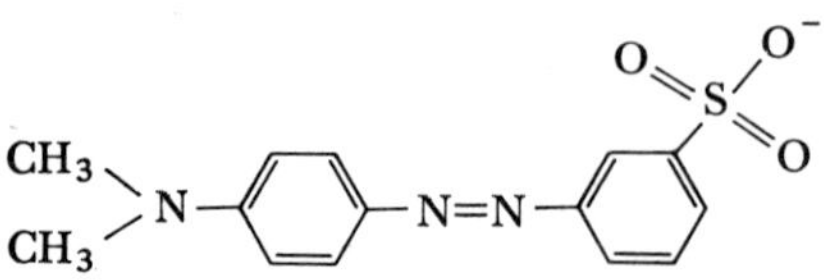 *Meta*-sulphonate analogue	Strong

	Binding to serum albumin
7. 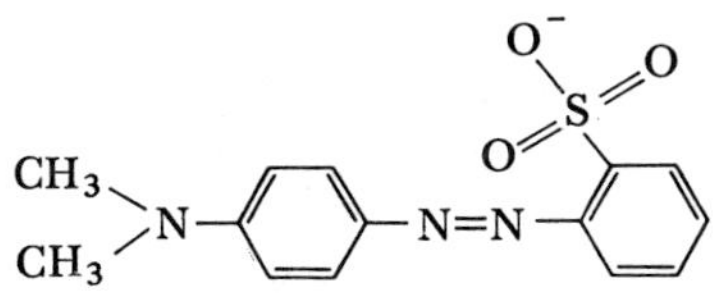 *Ortho*-sulphonate analogue	Nil
8. 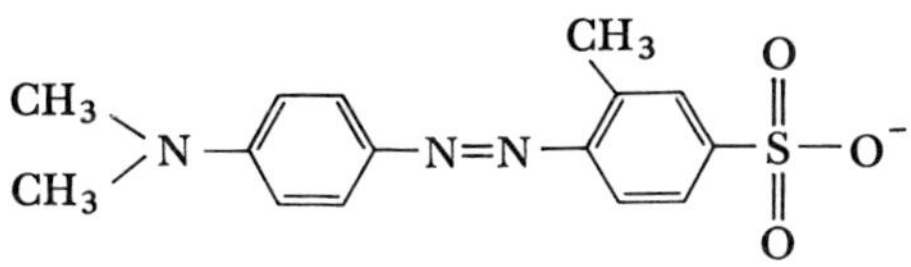 *Ortho*-methyl analogue	Strong
9. 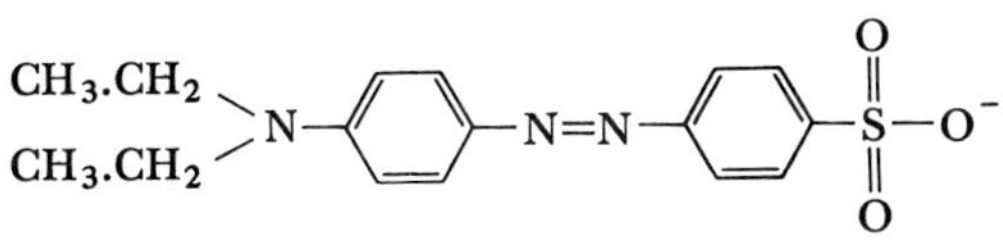 Diethyl-analogue	Moderate
10. 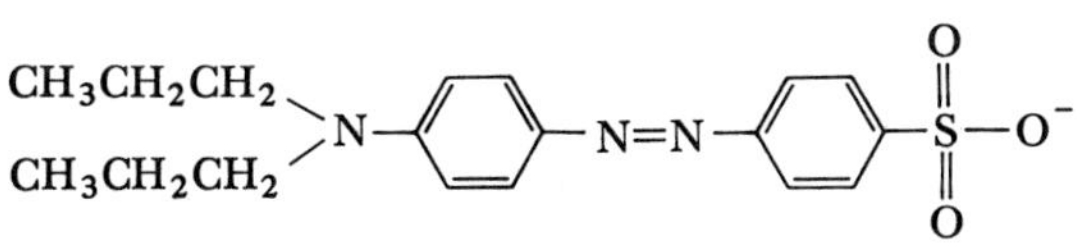 Dipropyl-analogue	Weak
11. 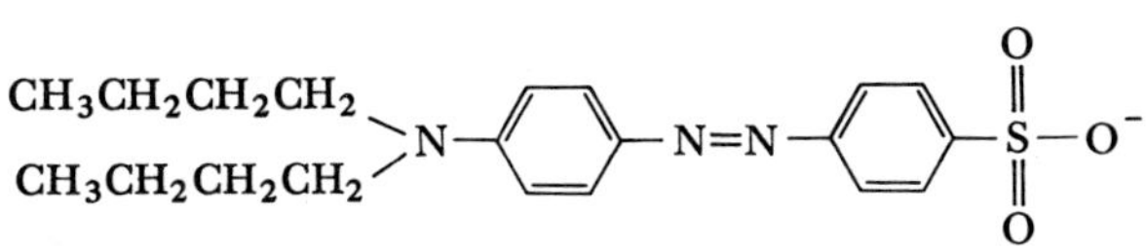 Dibutyl-analogue	Very weak

Figure 2.3. Protein binding of methyl orange analogues.

tives and less able to donate hydrogen for a hydrogen bond. Binding of methyl orange to iodinated-albumin was very much reduced.

The final picture of the binding site on the protein molecule is of a tyrosyl residue with an unionized hydroxyl about 12 Å from a strongly cationic $-NH_3^+$. Each serum albumin molecule possesses 18 tyrosyl residues, but few of these can be in the correct special relationship to a cationic group. Moreover, the two groups of the protein involved in the binding-site must both be in the correct ionized form before binding can occur. Changes in pH would be likely to alter the chances of finding a binding site on any particular albumin molecule. Studies *in vitro* with human serum albumin have shown that only one molecule of methyl orange binds to each protein molecule at the pH of blood plasma.

We have discussed protein-binding in some detail to illustrate the chemistry of drug-binding by serum proteins. The following *general* principles apply to most cases:

(i) Loose linkage of ionized drugs to serum proteins occurs non-specifically and is of little importance.
(ii) *Specific* linkage of particular drugs to individual serum proteins occurs in many cases.
(iii) Each specific linkage involves chemically precise interactions between the molecule of the drug and a *small* number of sites on each protein molecule.

In Table 2.8 we listed the percentage binding by serum proteins of a range of drugs. In fact, these were merely typical values at the usual therapeutic concentration of the drug in blood plasma, for the percentage of a drug bound to proteins varies with the concentration of the drug. This important conclusion follows from a simple mathematical treatment of the formation and dissociation of the drug–protein complex: Let D = concentration of free (unbound) drug in plasma

(PD) = concentration of drug–protein complex
K = dissociation constant of drug–protein complex
P = concentration of binding-protein
n = number of drug-binding sites per protein molecule
r = moles of drug bound per mole of protein (= PD/P)

Then, at equilibrium the rate of protein-binding of the drug equals the rate of dissociation of the drug–protein complex

$$\frac{((P \times n) - (PD)) \times D}{PD} = K$$

or,

$$\frac{PD}{(P \times n)} = \frac{D}{(K + D)}$$

so that,

$$r = \frac{(n \times D)}{(K + D)}$$

It follows that a plot of r against D will be a hyperbolic curve so that the moles of drug bound per mole of protein is dependent upon the concentration of *unbound* drug in the plasma. Now the *percentage* of drug present in the protein-bound form (%B) is given by:

$$B = \frac{PD}{(PD + D)} \times 100$$

$$= \frac{1}{1 + D/PD} \times 100 \qquad *$$

But we have show above that:

$$\frac{PD}{(P \times n)} = \frac{D}{(K + D)}$$

so that:

$$\frac{D}{PD} = \frac{(K + D)}{(P \times n)}$$

which can now be substituted in the previous equation* to give:

$$B = \frac{1}{1 + \dfrac{K}{(P \times n)} + \dfrac{D}{(P \times n)}} \times 100$$

In a clinical situation the concentration of plasma protein is constant, while both n and K will be dependent upon the particular drug under consideration. It follows then that both the term:

$$\frac{K}{(P \times n)}$$

and the term:

$$(P \times n)$$

are constants. Hence the only variable determining the value of (%B) is (D).

At *low* drug concentrations in plasma, if the binding affinity of the protein for the drug is high (i.e. K is low), most of the drug will be bound (%B approaching 100), but as more drug enters the plasma the available binding-sites on the serum proteins will become progressively saturated until a situation is reached where no further drug can be bound. If still more drug enters the plasma after this stage, it can only add to the pool of *unbound* drug, so that at very high drug concentrations, most will be unbound (%B reducing to a low value).

The implication of this reasoning is that a patient taking a single oral dose of a drug will, in the early stages, show a high percentage of protein-binding in the plasma as the drug begins to enter. The percentage binding will then *decrease* and reach a minimum at the highest plasma concentration of the drug, but will then begin to *increase* again as the drug is eliminated from the body. It follows, therefore, that values for the percentage protein-binding of a drug are without meaning unless the plasma concentration of the drug is specified.

It should also be noted that for drugs which bind strongly to serum proteins, it may be necessary to give very large doses to saturate completely all available binding-sites before the concentration of *unbound* drug in plasma reaches a sufficiently high level to allow the transfer of significant amounts to other compartments of body water. It should be noted as well, that only the *free* drug can be cleared by glomerular filtration in the kidneys, although some protein-bound drugs, such as penicillin, are readily cleared by tubular secretion.

The marked affinity of some drugs for plasma proteins is well illustrated by suramin, a prophylactic drug against trypanosomiasis, which persists in the blood for several weeks after a single dose, and by 3-hydroxy-2,4,6-triiodo-α-ethylhydrocinnamic acid which was once given to visualize the gall-bladder, but later abandoned when it was discovered to persist in the blood, bound to serum proteins for many years after a single dose.

Marked individual susceptibilities to different drugs have been occasionally traced to inherited differences in particular serum proteins which bind the drug. A good example is digitoxin where the clinically effective dose varies by a factor of over two, depending on the extent of its binding to an individual's serum proteins.

We have previously mentioned competition between two drugs for the same binding-sites on serum proteins, but drugs can also compete with the natural metabolites usually carried by serum proteins. One of the best known examples is the competition between sulphonamides and bilirubin, both of which bind to the same sites on albumins. This competition is of clinical importance in the treatment of bacterial infections in premature infants, who lack the ability to form bilirubin glucuronate and have high levels of serum bilirubin until the hepatic enzyme catalysing this conjugation has developed. Albumin has a high affinity for free bilirubin so that, even if the serum content of bilirubin is high, it is present almost entirely in combination with albumin. Sulphonamides displace bilirubin from albumin and preferentially bind to the albumin sites, so that the concentration of free bilirubin increases in serum, as well as in other body water compartments, such as CSF, by a process of equilibration. This leads to kernicterus, which can be either fatal or leave permanent damage to the nervous system.

(ii) Passage Across Membranes

Most drugs, particularly when administered orally or by *i.v.* injection, enter the blood, where many become bound to serum proteins. Only the unbound drug is able to pass into other parts of the body. To leave the blood plasma a drug has to cross membranes; in most cases these are the walls of capillary vessels.

Two main factors appear to determine drug transport across capillary membranes: lipid solubility and molecular size. Drugs with high solubility in lipids tend to leave capillaries at rates proportional to their lipid–water *partition coefficient*. For water-soluble drugs the major factor is molecular size; the lower the molecular weight, the faster the capillary transfer. However, water-soluble macro-molecules with M.W. in excess of about 60 000 (corresponding to a sphere of radius about 30 Å) cannot leave capillaries at all. The reason for this is not known, for electron microscopy has failed to reveal any pores in capillary walls.

The passage of a drug out of blood plasma will therefore be determined largely by its chemical properties and the *ratio* of free to protein-bound forms. While the free drug is by far the most important contributor to the extra-vascular drug pool, occasionally, it is possible for some protein-bound drug also to escape. If the binding-protein has a molecular weight less than about 60 000 it may pass through capillary walls and carry the drug with it. However, most serum proteins have molecular weights well in excess of this limit, while those few proteins which can pass, do so at very low rates. For example, a protein of M.W. about 40 000 passes through capillary walls several thousand times slower than a relatively small molecule, such as sucrose.

Table 2.10. Drug active transport

Organ	Process	Drugs transported
Kidneys	Tubular secretion	Penicillins, phenol red, tetraethylammonium salts
Liver	Bile secretion	Penicillin, bromosulphophthalein (BSP)
Choroid plexus	Transport from CSF to blood	Penicillin, phenol red, iodides, thiocyanates, various organic acids

The ease with which different unbound drugs can pass through capillaries is clearly very variable, but it is true to say that significant transfer can occur of *any* drug molecules in plasma which are not bound to proteins. The major exceptions to this general view are transfers from plasma into the brain; the capillary vessels of the CNS possess a cellular sheath, not seen on other capillaries, which greatly reduces permeability to all water-soluble compounds.

Cell membranes possess active transfer mechanisms for taking up and excreting different intermediary metabolites. It might be thought that, because most drugs are foreign compounds, no active transport mechanisms could exist for them within the body. While this is generally true, there are a number of important exceptions. Some are listed in Table 2.10. For each of these examples, the transport mechanism shows a high degree of steric specificity and structural analogues of compounds that may be transported will *competitively* inhibit the active transport mechanisms. *Non-competitive* inhibition results from oxygen deprivation or exposure to metabolic poisons, such as cyanides.

For the majority of drugs, however, active transport mechanisms are not found, so that penetration of cell membranes is by the process of simple diffusion. For *non-electrolyte* drugs, the diffusion rates for influx and efflux are identical, but the degree of "resistance" to diffusion offered by a cell membrane varies for different substances. As with the passage

pH	Relative proflavin concentration in cells after 4 min
8.5	5.6
7.9	1.4
7.4	0.4
7.0	0.1

Figure 2.4. Effect of pH on proflavin uptake by cells.

through capillary walls, lipid-soluble compounds penetrate membranes approximately in proportion to their lipid–water partition coefficient, through compounds of M.W. below about 5000 (corresponding to a sphere of radius 15 Å) penetrate faster than would be expected from their partition coefficient, while compounds of very high M.W. penetrate slower than expected.

For drugs that are weak acids or bases, cell membrane diffusion depends on the relative proportions of ionized and non-ionized forms; which in turn depends on pH. These effects are due, in part, to the large differences in lipid–water partition coefficients between the two forms of the drug. Ionized forms interact very weakly with the lipids of the cell membranes and so are absorbed poorly, or not at all. In most cases the diffusion of the drug is entirely that of the unionized form, so that the amount of ionized drug is irrelevant. This effect is demonstrated by the entry of proflavin at a fixed concentration into human cells in tissue culture (Figure 2.4).

(iii) CNS

Drugs can gain access to the tissues of the brain and spinal cord either from capillary circulation or from the CSF. Using radioactive drugs it has been shown that the penetration into the cerebral cortex is much greater than into the white matter and this corresponds to the relative blood supplies. As mentioned above, capillaries of the CNS have a unique cellular sheath that makes them much less permeable to water-soluble compounds than capillaries elsewhere in the body. This difference gave rise to the concept of the "blood-brain barrier" and it was once assumed that this was an absolute physiological phenomenon. It is now realized, however, that the permeability difference between the brain and other body tissues is a quantitative, rather than a qualitative one.

The second route of access for drugs to reach the brain and spinal cord is through the CSF, a fluid formed at the choroid plexus at the rate of about 0.3 ml/min in adults and with a total volume of about 200 ml. The CSF bathes the surfaces of the CNS then is removed by venous blood sinuses in the arachnoid villi. Some drugs can pass from blood plasma into the CSF, but, as we mentioned before (p. 126), some drugs are actively transported by the choroid plexus *out* of CSF into plasma. For drugs of this type, this is a way of eliminating drugs from the CNS, which may have entered either by direct administration into the CSF or via the capillary circulation.

It will be seen that the entry and removal of drugs from the CNS is a complex problem. For drugs coming via the blood plasma, usually only unionized unbound molecules can enter. The CSF contains only a trace of protein, so virtually no protein-binding of drugs occurs here, but the brain cells appear to contain protein receptors which can bind considerable amounts of particular drugs. The importance of ionization was well shown in studies with drugs of the sulphonamide series, where it was discovered that significant brain penetration occurred only for compounds with a pK_a greater than 7.4 (and so would be largely unionized at the pH of blood plasma). The final factor determining brain entry is lipid solubility. Compounds with a low lipid–water partition coefficient can enter only very slowly, even though they may occur in plasma as unionized unbound forms.

Table 2.11. Penetration of some drugs into the brain

Factor	Drug		Brain penetration (half-time, min)
Ionization		Per cent unionized at pH 7.4	
	Antipyrine	100	5.8
	Aniline	99	1.7
	Aminopyrine	99	2.8
	Sulphaguanidine	99	231
	Pentobarbital	83	4.0
	Thiopental	61	1.4
	Barbital	56	27
	Mecamylamine	2	32
	Salicylic acid	0.4	115
Protein-binding		Per cent bound to plasma proteins	
	Thiopental	75	1.4
	Pentobarbital	40	4.0
	Salicylic acid	40	115
	Aminopyrine	20	2.8
	Mecamylamine	20	32
	Aniline	15	1.7
	Antipyrine	8	5.8
	Sulphaguanidine	6	231
	Barbital	0	27
Lipid-solubility (of unionized drug)		Lipid–water partition coefficient	
	Mecamylamine	*c.* 400	32
	Thiopental	3.3	1.4
	Aniline	1.1	1.7
	Aminopyrine	0.2	2.8
	Salicylic acid	0.12	115
	Pentobarbital	0.05	4.0
	Antipyrine	0.005	5.8
	Barbital	0.002	27
	Sulphaguanidine	0.001	231
Ionization and lipid solubility		(Partition coefficient) times (per cent unionized)	
	Thiopental	200	1.4
	Aniline	110	1.7
	Aminopyrine	21	2.8
	Pentobarbital	4.2	4.0
	Antipyrine	0.5	5.8
	Mecamylamine	0.48	32
	Barbital	0.11	27
	Sulphaguanidine	0.01	231
	Salicylic acid	0.0048	115

In animal investigations it is possible to measure the half-time of brain penetration by a drug, and values range from less than one minute to several hours. It is of interest to compare these values for a series of drugs with the degree of ionization, protein-binding and lipid solubility. This is shown in Table 2.11.

It will be seen that the relative ease of penetration of these nine drugs cannot be predicted from a knowledge of only one of the three factors, but a combination of ionization and lipid-solubility gives a very reasonable estimate. If protein-binding is also taken into account, the two anomalies in the final part of the table can be resolved, for there is more unbound barbital than mecamylamine. Only the two drugs with the greatest half-time of brain penetration are ranked incorrectly, but this is probably due to the difficulty in assessing any accurate value of the lipid–water partition coefficient of sulphaguanidine, which is so low as to be almost unmeasurable.

The conclusions to be drawn from this table are primarily the importance of considering all factors together and that even though a drug may be largely protein-bound in plasma (such as thiopental) it still may enter the brain rapidly. Conversely, a high lipid-solubility (as with mecamylamine), or a high percentage of unionized molecules in plasma (as for sulphaguanidine), are no guarantees that a drug will enter the CNS with speed.

There are some important clinical implications to be drawn from a knowledge of drug transport into the CNS. In the treatment of meningitis, for example, penicillin is unsuitable for the following reasons. It is completely ionized in plasma, has low lipid-solubility and a high percentage binding to plasma proteins. Moreover, it is one of the few drugs actively transported *out* of CSF at the choroid plexus. In meningeal infections, drug transport into the CSF may be increased, but is an uncertain basis on which to base treatment of a serious condition. Sulphonamides are a more logical choice, but here great care is needed in selecting the best compound. Sulphonamides with a pK_a of less than 7.4 will not enter the CNS, while even some of those with a pK_a greater than 7.4 may not enter due to their poor lipid solubility and/or high protein-binding. For example, the pK_a of sulphaguanidine is about 10, but we have seen from Table 2.11 that it enters the brain very slowly indeed.

Conversely, drugs can be designed to exclude their penetration into the brain. For example, anticholinergic effects may be required peripherally but not centrally. Doses of atropine to produce a therapeutic effect at the periphery are readily taken up by the CNS, but this can be avoided by the use of atropine methyl sulphate, which is highly ionized, protein-bound, and scarcely penetrates the brain at all.

(iv) Foeto-Placental Barrier

In a later section of this chapter (p. 187) we discuss the undesirable side-effects of drugs and one of the most important of these is teratogenicity: the induction of foetal malformations. It is of great importance to understand the factors which determine the transport of drugs from the mother to the foetus.

In humans, fertilization occurs usually in the fallopian tubes. The zygote then begins to divide, until 72 hours later there is a blastocyst of 58 cells. Only 5 of these develop into the foetus, the others becoming the trophoblast. Implantation of the blastocyst into the uterine wall occurs on the 7th day and is followed by increased growth and development of the embryo, which at this stage is bathed by the maternal blood. Drugs may reach the embryo through the ambiotic fluid, but as growth progresses, only the foetal epidermis is in contact with this fluid. As the foetal vascular system develops, the foetal blood is prevented from mixing with the maternal blood by the membranes of the inter-villus spaces of the placenta, so that drugs (along with normal metabolites and nutrients) have to pass from the mother to the foetus across these membranes.

Early in pregnancy the placental membranes are about 25 μ thick, but this gradually decreases to about 2 μ at term, and is accompanied by a gradual increase in permeability. In terms of permeability, the placental membranes have properties very similar to those of capillary membranes elsewhere in the body, but differ by possessing several energy-coupled active transport systems. The latter lead to the existence and maintenance of significantly higher concentrations of various compounds in foetal than maternal blood, e.g. amino acids, inorganic phosphate, iodide, etc.

As may be readily appreciated, the investigation of foeto-placental transport of drugs is particularly difficult to investigate, especially in humans.

There are a number of analogies between drug passage into the foetus and into the brain, for the same three factors appear to be controls (percentage of unionized form, protein-binding, and lipid-solubility). The situation is complicated, however, by the existence of plasma proteins in foetal blood which combine with drugs passing through the placental membranes and so help to maintain a gradient of unbound drug. Where the maternal blood plasma drug concentration is constant, the fastest possible equilibration with the foetus is about 40 minutes, although most drugs require far longer than this, while it is improbable in practice that no change will occur in maternal drug level. Some drugs, such as tetracycline, apparently never reach equilibrium, but this is probably because most assay methods determine only "total" tetracyclines and the maternal blood contains a mixture of conjugates (mainly ionized) and free drug, while the foetal blood contains only the free drug. Normal foetal liver is unable to make glucuronates, due to the absence of an hepatic microsomal glucuronyl transferase (p. 19). Most other detoxication mechanisms are also absent. The failure of conjugates to pass from mother to foetus is also true for the steroids, where only free, unconjugated (unionized) forms may pass.

Equilibration times for some commonly used drugs are given in Table 2.12.

Table 2.12. Passage of drugs across the placenta

Drugs	Minimum equilibration time between foetal and maternal blood
Thiopental	*c.* 40 min
Trichloroethylene	40
Oxygen	40
Nitrous oxide	60 (?)
Penicillin	10 hours
Streptomycin	18
Morphine	Several hours
Phenothiazine tranquillizers	Several hours
Reserpine	Several hours
Tetracyclines	Never (?)
Erythromycin	Never (?)

There are a number of unexplained problems of foeto-placental transfer. Maternal serum albumin molecules pass into foetal blood at a very low rate, but the much larger antibody globulins transfer slowly yet faster than albumin. There is also conclusive evidence of intact erythrocyte passage in both directions, but this could conceivably follow vascular breaks in the placenta.

3. Elimination and Excretion

(i) General

We have already discussed in detail (p. 103) the way in which drugs and other foreign compounds may enter the body. We have also seen that most are converted in some degree to chemically modified metabolites and conjugates, often with enhanced water solubility. Active drug molecules are also removed from the circulation by binding to serum proteins or deposition in adipose tissues, from which they are only slowly released. The general problem of drug elimination is therefore one of dealing with the *free* molecules of drugs and their metabolites in the various compartments of body water.

Drugs acting internally are usually excreted in urine or in bile. For drugs leaving via the latter route, the faeces are the major form of excretion, but we have already mentioned the possibility of reabsorption via an entero-hepatic cycle, and drugs entering such a cycle may actually leave the body as conjugates in the urine, even though these were first formed in the liver and excreted in bile. A discussion has already been given of the more unusual routes whereby foreign compounds and their metabolites can be excreted, e.g. volatile compounds by the lungs. It must also be realized that the molecules of some drugs may be broken down to CO_2 and H_2O, at least partially.

Finally, we must mention the excretion of drugs which have little or no entry into body tissues, such as radiological contrast media for the gastro-intestinal tract, or topically applied skin preparations. The former are excreted unchanged in faeces, while most of the latter are lost in sloughed skin and hair.

(ii) Renal Excretion

An average man filters about 190 l daily (i.e. about 130 ml/min) of plasma water through the combined glomerular membranes of his kidneys. In usual circumstances between 1.0 and 1.5 l is excreted as urine, which is between 0.7 and 1.0 ml/min (assuming a constant rate of formation). Only free drug molecules, not bound to plasma proteins, are filtered and only drugs of very high molecular weight (in excess of about 70 000 can escape glomerular filtration.

Once a drug has been filtered at the glomeruli, its return into the blood with most of the filtered water across the tubular epithelium is based largely on the factors governing passage through biological membranes; except that protein-binding is not of importance here due to the very small amounts of protein in the glomerular filtrate. Only percentage ionization and lipid-solubility are significant. Polar compounds and ions are but little transported at the tubules and so are passed into the bladder. The only exceptions are compounds or ions for which a tubular energy-dependent active transport mechanism exists. Both proximal and distal renal tubules have active transport mechanisms to deal with the reabsorption of NaCl, glucose, amino acids, etc.

Substances not bound to plasma proteins that are filtered at the glomeruli, but not reabsorbed (nor secreted) by the tubules to any significant extent, can be used to measure the glomerular filtration rates (GFR) mentioned in opening paragraph of this section. Such an ideal substance would give the maximum values of about 130 ml/min and inulin or creatine approach this value. The GFR can therefore be regarded as the volume of plasma water cleared of all this ideal substance in one minute. At the other extreme is a compound such as glucose which fails to appear in the urine at all, despite being entirely filtered by the glomeruli. Here the concept of *renal clearance* is required, which is defined as the volume of plasma water cleared of a substance per minute by the entire mass of both kidneys, so that both glomeruli and tubules are taken into account.

The renal clearance (or, more simply, *clearance*) of a substance like glucose would be 0 ml/min while either inulin or creatinine would have a clearance of about 130 ml/min (= GFR), but compounds filtered at the glomeruli as well as being actively secreted by the tubules would have higher values. The average blood flow to the kidneys is about 1300 ml/min, or approximately 650 ml/min of plasma, so that 650 ml/min is about the maximum clearance that any compound can show.

Several drugs approach this value, including penicillin and *p*-aminohippuric acid. Most drugs excreted by the urinary route have values less than the GFR, while drugs with a clearance approaching zero have to be eliminated by some route other than urine.

If a drug is excreted from the body almost entirely in urine, then it is possible to deduce from its renal clearance the rate of elimination from the body. The only other information required is the volume of distribution (V_d) of the drug, i.e. is the drug confined to plasma, or present in all extracellular fluids, or distributed equally throughout body water? The half-time of elimination ($t_{\frac{1}{2}}$) is given by the equation:

$$t_{\frac{1}{2}} = \frac{V_d}{\text{clearance}}$$

For the average man, some examples of rates of elimination of drugs excreted only by the kidneys and distributed in different body water compartments are given in Table 2.13.

Table 2.13. Rates of elimination of compounds excreted principally via urine

Compound	Species	Elimination ½-life (min)
Tolazoline	Dog	120
p-Aminohippuric acid	Human (adult)	43
	Human (infant)	160

It is important to realize that while only free drug molecules are filtered at the glomerulus, protein-binding has little or no effect on tubular secretion; both free and bound forms of the drug being secreted by active transport mechanisms. The following is a partial list of some drugs excreted by tubular secretion:

- *p*-aminohippuric acid
- chlorthiazide
- hexamethonium salts
- histamine
- iodopyracet
- nicotinamide (partially as the *N*-methyl metabolite)
- penicillins
- probenecid
- tetraethylammonium salts
- tolazoline

As would be expected for an energy-dependent process, tubular secretion is non-specifically blocked by various metabolic poisons, such as cyanide. More interesting is the competition between various drugs for tubular secretion, so that the one may block (and so prolong the action) of another. Not only tubular secretion but also tubular *absorption* can also be affected by another drug. Most of these inhibitions are relatively specific for a particular class of compound and do not affect other drugs passing (either way) through the tubules:

Tubular *secretion* of penicillins is blocked by probenecid, of tetraethylammonium salts by mersalyl, of *p*-aminohippuric acid by malonate.

Tubular *absorption* of uric acid is blocked by probenecid, of salicylates by bicarbonate.

The blocking of tubular absorption of salicylates by bicarbonates is not strictly the same type of competition as in the other examples, where there is probably competitive uptake of the two drugs at the same tubular sites. The effect of bicarbonate, however, is to alter the pH of the fluid in the distal tubules and so change the proportions of ionized and non-ionized salicylate presented for transport. As only non-ionized forms are absorbed, a decrease in this fraction will reduce tubular absorption of salicylate. Urine shows one of the greatest ranges of pH for any body fluid, with normal limits between about pH 4.5 and 8.0. The administration of bicarbonate leads to alkalinization of the urine, so that its pH may change from, say 7.4 to 8.0. Let us consider the effects of such a change on the ratio of ionized to non-ionized forms of a weak acid, such as salicylate.

From the Henderson–Hasselbalch equation we know that:

$$\mathrm{pH} = \mathrm{p}K_\mathrm{a} + \log \frac{(\text{ionized})}{(\text{non-ionized})}$$

so that at pH 7.4, salicylate with a $\mathrm{p}K_\mathrm{a}$ 3.0 will have the ratio:

$$7.4 = 3.0 + \log \frac{(\text{ionized})}{(\text{non-ionized})}$$

therefore

$$4.4 = \log \frac{(\text{ionized})}{(\text{non-ionized})}$$

therefore

$$\frac{(\text{ionized})}{(\text{non-ionized})} = 25\ 000$$

Similarly at a urine pH of 8.0:

$$8.0 = 3.0 + \log \frac{(\text{ionized})}{(\text{non-ionized})}$$

therefore

$$\frac{(\text{ionized})}{(\text{non-ionized})} = 100\ 000$$

Hence the ratio of the relative proportions of ionized to non-ionized forms of salicylate in urines with pH 8.0 and 7.4 is:

$$\frac{100\ 000}{25\ 000} = \frac{4}{1}$$

Conversely, if the urine becomes acid, tubular absorption of salicylate (or other weak acids) is increased. For example, at pH 5.0:

$$5.0 = 3.0 + \log \frac{(\text{ionized})}{(\text{non-ionized})}$$

therefore

$$\frac{(\text{ionized})}{(\text{non-ionized})} = 100$$

This example shows that changes in urinary pH of only a few units can have dramatic effects on the excretion of ionizable drugs. A practical application is in the treatment of poisoning by drugs of this type; *i.v.* sodium bicarbonate solution can greatly increase the urinary clearance of salicylate, barbiturate, etc. For example, the effect of such treatment on the clearance of phenobarbital is an increase from about 5 ml/min to 30 ml/min, which for a drug in extra-cellular fluids means a reduction in the elimination half-time from 1690 minutes to 277 minutes.

The renal tubules of infants, particularly of premature babies, are only partially active, especially for secretion. Glomerular filtration is also imperfect. The inulin clearance, which should be a measure of GFR, is often much lower than expected and suggests that the glomeruli are not completely removing the inulin, or that some is being absorbed by the tubules. Similarly, in such infants the elimination half-time of *p*-aminohippurate is about four times the adult value, so that its tubular secretion is clearly much reduced.

(iii) Biliary Excretion

As for urine, it is *ionized* drugs, and their metabolites, that are mainly excreted into bile by cells of the liver, though certain non-ionized drugs are also found. Most organic compounds in bile that are unionized at the pH of the intestines are re-absorbed into the portal blood, and so enter an entero-hepatic cycle. The major importance of bile as a route for drug excretion is therefore for drugs which are not re-absorbed in the lower intestines and so pass out of the body in the faeces. Most drugs in this category are weak organic acids and bases, e.g. penicillin, fluorescein, bromosulphophthalein (BSP), quaternary ammonium compounds.

The rate of drug excretion into bile depends primarily on its concentration in blood plasma, but almost every drug that passes into bile shows a definite "cut-off" concentration, after which further increase does not affect the rate of excretion into bile. For BSP this gives a maximum secretion rate into bile of 2.5 mg/min (dog).

As for tubular secretion in the kidneys, the degree of protein-binding of a drug is of little significance compared with its biliary excretion. At usual doses, BSP is bound to plasma proteins to a large extent, but is readily removed and excreted by the liver cells.

There is no doubt that excretion of most drugs into bile is via energy-dependent active transport mechanisms. Drugs are excreted *against* high concentration gradients, sometimes as high as 50-fold, but can be blocked by metabolic poisons, such as cyanide or 2,4-dinitrophenol. As with the kidney tubules, two drugs sometimes compete for biliary excretion, so that the one that is preferentially secreted by the liver cells prolongs the elimination half-time of the other, and also its biological effects. Any two of the following quaternary ammonium compounds, whose principal route of elimination from the body is via the bile, will compete for biliary excretion:

procaine amide ethobromide
mepiperphenidol
benzomethamine
oxyphenonium
N'-methylnicotinamide
tubocurarine.

However, neither tetraethylammonium salts or decamethonium salts are excreted in bile and so do not interfere with the excretion of any of the above drugs.

Important *normal* constituents are the bile acids, such as taurocholate. A number of drugs appear in bile as conjugates with bile acids; for example, strychnine and quinine, although for both these drugs the principal route of elimination is in urine.

The significance of biliary excretion for many drugs is low, though several antibiotics, which appear in urine to some degree, are present in bile at high concentrations, e.g. penicillin, streptomycin, tetracycline. Both natural and synthetic steroid hormones are readily excreted in bile as conjugates, but are largely reabsorbed from the intestines, while the small amounts escaping to the lower gut tend to be decomposed by bacteria.

Generally speaking, almost every drug, or its metabolite, capable of occurring in an ionized form is excreted in bile to some extent, while for the few mentioned above it is of primary importance.

Diseases of the liver and gall bladder, or of the associated blood vessels can lead to impaired drug excretion in bile; which is why the measurement of hepatic clearance of compounds such as BSP make a sensitive test of hepatic function and integrity.

4. Biochemical Factors Influencing Drug Action

We began this section by introducing the concept of drug receptors and by pointing out, that, to have an action, a drug must be able to reach its receptors and remain in combination with them for a time sufficient to allow the effect to ensue. We feel we should close this section by attempting to summarize some of the biochemical factors influencing drug action. From the clinical viewpoint, *most* drugs in *most* patients give expected responses, but occasionally any of the following may result:

(i) The drug may have an abnormally *prolonged* effect
(ii) The drug may produce *little or no* effect
(iii) The drug may produce an *unexpected* effect.

Let us consider these possibilities separately from the following factors discussed earlier:

(a) It is the concentration of *free* drug molecules in the region of the receptor that determines the effects.
(b) The concentration of drugs in the body is a balance between rate of *administration* and rate of *elimination*.
(c) Drugs can be eliminated from the body by *excretion* in urine, bile and other routes.
(d) Drugs are *metabolized* to related compounds, most of which are devoid of biological activity, although a few are more active than their parent substance (see p. 160)
(e) The distribution of drugs throughout body water usually depends on their degree of *ionization, lipid-solubility* and *binding by plasma proteins*.

We have already mentioned the following examples of drugs having abnormally prolonged actions:

(1) *Competitive transport*: Probenecid given with penicillin blocks tubular secretion of the latter; any two drugs in the quaternary ammonium series given together will potentiate the effects of one due to inhibition of its biliary excretion.
(2) *Competitive protein-binding*: The concentration of active, free warfarin in plasma is increased if chloral hydrate is given simultaneously, for a metabolite of the latter displaces warfarin from plasma proteins (see p. 46).
(3) *Defective metabolism and excretion*: Infants, especially premature babies, or unborn foetuses, lack many of the enzymes required to metabolize drugs, while renal function is not fully developed, so that many drugs have a prolonged effect.

Several other types of abnormally prolonged drug actions in humans are known. A popular muscle relaxant is succinylcholine, which is rapidly inactivated in normal people by the cholinesterase ("Pseudocholinesterase") of liver and plasma. The drug is usually administered by *i.v.* infusion; however, the plasma concentration is greatly reduced in only a few minutes. A small number of patients react abnormally to succinylcholine and develop prolonged muscular paralysis and apnoea. It can be shown that the rate of metabolism of succinylcholine by these patients is very slow and that drug concentration in plasma falls at a very low rate. This defect, which is inherited as an autosomal autonomous trait, is in the cholinesterase enzyme. The normal isozyme is absent from liver and plasma and is replaced by another with defective capacities to catalyze the hydrolysis of succinylcholine. A measurement of the defective enzyme is the so-called "*dibucaine number*"; which is defined as the percentage inhibition of cholinesterase by dibucaine, under arbitrary standardized conditions. In normal persons, the dibucaine number averages about 78, with a range of ± 8, but in patients showing abnormal sensitivity to succinylcholine, the number is under 20. It is evident that the inhibitory effect of dibucaine on the defective isozyme is much less than on the normal form. Only persons homozygous for the defective isozyme show the abnormal drug sensitivity; heterozygotes (who have dibucaine numbers between 40 and 68) show a normal therapeutic response.

A second defective cholinesterase isozyme has been described, in which the dibucaine number is normal, but the "fluoride number" (an arbitrary measurement under standard conditions of the degree of inhibition of the isozyme by fluoride ions) is very low. Occasional individuals have been found who are heterozygous for *both* defective isozymes. The dibucaine-insensitive isozyme is normally inhibited by fluoride, just as the fluoride-insensitive enzyme is inhibited by dibucaine, so that individuals with *both* forms have dibucaine number and fluoride number in the moderately low range, yet show abnormal sensitivity to succinylcholine, as in *homozygotes* for either of the two defective isozymes.

Isoniazid is a widely used drug for the treatment of tuberculosis. It is metabolized by the liver to an acetylated metabolite and, to a much smaller extent, a pyridine carboxylic acid. The rate of acetylation determines the rate of elimination from the body, so that blood concentrations of isoniazid are an indication of the *coenzyme A-acetyl transferase* activity in the liver. Studies in large groups of patients have shown that plasma concentration of isoniazid several hours after a single administration is high in about 50% of people, but very low in the remainder. The activity of the acetyl transferase enzyme in liver biopsies from patients of these two types has revealed very much greater activity in the livers of "fast" metabolisers of isoniazid. In this case, however, no differences could be found in the properties of the enzymes from "fast" and "slow" metabolizers; so that, unlike cholinesterase, the abnormal response to isoniazid is *not* due to inheritance of defective isozymes, but to the liver's ability to form the acetyl transferase enzyme. This ability, the nature of which is unknown at the molecular level, is determined by two alleles, so that a "slow" metabolizer of isoniazid is homozygous for one allele, whereas a "fast" metabolizer is either homozygous for the alternate allele, or heterozygous. The mode of inheritance appears to be as an autosomal recessive trait.

The principal metabolic pathway for several other drugs also involves acetylation, and all the following are excreted mainly as acetyl esters:

sulphamethazine
sulphanilamide
hydralazine
p-aminobenzoic acid
phenelzine (β-phenylethylhydazide).

Each of these compounds has now been examined with "fast" and "slow" patients (for isoniazid) and the surprising discovery made, that for two of these drugs (sulphanilamide and *p*-aminobenzoic acid) there is no difference in rate of metabolism between the two types of patients, although the other three are all metabolized slowly by patients who metabolize isoniazid slowly and fast by those who metabolize it fast. The most likely explanation is that hepatic acetyl transferase is not involved in acetylation of sulphanilamide and *p*-aminobenzoic acid, though it is in acetylation of the other drugs. This view is confirmed by studies *in vitro* with liver homogenates from "fast" metabolizers, where no acetylation of sulphanilamide or *p*-aminobenzoic acid could be detected, though the other compounds were metabolized well.

The clinical significance of low hepatic acetyl transferase is that the half-time of elimination of a drug requiring acetylation by this enzyme is greatly increased. Thus, after a standard dose of isoniazid (40 mg/kg *p.o.*), mean plasma concentration after 6 hours in "fast" metabolizers is only about 1.6 μg/ml, but is about 7 μg/ml in "slow" metabolizers. One of the commonest, and most serious, side-effects of isoniazid is peripheral neuropathy, and this develops with much greater frequency in "slow" than in "fast" metabolizers. Fortunately, this side-effect can be prevented by simultaneous administration of pyridoxine. Serious side-effects to phenelzine are also found with greater frequency in "slow" metabolizers, but not, somewhat surprisingly, for other drugs requiring hepatic acetyl transferase.

Whereas about half the population of Europe and North America, including Negroes, are "slow" metabolizers of isoniazid (the relative frequencies of the "slow" and "fast" alleles are about 0.7 and 0.3), in Japan only about 10% of the population is "slow".

Abnormally prolonged action has been reported for several other drugs (e.g. bishydroxycoumarin, diphenylhydantoin, nortryptyline, etc.) in particular patients. In each case, investigation has revealed a genetic defect in a particular enzyme required to inactivate the drug *in vivo* and, in most cases, the defect is associated with markedly higher incidence of serious side-effects.

The converse situation, where usual therapeutic doses of a drug have little or no effect on particular patients, has also been found. One obvious cause is malabsorption of a drug given orally, usually due to gastro-intestinal disorders, while we have already discussed the increased rate of elimination of salicylates or barbiturates following bicarbonate. By analogy, any disease process that leads to alkalinization of urine would be expected to block tubular absorption of such drugs and so decrease their residence time in the body. Of greater interest are a variety of inherited forms of specific drug resistance.

We have discussed the inheritance of a defective cholinesterase isozyme that, in persons homozygous for this trait, leads to inability to metabolize succinylcholine. The incidence of such persons in Europe and North America is about 1 in 3000, but a much rarer state was discovered during investigations of the condition. Occasional individuals have been found to possess a further cholinesterase isozyme with an activity about three times that of the common isozyme. In such persons, succinylcholine is hydrolyzed so fast that intravenous infusions of usual doses have almost no muscle relaxant action.

A similar example is resistance to warfarin or related anticoagulants. Individual patients have been described who require up to 20 times the usual dose to maintain the desired increase in prothrombin clotting time. This drug resistance is inherited as an autosomal dominant trait.

The molecular basis of warfarin-resistance is uncertain. The drug is normally about 98% bound to plasma proteins, so that only 2% of the drug in blood is in an active form. It has been suggested that resistant patients may possess increased protein binding capacity, so that plasma contains virtually no free drug. However, the few warfarin-resistant patients who have been studied appear to have normal free drug levels in plasma. Other possibilities under investigation are that resistant patients metabolize and eliminate warfarin more rapidly than normal, which would make the condition analogous to succinylcholine resistance mentioned above or have a defect in the competitive site for warfarin and vitamin K in biosynthesis of clotting factors. As warfarin-resistance is exceedingly rare, it is most difficult to investigate. It is, however, of more than theoretical interest, for warfarin is the most successful and widely used rodenticide, but resistant rats have begun to appear in several regions. The animal condition is probably similar to that in humans.

Vitamin D-resistant rickets is a similarly poorly understood condition. It is familial; inherited as a sex-linked dominant trait. Serum calcium concentration is normal, but serum phosphate is low, even in members of the family who do not show bone changes. A child with this condition resembles one with vitamin D-deficiency rickets, but is otherwise in good health with normal muscles. Often no evidence of dietary vitamin D deficiency is found and administration of usual therapeutic doses of calciferol (1500 to 3000 IU daily) are without effect. Very high doses will, however, produce a therapeutic action and the rickets can be completely reversed in time. Occasionally it is necessary to give up to 1000 times the usual dose and renal damage occurs in almost all patients, sometimes to a severe extent.

Only recently has the mode of action of vitamin D at the molecular level come to be understood. It is clear that there are a number of tissue-specific metabolites of calciferol and it seems possible that the defect in refractory rickets is of an enzyme catalyzing formation of one of these metabolites. Presumably, high doses needed to produce an effect, supply a metabolite with only weak activity in the place of a potent specific metabolite, so very large quantities are needed.

An interesting form of drug resistance is insensitivity of many Negroes to mydriatic agents. Neither ephedrine-type compounds or atropine have much effect on the pupil when applied to conjuctiva, though all other actions of these drugs occur in Negroes following administration by alternative routes. A generalized atropine-resistance has been described for rabbits, which appear to be unique in possessing a plasma *atropine acyl hydrolase*, which rapidly destroys the drug. Mydriatic-resistance in Negroes does not appear to be a phenomenon of this type, for atropine and ephedrine act on the pupil by different mechanisms. Atropine paralyses the oculomotor endings (and also induces cycloplegia—loss of accommodation), while ephedrine is adrenergic and does not induce cycloplegia. It seems improbable that Negroes have some specific biochemical difference in ocular transport or metabolism of two drugs acting on different receptors, so that resistance to mydriatic drugs is probably due to non-specific structural differences in tissues of the eye.

Several other types of drug-resistance, or poor response to usual doses are known. Juvenile pernicious anaemia is an inherited defect in formation of the gastric mucoprotein (known as "intrinsic factor") required for absorption of vitamin B 12. There is no response to normal doses of oral vitamin, for it is not absorbed, but huge oral doses can be effective. Parenteral administration is more usual.

All the previous examples have been of *constitutional* factors which decrease responses to particular drugs, but acquired resistance is also well known. Infections resistant to antibiotics are discussed elsewhere (p. 268), while gradually increasing resistance to drugs is common. We have already mentioned sensitivity in infants to many drugs, due to lack of drug-metabolizing enzymes. These enzymes, however, are induced by exposure to drugs, so that during the course of treatment, elimination half-time of a drug gradually decreases and increased dosage may be required to maintain therapeutic effects. To a lesser extent the same phenomenon occurs in adults, who may have only low levels of hepatic microsomal oxidases at the beginning of treatment, but acquire high activity during its course. As these oxidases are of low specificity, a patient who has received drug therapy in the recent past may be able to metabolize and clear a second drug much more easily than patients who have not had recent exposure to other drugs.

Psychotropic drugs probably present a special case. There is little doubt that ethanol, amphetamines, heroin, etc., are taken by addicts in doses that would have striking effects in non-addicts, yet which in them produce far milder results. This may be partially due to the addict's increased capacity to metabolize and clear the drug, but other physiological and psychological factors are certainly involved.

Finally, we come to the problem of drugs inducing unexpected effects, unrelated to their therapeutic actions. We will consider later the biochemical basis of those drug side-effects which are common to many classes of pharmaceuticals and mention here only a few relatively specific unexpected actions of some common drugs.

Inheritance of haemoglobins differing in structure from normal haemoglobin A is well described. All haemoglobin types are oxidized to methaemoglobin by a wide range of agents and about 1% of haemoglobin present in blood of normal people is in this form. Many drugs increase slightly the amount of methaemoglobin, even in normal people, but induce a striking, and sometimes fatal, methaemoglobinaemia in susceptible patients. These drugs are listed below:

nitrites and nitrates
chlorates
quinones
aniline
acetanilid
acetophenetidin
sulphonamides
various aromatic nitro—or amino—compounds.

Two quite different types of drug-induced methaemoglobinaemia can be recognized. The first occurs in patients with unusual haemoglobins particularly susceptible to oxidation to methaemoglobins; such as haemoglobins M or H. The second group have normal haemoglobin A, but have an inherited defect in the red cell enzyme, *$NADH_2$-methaemoglobin oxido-reductase*. A number of different structural defects in the enzyme molecule appear possible and are controlled by different abnormal alleles. Patients homozygous for defective enzyme show very low methaemoglobin reductase activity in their red cells and react seriously if exposed to the previously mentioned drugs. Heterozygotes with both normal and defective alleles have reduced red cell enzyme activity, but do not usually react to these drugs. The abnormal haemoglobins are inherited as autosomal dominant traits, while the defective enzyme is an autosomal recessive trait.

A similar phenomenon is "primaquine-sensitivity", which is discussed in detail elsewhere (p. 296). Susceptible patients, who are usually Negroes, have an inherited defect in red cell *glucose-6-phosphate dehydrogenase* and develop an haemolytic anaemia when exposed to primaquine, or any of a long list of other drugs.

Another abnormal drug response is porphyria. Acute porphyria on exposure to drugs is sometimes fatal and appears to be caused by a sudden induction of very high levels of *delta-aminolevulinic acid synthetase* in hepatic mitochondria; which results in uncontrolled formation of porphobilinogen and its metabolites. Susceptibility is inherited as an autosomal dominant trait and occurs even in heterozygotes. The nature of the molecular defect is unclear and presumably lies in the repression mechanisms for the gene controlling formation of the enzyme protein. Even without exposure to drugs, persons with this trait have higher than normal synthetase activity in liver biopsies. Exposure to any drug in the following list results in further marked de-repression of enzyme synthesis:

diallylbarbiturate
allylisopropylbarbiturate
allylmethylbutylbarbiturate
allylisopropylacetylurea
chloroquine
sulphonamides
aminopyrine
hexachlorobenzene

It was once believed that only drugs containing an allyl group were implicated in porphyria, but this is now known to be untrue. Interestingly, the enzyme of bone marrow is not affected, while half-life of the hepatic isozyme is the same before and after induction.

Other unexpected effects of particular drugs, such as allergy, carcinogenesis, teratogenicity, etc. are described in other sections.

III. MODE OF ACTION OF DRUGS

1. Drug Receptors

We have discussed the concept of macro-molecular receptors for drugs and shown that action of a drug usually depends upon its ability to be transported from the site of administration to its receptors, which may be anywhere in the body. We also mentioned a small group of drugs which appear to act on small molecules or ions, and not to require combination with macro-molecules.

The investigation of drug receptors is the principal interest of biochemical pharmacology, but in many cases is a very difficult study. The task is made somewhat easier by the use of radioactively labelled drug molecules, which appear to be transported in the body and to have the same effects as unlabelled drugs. A radioactive drug will be carried to the site of action and its chemical interaction with its receptors investigated. The receptor–drug complex is a hypothetical entity, very analogous to the enzyme–substrate complex required by the Michaelis–Menton hypothesis of enzyme action. Like the enzyme–substrate complex, the receptor–drug complex may be transient, unstable and almost impossible to isolate, except in a limited number of cases.

The combination of a drug with a particular macro-molecule is no guarantee that the molecule is its receptor. We have already seen that many drugs combine with plasma proteins, while others combine with relatively inert tissue proteins, such as collagen. Despite the fact that these interactions are often highly specific at the molecular level, they are in no sense drug actions. Theoretically, interactions of drug molecules with macro-molecules other than the specific receptor are irrelevant to an understanding of drug action, though they may be of importance to transport, metabolism or excretion.

Similarly, we have seen that most drugs are metabolized in the body, particularly by the hepatic microsomal enzyme complex. Metabolism implies the formation of enzyme–drug complexes as intermediaries of the catalytic process, but these macro-molecular complexes are again not receptor–drug complexes.

To appreciate the various types of drug action at the biochemical level it is necessary to consider the various chemical types of drug receptors occurring in the human body.

Table 2.14. Drug receptors

Type of macromolecule	Varieties	Drug interaction
1. Proteins	Enzyme proteins	(i) Inhibitor (ii) Substrate
	Non-enzyme proteins	(i) With gene repressor (ii) With membrane lipoproteins (iii) With contractile proteins
2. Nucleic acids	DNA	(i) Alkylation (ii) Base changes (iii) Viruses
	RNA	(i) Alkylation (ii) Base changes (iii) Viruses
3. Lipids	Lipoprotein complexes	(i) Changes permeability

A simple classification is shown in Table 2.14. This list should not be regarded as complete, nor as mutually exclusive. New types of receptors may well be discovered, while many drugs act simultaneously on several types of receptor, e.g. mechlorethamine reacts with proteins, RNA and DNA.

We will consider each type of drug receptor separately, giving examples of important drugs acting via this mechanism.

2. Proteins as Drug Receptors

An obvious way of altering body metabolism is to interfere with the activity of a particular enzyme. This can be achieved in two quite separate ways—by inhibiting the catalytic activity of the enzyme, or by offering it an unnatural substrate which is nevertheless acted on to yield a foreign metabolite. We will first consider drugs acting primarily as enzyme inhibitors.

A very large number of drugs of many different types are thought to act as inhibitors of specific enzymes. One of the best investigated are the cholinesterase inhibitors, as these drugs form a relatively stable complex with the enzyme.

Some examples of other drugs combining with enzymic or non-enzymic protein receptors are listed in Table 2.15.

Table 2.15. Some protein drug receptors

Drugs	Receptors	Net effects
MAO-inhibitors (e.g. phenelzine, isocarboxazid, tranylcypromine, pargyline, etc.)	Monoamine oxidases (MAO) of brain, liver, and other organs	Delayed metabolism of natural amines such as noradrenaline and serotonin and potentiation of their actions (also of phenylethylamine drugs and food amines, such as tyramine from cheese).
Disulfiram (anatabuse)	Aldehyde dehydrogenase of liver	Competes with NAD^+ for binding-site on enzyme so that activity is reduced and acetaldehyde (produced from ethanol) is poorly metabolized to acetate.
Anti-folate drugs (e.g. aminopterin methotrexate) pyrimethamine, chlorguanide)	Dihydrofolate reductase of cancer tissues (and others)	Compete with dihydrofolic acid for conversion to active THFA coenzymes. The drugs bind tightly to the active site of the reductase.
Sulphonamides	PABA conjugase of micro-organisms	Biosynthesis of folic acid is blocked binding competitively to active site.
Penicillins	Enzyme of cell wall glycoprotein biosynthesis	Biosynthesis of cell wall components is impaired in micro-organisms.

Table 2.15—continued

Drugs	Receptors	Net effects
Cycloserine	Enzyme catalysing incorporation of D-alanine into glycopeptides	Biosynthesis of cell wall components is inhibited in micro-organisms.
Steroid hormones and analogues	Nuclear and cyto-plasmic binding proteins	Stimulation (or inhibi-tion) of metabolism.
Anti-sera (e.g. γ-globulin, tetanus anti-serum, etc.)	Pathogenic micro-organisms; protein of cell walls or of viruses	Formation of antibody–antigen complex with death of organism.
Retinol	Proteins of the eye (opsin) and other tissues	Reversal of deficiency symptoms.
B Vitamins	Numerous tissue enzymes	Converted to enzyme cofactors.
Allopurinol	Xanthine oxidase	Reduced formation of uric acid.
5-azaorotic acid	Orotidine-5′-phosphate pyrophosphorylase	Reduced synthesis of pyrimidine nucleotides and, hence, of nucleic acids.

Cholinesterases occur widely in animal tissues and catalyse the hydrolysis of the neurotransmitter, acetyl choline. Two cholinesterases are usually distinguished. The first, which is called "acetylcholinesterase", will catalyse only the hydrolysis of acetylcholine, but the second, "pseudocholinesterase", will act on a wide range of choline esters. Both types of cholinesterase belong to the "serine enzymes", in that their active site contains a serine residue essential for catalytic activity. The amino acid sequence of the cholinesterases around the active site appears to be:

. Phe . Gly . Glu . *Ser* . Ala . Gly . Ala .

The serine enzymes include many other important body enzymes, such as muscle phosphorylase, phosphoglucomutase, alkaline phosphatase, trypsin, chymotrypsin, and thrombin.

Many cholinesterase inhibitors will inhibit all or most of the serine enzyme group. Only a few are specific for the cholinesterases. The mechanism of catalysis by any member of the serine enzymes is basically the same; the substrate reacts with the hydroxyl group of the serine residue at the active site to yield an acylated enzyme. The latter is often relatively stable and only slowly spontaneously hydrolyzes back to the free enzyme. The sequence of events in the hydrolysis of acetyl choline by a cholinesterase is shown in Figure 2.5.

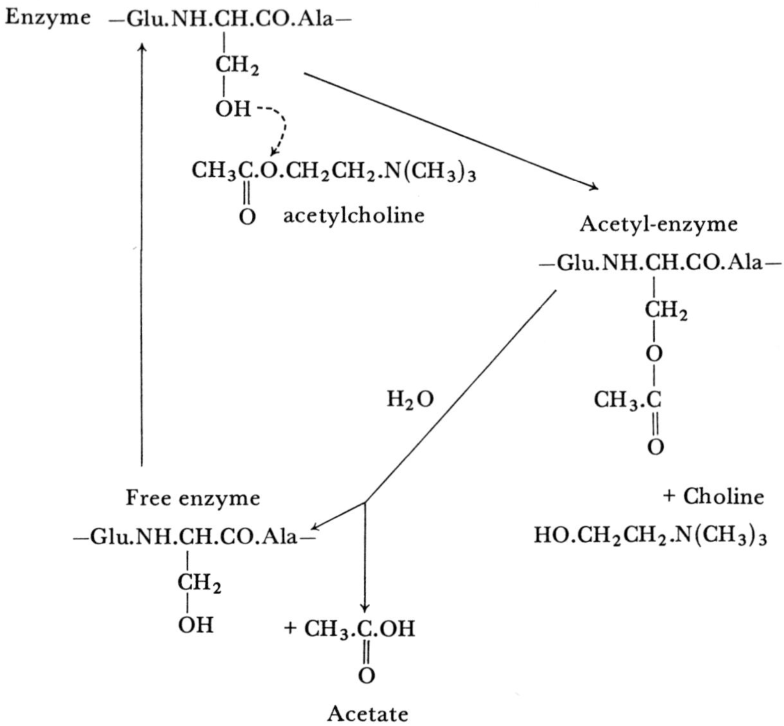

Figure 2.5. Mode of action for cholinesterases.

Substrate activation by the cholinesterases is by a second amino acid residue, thought to involve the electron-rich imidazole ring of a histidine, located close to the serine residue. The primary reaction of the substrate with the enzyme involves electron donation from the histidine residue to the carbonyl of the substrate, which is then transferred to the nearby serine hydroxyl shown in Figure 2.6. Specificity of the reaction involves a third group, still unidentified, located about 7 Å from the serine residue. The histidine and the serine are both concerned with the formation of the acyl enzyme and so may be considered as a single "esteratic" site of the enzyme. The third group forms an ionic bond with the cationic N of choline esters and so is a chemically distinct second active site (usually called the "anionic site"). Presumably the sterochemistry of these sites on acetylcholinesterase is such that esters of choline other than acetyl are unable to interact, while the spacial arrangements of the pseudocholinesterases permits some latitude in the nature of the choline ester.

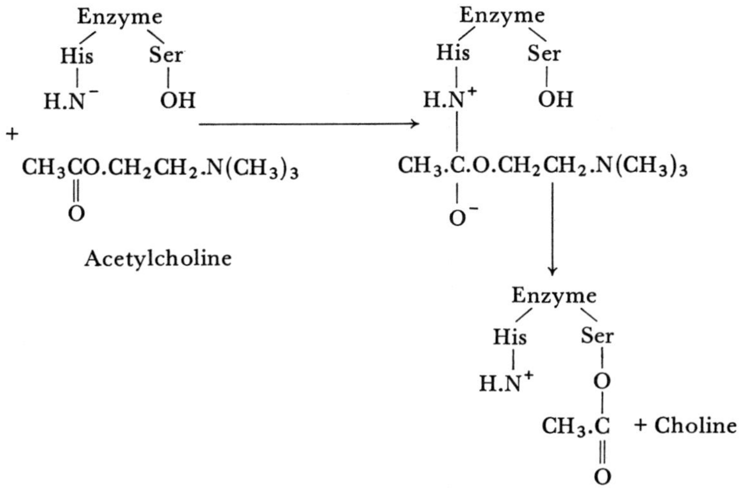

Figure 2.6. Role of histidine in cholinesterase action.

It is obvious that cholinesterases may be inhibited by chemically blocking the serine hydroxyl at the active site. A variety of organo-phosphates are known which will interact with cholinesterases to yield a phosphorylated enzyme, analogous to the acetyl enzyme that is formed during the hydrolysis of acetyl choline. While the acetyl enzyme is fairly rapidly hydrolyzed at the pH of the body, the phosphorylated enzyme is very stable, so that once formed no free enzyme can be liberated. Figure 2.7 illustrates the formation of the enzyme phosphate by interaction with diisopropylfluorophosphate (DFP).

It seems likely that this reaction does not require the participation of the anionic site, for DFP has no positively

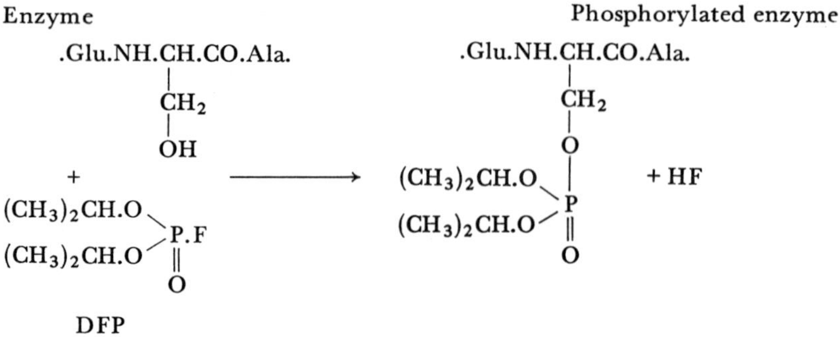

Figure 2.7. Cholinesterase inhibition (by DFP).

charged group with which it could form an ionic bond. Moreover, DFP (and related organophosphorus compounds) will phosphorylate all members of the serine enzyme group, most of which do not possess any anionic sites for substrate interaction. Due to its wide ranging effects on many enzymes, DFP (dyflos) has found little therapeutic application, but a wide range of *specific* anti-cholinesterases are known (physostigmine, neostigmines, etc.) and are used in the treatment of myasthenia gravis and to increase the activity of involuntary, or voluntary, muscles; also as miotics.

The interaction of neostigmine with a cholinesterase is shown in Figure 2.8. As with DFP, reaction occurs between the drug and the serine hydroxyl to yield, in this case, a dimethyl-

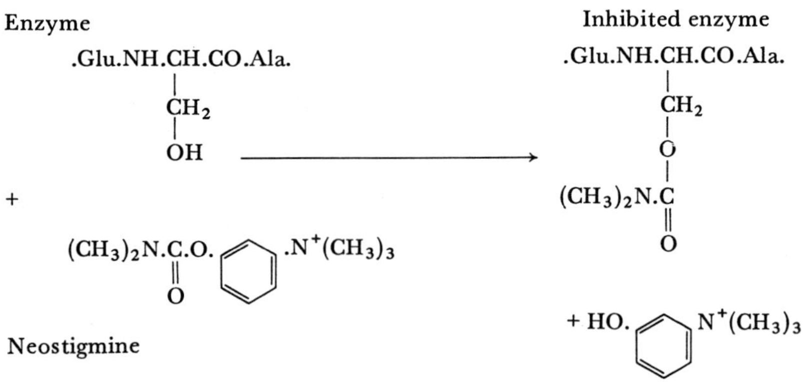

Figure 2.8. Cholinesterase inhibition (by neostigmine).

carbamylated enzyme, which is stable and resistant to hydrolysis back to the free form. It is likely that both specific and non-specific inhibitors of cholinesterase react initially with the histidine imidazole before transferring to the serine hydroxyl, but only the *specific* inhibitors form an ionic bond to the anionic site of the enzyme. All drugs in this category must therefore possess a positively charged grouping.

The anti-cholinesterases are a group of drugs with a receptor of clearly defined function, and their chemical combination with their receptors inhibits the break-down of acetylcholine. The net result is the prolonged action of this neurotransmitter within the body. A second group of drugs are also concerned with acetylcholine, but their function is to prevent it acting, and so is the exact opposite of the anti-cholinesterases.

The receptors for the acetylcholine antagonists are the natural receptors for acetylcholine itself, and so are distributed throughout the body in many tissues. Acetylcholine is released from cholinergic neurones and its receptors occur in the end-plates of skeletal muscles, in secretory glands, in smooth muscles, in autonomic ganglia, and also in certain parts of the CNS. Precisely what these acetylcholine receptors are, is still unclear, but they are not simply cholinesterase enzyme molecules, even though their distribution in tissues is very similar. It seems probable that the action of acetylcholine on any tissue is determined by the competition between the uptake by receptors and the uptake by cholinesterases for destruction. Most acetylcholine antagonists block uptake by the receptors, but have no effect on cholinesterases.

The most likely hypothesis is that the acetylcholine receptors are proteins of various membranes, but lack enzyme activity. The effect of acetylcholine is to greatly increase membrane permeability to Na^+, so that it depolarizes nerves by allowing an influx of Na^+. As the acetylcholine is destroyed, the previous concentration gradient is restored by the "sodium pump". Presumably, interaction of acetylcholine with the membrane protein results in altered conformation, so that permeability is increased. Alternatively, acetylcholine could combine with, and inhibit, some protein component of the "sodium pump" mechanism. In either case, an acetylcholine antagonist must be able to prevent the uptake by the receptor, yet not itself mimic the biochemical actions of acetylcholine.

A consideration of the structure of acetylcholine Figure 2.9 shows that it is

(a) a quaternary ammonium compound with three methyl groups
(b) an *acetyl* ester
(c) an *ester* with 2 carbons (methylenes) between the esterified hydroxyl and the quaternary N.

We have already seen that the interaction of acetylcholine with one type of protein (the cholinesterases) is highly specific and involves two distinct sites. Similar specificity is therefore to be expected in its interaction with its other protein receptors. This can be investigated by studying various structural analogues. The potency of these in a number of *in vitro*

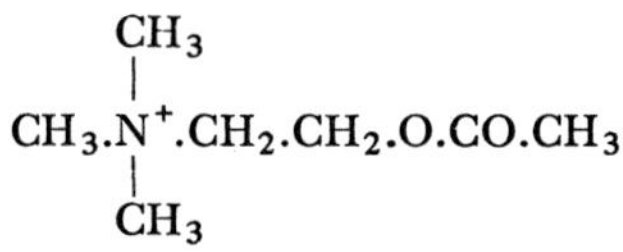

Acetylcholine

Pentamethonium salts

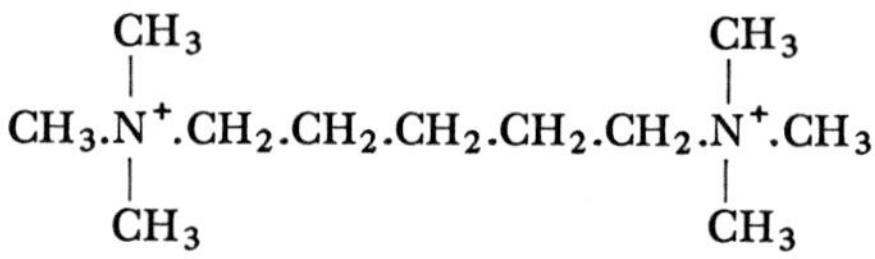

Decamethonium salts

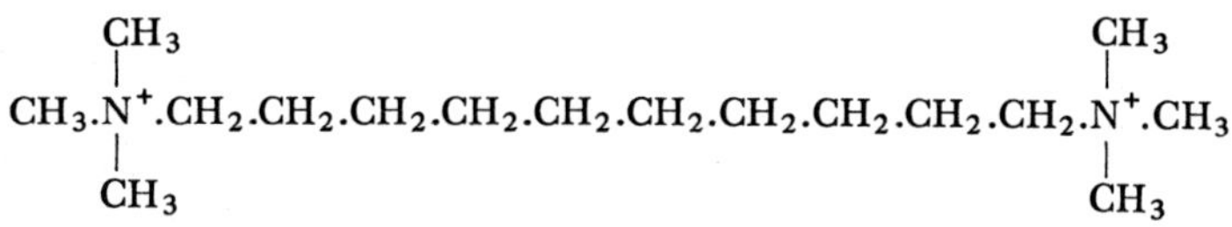

Succinylcholine

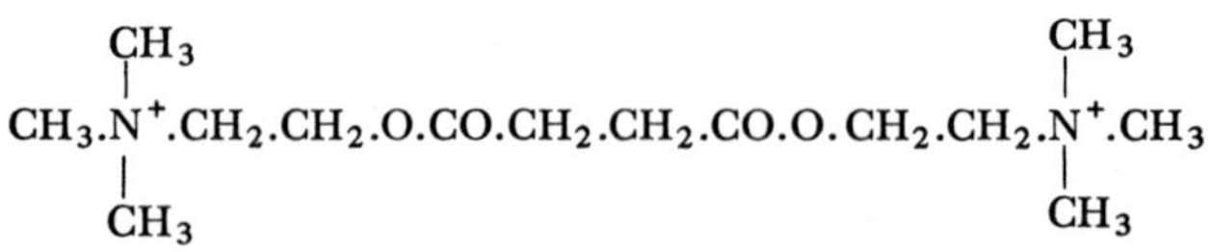

Figure 2.9. Comparison of chemical structures of acetylcholine and anti-cholinergic compounds.

systems, such as contraction of perfused heart muscle, can be readily determined. The following conclusions have been reached from such studies:

(a) Replacement of the methyl groups on the quaternary N atom leads to a gradual loss of activity. The dimethyl compound is fairly active, the monomethyl compound much less so, while the unsubstituted primary amine is almost inert. A monoethyl-dimethyl derivative is quite active, but the diethyl-monomethyl compound is very weak. None of these compounds is very effective as acetylcholine antagonists, although the triethylanalogue has some stimulant, rather than inhibitory, properties on heart muscle. Removal of the N atom results in a total loss of all activity.

(b) Unesterified choline is almost inert, but various esters other than acetyl prove to have either moderate activity, or to be antagonists.

(c) The ketone analogue of the ester linkage has moderate activity, as does the ether analogue. Compounds with more than 2 carbons between the N and the ester bond are only weakly active, as are *n*-paraffinic derivatives of varying chain length.

The best model to explain the activities of these analogues is that the receptor protein (like the cholinesterases) possesses two sites which interact with acetylcholine. One of these is an *anionic site*, with a negatively charged group, which forms an ionic bond with the quaternary N atom. The effects of ethyl substitution suggests that the N atom may be held in the correct spacial relationship by two of the methyls fitting into "grooves" or "holes" in the receptor around the anionic group.

The second site is 7 Å away and forms some type of link with the carbonyl group of the acetyl; this is unlikely to be a strong interaction with a serine hydroxyl, as in the cholinesterases, for there is no evidence that the acetylcholine receptors break the ester linkage. A hydrogen bonding seems the most likely.

Further evidence concerning the acetylcholine receptors comes from a study of antagonists, for these obviously compete for the same receptor molecules. The three compounds in Figure 2.9 are all active antagonists. Quite marked differences have been found in the biological effects of these antagonists.

Pentamethonium compounds effectively block the action of acetylcholine at *autonomic ganglia* (and this and related compounds are often referred to as "ganglion-blocking agents"), but have little effect at the *muscle-end plate* receptors. Conversely, the longer molecules of decamethonium salts or succinylcholine are efficient muscle-relaxants, but do not significantly block autonomic ganglia. By examining the activity of various analogues of these antagonists, the following further conclusions can be drawn about the molecular architecture of acetylcholine receptors.

(a) While the structure of the receptors in ganglia and muscle end-plates must be basically similar, the surrounding parts of the protein must be significantly different.
(b) In autonomic ganglia the acetylcholine receptors must possess a *second anionic site*, with a structure very similar to the first, about 10 Å away, but in approximately a straight line across the carbonyl-binding site. In this way pentamethonium ions bind to the two anionic sites of the receptor and, in so doing, lie across the carbonyl-binding site. Both ends of the molecule of antagonist must bind to the receptor, so that compounds of the decamethonium type cannot attach to the ganglion receptors.
(c) In contrast, the acetylcholine receptors of the muscle end-plates must also possess a second anionic site, but this must be more than 10 Å from the anionic site required for acetylcholine binding, for pentamethonium ions do not bind. The quaternary N atoms of both decamethonium ions and succinylcholine are about 19 Å apart in linear conformation, so that the second receptor could be about this distance from the first, and again, the molecule of antagonist is probably bound to the end-plate receptor such that it lies across, but not bound to, the carbonyl binding-site.

A further well-known acetylcholine antagonist is atropine, together with its related alkaloids (hyoscine, scopolamine, etc.). At low doses it is an effective inhibitor of muscle end-plate response, but at high doses it can also block autonomic ganglia. Its action is also to compete with acetylcholine for the receptor molecules, for as with the other type of antagonist, there is no evidence that it in any way interferes with the biosynthesis or release of acetylcholine. The nature of the binding of atropine to either site is unclear, for it is a complex molecule with both primary alcohol and ester groups, as well as a basic tri-substituted N. Presumably the latter can combine with the anionic site of the acetylcholine receptors and the carbonyl of the ester link with the second, for the separation is about right. However, from this reasoning it would follow that atropine should be a good ganglion-blocker and a poor end-plate inhibitor, whereas the reverse is true. It is therefore likely that some other group of the atropine molecule, perhaps the primary alcohol, is involved in interactions with the acetylcholine receptors at sites not involved in interactions with any of the previously mentioned molecules.

The mechanisms and kinetics of enzyme inhibitors can be differentiated between reversible and irreversible inhibition. We have seen that many drugs act as enzyme inhibitors by combining with active sites, yet the drug molecules escape catalytic change. Theoretically, it would be expected that most of these inhibitors are of the competitive type, so that for an enzyme (E) catalyzing the reaction:

$$A \overset{E}{\rightleftharpoons} B$$

we would have the equilibrium:

$$A + E \rightleftharpoons AE$$

and the second equilibrium:

$$AE \rightleftharpoons E + B$$

so that, in the presence of the drug inhibitor, there would be a third equilibrium:

$$D + E \rightleftharpoons DE$$

so that the rate of catalysis would be limited by the amount of enzyme not inactivated by combination with the drug. However, the binding affinity of most drugs for the enzymes which they inhibit is usually very great, so that even where there is excess enzyme, almost all drug molecules are enzyme-bound. Under physiological conditions they cannot be removed, though in the laboratory they often can by prolonged dialysis *in vitro*.

1. Chlorguanide (antimalarial, active *in vivo*, but not *in vitro*).

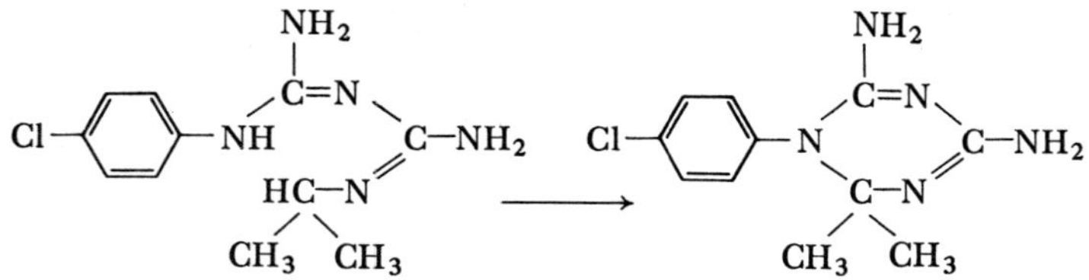

Triazine
(active metabolite)

2. Iproniazid (intended as anti-tubercular drug, but found to be MAO-inhibitor)

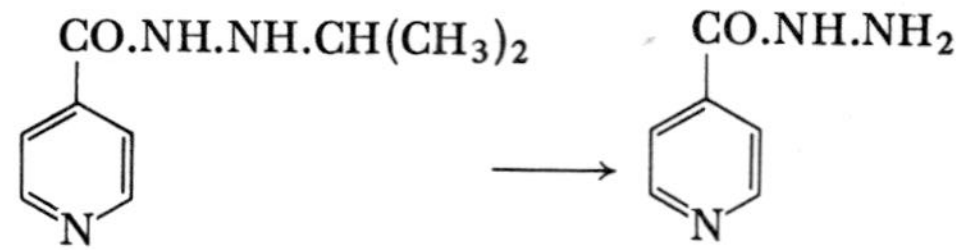

Isoniazid
(active metabolite)

3. Parathion (insecticide and cholinesterase-inhibitor)

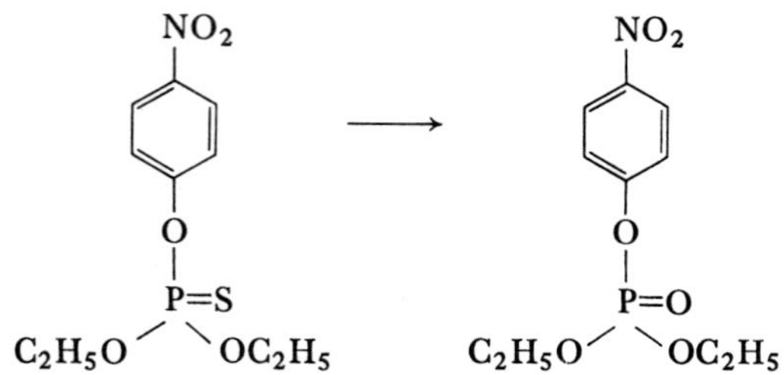

Paraoxon
(active metabolite)

4. (Slow acting phenacetin analgesic and antipyretic, probably requiring conversion to paracetamol before effective)

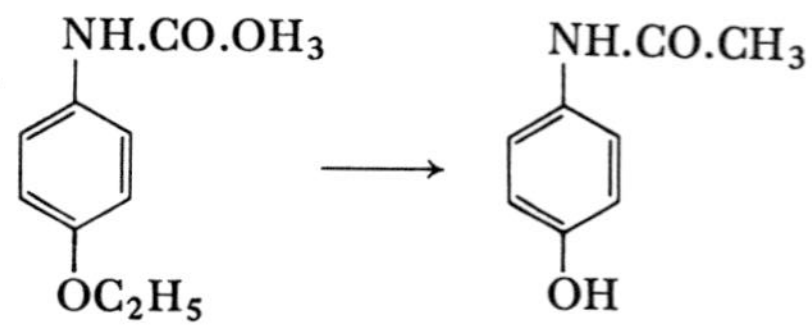

Paracetamol
(active metabolite)

Figure 2.10. Some drugs requiring metabolic activation.

The overall result is that, while *theoretically* reversible, enzyme inhibition by drugs is usually *in practice* irreversible (and is sometimes called "pseudoirreversible"). For these reasons, mathematical treatments of the kinetics of drug inhibition for most enzymes follow the equations for irreversible inhibitors.

We have made the point that very little, if any, of the drug is converted by the enzyme it inhibits to some other substance, i.e. these drugs cannot serve of substrates. For most drugs of this type (e.g. MAO-inhibitors, cholinesterase-inhibitors, sulphonamides, etc.) the number of drug molecules converted by the enzyme is less than 1%. This brings us, therefore, to a separate class, that also has specific enzymes as its receptors, but which is acted upon by these enzymes and converted to a metabolic analogue. Drugs of this type fall into two distinct sorts. The first we have already met before (p. 43) and are drugs which require metabolic activation before they can bring about their actions. This may be considered a further type of "lethal synthesis". Examples are given in Figure 2.10. In each case the drug administered acts on a single "receptor", i.e. the enzyme which converts it to an *active* metabolite. Similarly, the metabolites have different enzyme receptors to produce the biological effects seen when the drug is administered. Strictly, the first *receptor* of a drug requiring metabolic activation may be *totally unrelated* to its biological actions (Table 2.16).

It will be noted that these drugs all require metabolic activation before producing a biological effect. It is an interesting academic exercise to separate the action of a parent drug from those of its metabolite. Where microsomal oxidases

Table 2.16. Receptors for drug metabolites

Drug	First receptor	Metabolite	Second receptor	Biological actions
Chlorguanide	Hepatic oxidases (cyclization)	Triazine	Dihydrofolate reductase	Folate antagonist for malaria parasite
Iproniazid	Hepatic oxidases (*N*-dealkylation)	Isoniazid	MAO	Anti-depressant
Parathion	Hepatic oxidases (desulphuration)	Paraoxon	Cholinesterases	Potentiation of acetylcholine effects
Phenacetin	Hepatic oxidases (de-ethylation)	Paracetamol	CNS proteins (?)	Analgesic and anti-pyretic

are concerned, this can be attempted in animals by β-diethylaminoethyl diphenylpropyl-acetate (DAEDPA; more usually known as SKF 525A):

$$CH_3.CH_2.CH_2.C(C_6H_5)_2.CO.O.CH_2CH_2.N(C_2H_5)_2$$

At a dose of 15 mg/kg *i.p.*, this compound will virtually eliminate microsomal oxidation in the liver of a rat for several hours. Its use has shown, for example, that phenacetin itself has some moderate anti-pyretic activity, but significantly less than when freely metabolized to paracetamol.

Many new drugs appear to fall into the type requiring metabolic conversion to some other substance before the desired biological actions are seen. Many so-called "analogue

Table 2.17. Analogues of nucleic acid bases

Drug	Natural compound
6-Mercaptopurine	Adenine
8-Azaguanine	Guanine
5-Fluorouracil	Uracil
5-Fluorodeoxyuridine	Deoxyuridine
5-Azaorotic acid	Orotic acid

inhibitors" used to impede cell growth and division in cancer chemotherapy act in this way. We will consider the following analogues of purine and pyrimidine bases, all of which have been used clinically (Table 2.17).

6-Mercaptopurine is one of the earliest drugs in this category and was developed empirically as a potential inhibitor of nucleic acid biosynthesis. It proved to have anti-tumor activity in laboratory animals and, in clinical use, produced temporary remissions in about half the cases of childhood acute lymphatic leukaemia in which it was tried. From laboratory studies, it was clear that the compound inhibited biosynthesis of nucleic acid, for the incorporation of ^{14}C from such precursors of purine and pyrimidine bases as labelled formate or glycine (see Table 2.2) was much reduced. However, the incorporation of preformed adenine into nucleic acids was unaffected, though the utilization of hypoxanthine (6-hydroxypurine) was greatly reduced. To appreciate the significance of this result, one must consider the pathways of biosynthesis of nucleic acid precursors. We will summarize the principal facts:

(1) Nucleic acids are formed by the polymerization of nucleoside triphosphates, with the elimination of inorganic pyrophosphate.
(2) DNA is formed almost exclusively from:
 thymidine triphosphate (a deoxyribonucleotide) (TTP)
 deoxyguanosine triphosphate (dGTP)
 deoxyadenosine triphosphate (dATP)
 deoxycytidine triphosphate (dCTP)
(3) Most types of RNA are formed almost exclusively from:
 uridine triphosphate (a ribonucleotide) (UTP)
 guanosine triphosphate (GTP)
 adenosine triphosphate (ATP)
 cytidine triphosphate (CTP)
(4) Purines are made in the body from smaller molecules (glycine, formate, CO_2, etc.) as *pre-formed* nucleoside monophosphates, *not* as free bases.
(5) Pyrimidines also are formed in the body from smaller molecules (carbamyl phosphate and aspartic acid), but the basic pyrimidine ring is formed (as orotic acid) *before* the formation of a nucleotide. The other pyrimidine bases found in nucleic acids are all formed from the orotic acid *residue* of orotidine-5′-monophosphate, *not* as free bases.
(6) Deoxyribonucleotides are formed from ribonucleotides.
(7) Purine-containing nucleotides are all formed from inosinic acid (hypoxanthine ribotide).
(8) The pathways of nucleic acid biosynthesis may therefore be represented as in Figure 2.11.
(9) If free purine and pyrimidine bases are introduced into an organism, some may be converted to nucleoside phosphates by the "*salvage mechanisms*". Animal cells can utilize free adenine and hypoxanthine, but guanine is poorly used. Orotic acid (which is formed within tissues in the free state) is well used, while uracil is used to some extent. The "salvage route" involves reaction with PRPP (5-phosphoribosyl-1-pyrophosphate) and is catalyzed by *nucleotide pyrophosphorylase*. It is not

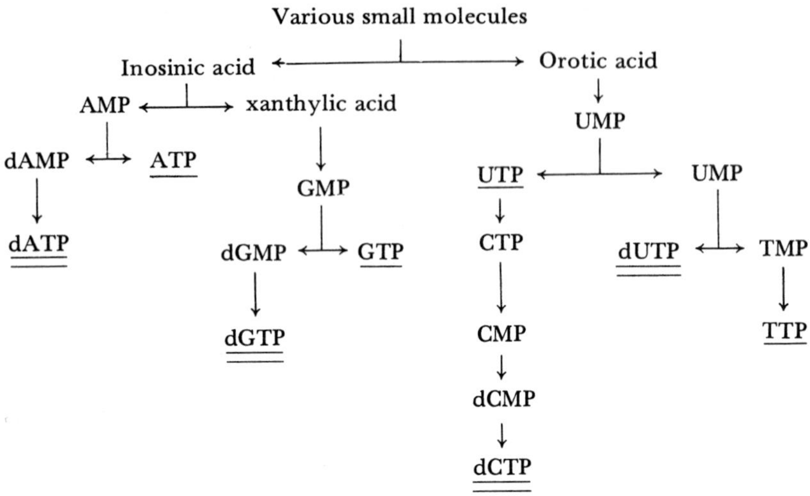

Nucleoside triphosphates underlined *once* are incorporated into RNA and those underlined *twice* into DNA.

Figure 2.11. Biosynthesis of nucleic acid precursors.

clear whether the same enzyme can act on both purines and pyrimidines, but *specific* pyrimidine pyrophosphorylases are known.

It will be seen that the decrease in hypoxanthine utilization induced by treatment with 6-mercaptopurine could be explained as a *competition* between hypoxanthine and its analogue for conversion by nucleotide pyrophosphorylase to the corresponding nucleotide, but this would not explain the continued utilization of adenine, which requires same enzyme. Detailed investigations revealed that, under physiological conditions, there is no significant impairment of conversion of hypoxanthine to inosinic acid.

It was found that the pyrophosphorylase acts on 6-mercaptopurine to convert it to *6-mercaptopurine ribose-5'-phosphate*, the sulphur analogue of inosinic acid. The enzyme may therefore be regarded as a primary drug receptor, although it is *not* involved in the principal biological action; only in conversion of the drug to an active metabolite.

The conversion of inosinic acid to AMP requires two enzymes; the first catalyzes the condensation of inosinic acid with aspartic acid to yield adenylosuccinate, which is then hydrolyzed to fumaric acid and AMP by the second enzyme. The latter appears to be completely inhibited by 6-mercaptopurine nucleotide. Inosinic acid is also converted to GMP, via xanthylic acid. This requires oxygenation at carbon-2; a reaction catalyzed by an NAD^+-requiring oxidoreductase. This enzyme also appears to be competitively inhibited by 6-mercaptopurine nucleotide, although probably less so than adenylsuccinase.

The net result is decreased formation of both AMP and GMP, which are essential precursors of RNA and DNA. It may be thought that 6-mercaptopurine could be incorporated into nucleic acids and so alter the "information" carried by their base sequences. There is some evidence for conversion of 6-mercaptopurine nucleotide to its di- and tri-phosphates; but very little, if any, appear to be incorporated into nucleic acids. This may be due to the fact that nucleic acid formation can only occur when all four trinucleotides are available, and in the presence of 6-mercaptopurine, one or more become limiting.

Aside from their role in nucleic acids biosynthesis, adenine-containing nucleotides are concerned in many essential enzyme cofactors (coenzyme A, NAD, FAD, etc.) and there is some evidence that, particularly at high doses, 6-mercaptopurine nucleotide may be used to produce analogues of these coenzymes.

Finally, there is now no doubt that not only are the rates of formation of AMP and GMP decreased following treatment by 6-mercapto purine, but that the *overall formation of purines* is also less. At first sight it might be expected that purine biosynthesis would be either increased to compensate for the lack of nucleic acid precursors, or unaffected below the level of inosinic acid. However, most metabolic sequences in the body are the subject of "*end-product*" *inhibition* (or activation) when the product of the sequence alters the activity of one enzyme earlier in the pathway. Purine biosynthesis is regulated by AMP concentration: a decrease stimulates increased formation, while an increase reduces it. In the presence of 6-mercaptopurine, AMP concentration is reduced; but 6-mercaptopurine nucleotide cannot be distinguished by the control mechanisms from AMP, so that the *combined* concentrations of AMP and 6-mercaptopurine nucleotide act as an end-product inhibitor on

5-phosphoribosylpyrophosphate aminotransferase (the regulatory enzyme of the sequence).

It will be seen that the nature of the enzyme receptors of 6-mercaptopurine nucleotide (the active metabolite of 6-mercaptopurine) are *multiple*, so that the *net* effect of the drug (selective metabolic poisoning of cancer cells through inhibition of nucleic acid biosynthesis) is achieved by several distinct biochemical actions.

The second purine analogue mentioned, 8-azaguanine, could be expected to produce effects similar to 6-mercaptopurine, but a number of significant differences have been found. When first tested in laboratory animals its initial actions were to inhibit *protein* biosynthesis but, unlike 6-mercaptopurine, decreased formation of nucleic acids came much later, even at high doses. It was clear that 8-azaguanine could not be inhibiting the biosynthesis of purines or their nucleotides. A comparison of the chemical structures of the two purine analogues (see Table 2.2) shows that 6-mercaptopurine is altered in the 6-membered ring, which is involved in numerous metabolic transformations among the various purine nucleotides, while 8-azaguanine is altered in the 5-membered ring, which apart from the linkage to ribose is not involved in reactions.

Using ^{14}C-labelled compound, the incorporation of 8-azaguanine into nucleic acids was demonstrated for a wide range of tissues in laboratory animals. By far the greatest incorporation was into *messenger*-RNA, or *soluble*-RNA (depending on the tissue and the species), with little or none into DNA. To be incorporated into nucleic acids, the compound must first be converted to the corresponding nucleotide (8-azaguanylic acid), (presumably by *nucleotide pyrophosphorylase*) then further phosphorylated to the di- and tri-phosphates. The latter is then incorporated into RNA and the triplet sequences in which it is included probably are "mis-read" or "nonsense" so that protein biosynthesis is impaired.

Significantly, 8-azaguanine is relatively ineffective against cells possessing a high concentration of *guanase*, which catalyses deamination to 8-azaxanthine, which is inactive.

We have also mentioned three *pyrimidine* analogues used in cancer chemotherapy. Free pyrimidines, or pyrimidine nucleosides, are readily taken up by many tissues, particularly those in rapid growth, and converted to the corresponding tri-phosphates for incorporation into nucleic acids. The free pyrimidine analogue, 5-fluorouracil, is rapidly converted to 5-fluorouridine, then to the mono-, di- and tri-phosphates. Little conversion to the corresponding deoxyribose forms occurs and 5-fluoruracil, labelled with ^{14}C, is incorporated (like 8-azaguanine) almost entirely into RNA, rather than DNA, and *protein* biosynthesis is inhibited, before that of *nucleic acids*.

By contrast 5-fluorodeoxyuridine does not appear to be significantly incorporated into *nucleic acids* and its primary action is seen as an inhibition of DNA *biosynthesis*. This nucleoside analogue is phosphorylated *in vivo* by *uridine kinase* to 5-fluorodeoxyuridine monophosphate, which appears to be the active metabolite and a potent inhibitor of *thymidylic acid synthetase*. This enzyme catalyses the conversion of dUMP to TMP and requires a tetrahydrofolate cofactor. The enzyme is also inhibited, though to a lesser extent, by 5-fluorouridine (formed *in vivo* from 5-fluorouracil).

Both compounds also produce a general "*end-product inhibition*" pyrimidine biosynthesis by suppression of *aspartate transcarbamylase*; the regulatory enzyme of the biosynthetic sequence (in a manner similar to the action of 6-mercaptopurine nucleotide on purine formation).

The final base analog we will mention is *5-azaorotic* acid: an inhibitor of nucleic acid biosynthesis, that is not significantly incorporated in either RNA or DNA. As would be expected from the key position of orotic acid in pyrimidine nucleotide formation, any drug interfering with the utilization of this intermediary would have wide-ranging effects on all kinds of nucleic acids. Unlike the other analogues so far discussed, 5-azaorotic acid does not appear to require metabolic activation before acting, and is *not* converted to 5-azaorotidine-phosphate in any significant amounts. The conversion of orotic acid to UMP requires two enzymes, *orotidine-5'-phosphate pyrophosphorylase*, which catalyses the condensation of orotic acid with PRPP; then a *decarboxylase*, which converts orotidine-5'-phosphate to UMP. The first enzyme is strongly inhibited by 3-azaorotic acid, so that the biosynthesis of both pyrimidine *ribo*nucleotides and *deoxyribo*nucleotides are blocked.

Summarizing the effects of these five analogue drugs (and their metabolites) (Table 2.18) for the wide range of receptors and enzymes involved, only one of these five drugs acts directly;

Table 2.18. Primary and secondary receptors for nucleic acid-base analogues

Drug	Primary enzyme receptors	Metabolites formed	Secondary enzyme receptors for metabolites
1. 6-Mercaptopurine (6MP)	Nucleotide pyrophosphorylase	6MP nucleotide	(a) Adenylsuccinase (b) 5-Phosphoribosyl-pyrophosphate amidotransferase
2. 8-Azaguanine (8AG)	Nucleotide pyrophosphorylase *then* Nucleoside mono phosphate kinase *then* Nucleoside diphosphate kinase *then* RNA polymerase	8AZ nucleotide *then* 8AZ nucleoside diphosphate *then* 8AZ nucleoside triphosphate *then* 8AZ-containing RNA	
3. 5-Fluorouracil (5FU)	Uridine pyrophosphorylase *then* Nucleoside mono phosphate kinase *then* Nucleoside diphosphate kinase *then* RNA polymerase	5-Fluorouridine *then* 5-Fluoro UDP *then* 5-Fluoro UTP *then* 5FU-containing RNA	(a) Thymidylic acid synthetase (b) Aspartate *trans*-carbamylase
4. 5-Fluorodeoxyuridine	Uridine kinase	5-Fluorodeoxyuridine	(a) Thymidilic acid synthetase (b) Aspartate *trans*-carbamylase
5. 5-Azaorotic acid	Orotidine-5′ phosphate pyrophosphorylase	None	–

the others require *in vivo* conversion to active metabolites to produce any effect. For 8-azaguanine and, to a lesser extent, 5-fluorouracil, the metabolites are converted into a "*false*" *RNA*, and so can hardly be said to "react with" second receptor. However, 6-mercaptopurine and 5-fluorodeoxyuridine (and 5-fluorouracil, to some degree) act *through* their metabolites by inhibiting specific enzymes.

3. Nucleic Acids as Drug Receptors

We have just seen that certain drugs interfere with the biosynthesis of nucleic acids by inhibiting enzymes concerned with the formation of various precursors, while both 8-azaguanine and 5-fluorouracil become incorporated into RNA molecules, after metabolic activation. Only in a trivial sense could nucleic acids be said to be the receptors for these drugs. There is another type of drug, however, which chemically combines with nucleic acid molecules and its biological action is dependent upon this interaction.

The two base-analogues mentioned above supply the body with a *pre-formed* unnatural base, which is nevertheless still incorporated into nucleic acid molecules and so alters their "information" content. Alkylating agents are drugs which act on the bases of nucleic acid molecules *already present* in the body. The commonest drugs in this class are Melphalan, nitrogen mustard, cyclophosphamide, busulphan (Myleran), and

Type of grouping	Active forms	Drug examples
1. Nitrogen mustard $R'R''{>}N.CH_2.CH_2Cl$	$R'R''{>}N.CH_2.CH_2{+}$ and $R'R''{>}\overset{+}{N}{<}(CH_2)_2$ (ring)	Nitrogen mustard; melphalan; cyclophosphamide
2. Ethylenimine $(CH_2)_2{>}N.R$ (ring)	$^{+}CH_2.CH_2.N^{-}.R$	Thio-TEPA
3. Alkyl sulphonate $R'.CH_2.O.SO_2.R''$	$R'.CH_2^{+}$ and $R''.SO_3^{-}$	Busulphan

Figure 2.12. Some alkylating agents.

thio-TEPA. Structures are given in Table 2.5. Within the body, each of these compounds exists as an equilibrium mixture of a variety of molecular species, some of which are highly reactive. Figure 2.12 illustrates the active forms of chemical groupings present in these drugs.

All alkylating agents interact with several types of cell components, but it is their effect on DNA which appears to be the basis of therapeutic action in cancer. It has been shown by giving the drugs labelled with ^{14}C that they are taken up by the nucleus of cancer cells, and, to a lesser extent, by the nucleic of

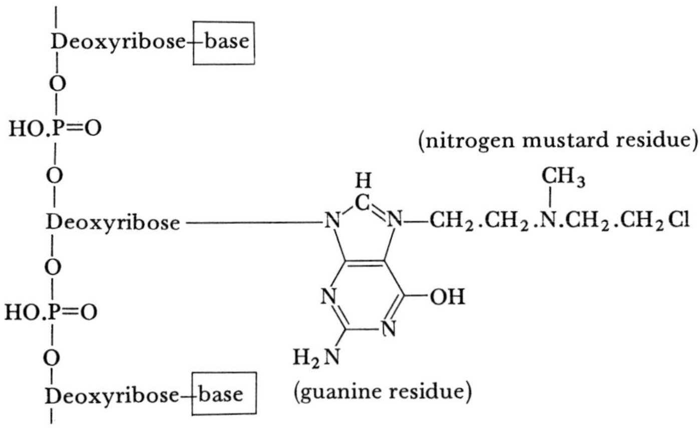

Figure 2.13. Combination of nitrogen mustard with a nucleic acid base.

cells of various normal tissues. If the DNA is extracted from the cells of an animal treated by an alkylating agent, the drug is found to be chemically combined with the nucleic acid by covalent bonds. Different parts of the DNA molecule appear to be attacked by the drugs, but by far the commonest is alkylation of the N atom at position 7 of guanine residues. The result is a DNA molecule with the partial structure shown in Figure 2.13.

By several ways alkylation may disrupt the normal functioning of DNA molecules:

(i) The presence of an alkylated guanine residue on a DNA molecule undergoing replication leads to a mis-matching of bases on one of the two daughter molecules. Alkylation at N-7 of guanine increases acidity at N-1, so that instead of pairing with cytosine, alkylated guanine pairs with thymine. This changes the information content of the base triplet. For example, if the original triplet was C.A.A., which codes for glutamine, alkylation of the opposite strand of DNA would change this triplet to T.A.A., which is a nonsense or "terminate" coding. The effect of a single alkylation of guanine is shown diagramatically below.

Part of original DNA double helix (bases shown as initials)

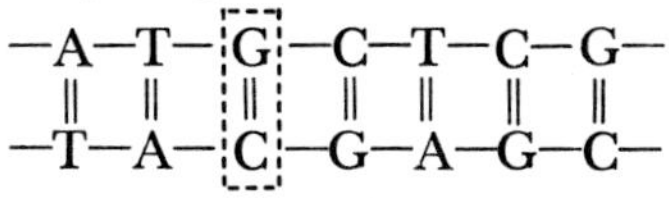

chain α

chain β

Molecule after alkylating agent has changed *one* guanine residue:

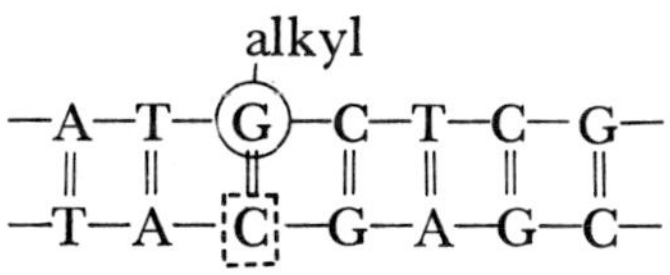

chain α

chain β

Replication would give rise to two daughter molecules:

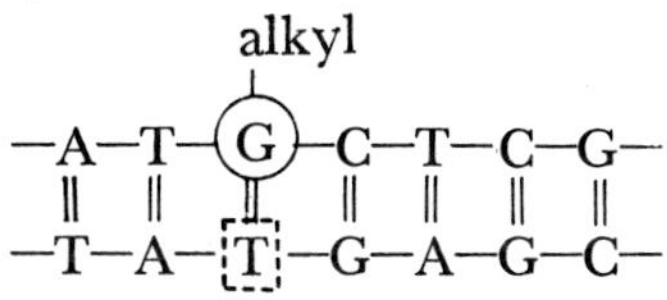

chain α

chain α—1

and

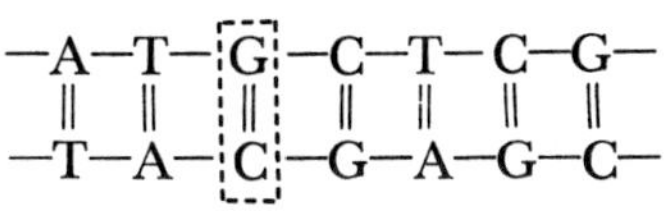

chain β—1

chain β

Further replication of the DNA derived from the alkylated chain will give rise to two further daughter molecules, *both* with altered base sequences:

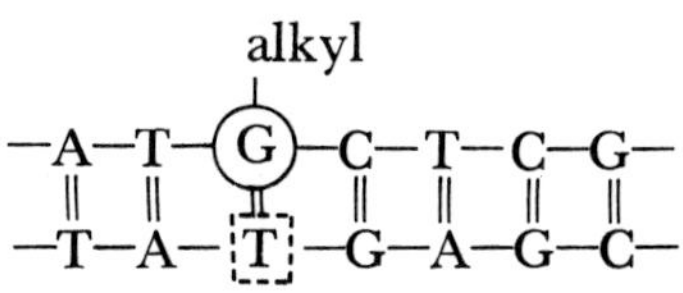

chain α

chain α—2

and

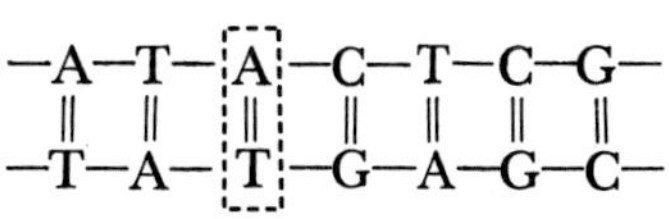

chain α—3

chain α—1

In this manner, the altered triplet will continue to be perpetuated, even on DNA chains not containing the alkyl group.

(ii) The presence of the alkyl group on a DNA molecule leads to difficulties during replication and there appear to be nuclear enzymes capable of removing the defective nucleotide. This will result in a break in one chain of the molecule, so that at the next replication, the break will be perpetuated and two smaller molecules will be formed. This is shown below. The original double helix of DNA is the same as in the previous example and the guanine residue in chain α is alkylated, then deleted at the next replication to give a molecule with the structure:

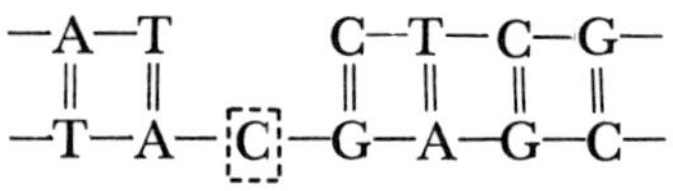

chain α (split)

chain β

On replication, this will produce *three* daughter molecules:

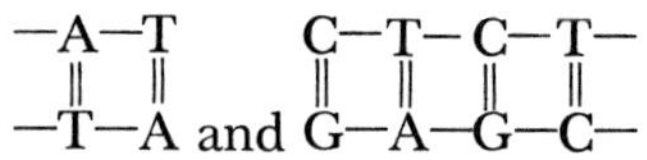

from chain α

and

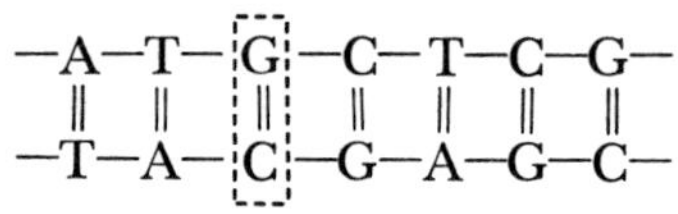

chain β—1

chain β

If the daughter molecules from the broken chain α are very small, they will probably be destroyed, but if they are of a reasonable size they will continue to be replicated and pass on false genetic information.

(iii) Most alkylating drugs in common use contain two (or more) reactive groups, so that it is possible for one molecule of drug to combine with *two* guanine residues. It will be recalled that the DNA double helix has bases in the centre of the molecule with the bases of opposite strands linked by hydrogen bonds. If there are two guanine residues close together, but on opposite strands, these can be linked by an alkylating agent. An example is shown below:

Original molecule

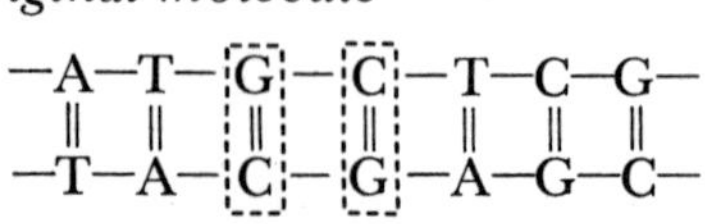

After alkylation

```
—A—T—(G)—[C]—T—C—G—
  ‖   ‖   alkyl   ‖   ‖   ‖
—T—A—[C]—(G)—A—G—C—
```

Such a cross-linked DNA molecule cannot be replicated, for the two strands are now joined by covalent bonds. Should the body manage to eliminate the alkylated nucleotides, *both* strands of the molecule will be broken.

(iv) The final possibility is for the alkylating agent to cross-link one strand of the DNA molecule to a nearby protein. It is known that nucleic acids occur in close association with the acidic proteins known as histones, which are involved in the biological function of DNA. Histones possess large numbers of acidic amino acid residues, such as aspartate or glutamate, and the free carboxylic acid grouping of these residues can react with an alkylating agent. A DNA molecule with one strand cross-linked to a protein cannot be replicated and, again, if the alkylated nucleotide is deleted, the molecule will be broken into two smaller units.

Treatment by an alkylating agent usually leads to the uptake of about one molecule of drug per 10^8 molecular weight of DNA, which is about one molecule per 150 000 base pairs. As the genetic code is in triplets, this is equivalent to about 50 000 amino acid residues. As most body proteins probably contain between 500 and 1000 amino acid residues, it seems likely that only about one or two DNA molecules in every 100 are alkylated by the drug, yet this is sufficient to disrupt metabolism and kill the cell.

Of the alkylating agents so far mentioned, most appear to act on nucleic acids directly and do not require any prior metabolic activation. An exception is cyclophosphamide, which will not alkylate nucleic acids *in vitro*. The reason is that the cyclophosphamide molecule cannot ionize to its active alkylating form while the cyclic phosphate ester ring is intact. Hydrolysis breaks the ring and ionization then occurs. *In vivo*, this hydrolysis is catalyzed by phosphatase enzymes, so that, as with so many other drugs, it is the metabolite and not the parent drug which brings about the biological action.

Nucleic acids are only one of the receptors for alkylating drugs, although they are probably the most important in determining their actions. A more specific interaction with nucleic acids is given by the antibiotic, actinomycin D (structure in Table 2.3), which also chemically links to the N-7 of guanine residues, mainly of DNA, but also of RNA. The principal action of this drug is thought to depend on the reaction with DNA. Molecular models indicate that a molecule of actinomycin D can lie along the groove in the DNA helix. The DNA-actinomycin D complex cannot be used as a template for RNA biosynthesis by transcriptase enzymes, so that the action of the antibiotic is to inhibit RNA synthesis. The peptide chains of actinomycin D are involved in the reaction with DNA, so that if they are removed or modified, combination with DNA is either eliminated or reduced. The precise nature of the bonding to guanine residues by actinomycin C is uncertain, but may involve hydrogen bonds between the peptides and two or more guanine residues. The interaction is not a strong one, for it is insufficient to prevent normal replication of DNA.

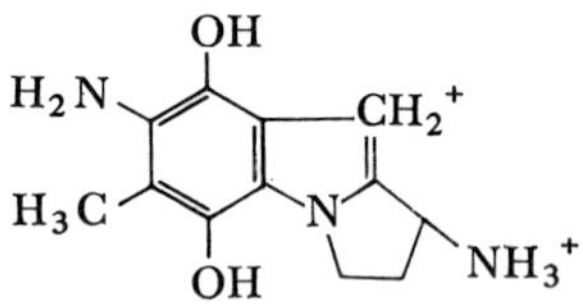

Figure 2.14. Reactive ion.

Another antibiotic which acts more or less specifically on DNA is mitomycin (structure, Table 2.3). It has little effect on DNA *in vitro*, but *in vivo* it combines strongly and irreversibly. This is strongly suggestive of an active metabolite and there is evidence for reduction to the corresponding *p*-diphenol, followed by loss of the methoxyl group and the amide to give the reactive ion shown in Figure 2.14. The ion then acts as an alkylating agent and cross-links two strands of DNA and so prevents them replicating. The mitomycin–DNA complex is also a poor substrate for RNA transcription, so that RNA biosynthesis is reduced. Again it is the N-7 of guanine residues which appear to be involved in the interactions.

A number of other antibiotics have a more or less specific inhibitory action of nucleic acid biosynthesis, e.g. griseofulvin and novobiocin, but they do not combine with nucleic acids, so their receptors are probably one or more of the enzymes involved in DNA biosynthesis, such as DNA-polymerase.

A further antibiotic to act via interaction with nucleic acids in streptomycin (structure in Table 2.3), which has been shown to bind tightly and specifically to 30S ribosomes of sensitive bacteria. Micro-organisms which are not affected by streptomycin possess 30S ribosomes that do not bind this antibiotic. As the cationic molecule of streptomycin will bind to RNA even *in vitro*, it seems likely that the antibiotic's receptor is a particular type of ribosomal RNA. The biochemical action of streptomycin is of a rather unusual type, but is well illustrated by the results of various *in vitro* studies of protein synthesis in cell-free systems from sensitive bacteria. The amino acid sequence of a particular protein is determined in ribosomes by specific matching of triplet base codons of messenger-RNA with anti-codons of transfer-RNAs carrying amino acid residues. In *in vitro* experiments a synthetic *m*-RNA was added (poly-uridylic acid–poly-U) and, in the absence of streptomycin, poly-phenylalanine was formed, for the triplet U.U.U. codes *only* for this amino acid. However, in the presence of streptomycin, under exactly the same conditions, proteins were formed containing significant amounts of leucine, isoleucine and serine, together with phenylalanine. The triplet base codes for these amino acids are U.U.G., A.U.U., and U.C.U., so the most likely explanation of the streptomycin effect is that it permits errors in the codon–anticodon matching, thus allowing the entry of *t*-RNA molecules with an anti-codon triplet slightly different to the correct molecule into the polymerization site of the ribosome.

In a similar study, polycytidylic acid was used as a synthetic messenger and, in the absence of streptomycin, only poly-proline was formed, because the triplet C.C.C. codes for proline. In the presence of streptomycin, significant amounts of histidine (C.A.C.) and serine (U.C.C.) were incorporated into the protein. The mode of action of streptomycin, therefore, involves binding to RNA receptors of bacterial ribosomes such that *m*-RNA is "mis-read" with a high frequency, so that the organism manufactures proteins with incorrect amino acid sequences, foreign to its metabolism.

In view of this disruptive effect of streptomycin, it might be thought surprising that certain mutant strains of bacteria are known which grow only in the presence of the antibiotic. Ribosomes from these organisms will incorporate amino acids into protein in the correct sequence only after streptomycin has been added; in its absence there is a high percentage of "mis-reading". For these organisms, there is an inherited defect in the ribosome structure at, or near, the binding-site for streptomycin, which is corrected by the presence of the antibiotic. It has been shown that sensitivity, resistance and dependence for streptomycin are all determined by alleles of the same genetic locus.

A second type of streptomycin resistance of a rather general type is shown by a wide range of bacterial mutants, which will often show moderate growth in the presence of streptomycin in a synthetic medium, although there is no growth in its absence. For example, an arginine-requiring mutant of *E. coli* shows no growth in a medium lacking arginine, but will grow moderately well if streptomycin is added. The bacteria require arginine because they lack the enzyme ornithine transcarbamylase, which is required for its synthesis. Surprisingly, in the presence of streptomycin the missing enzyme appears. The explanation is that the mutant is making a protein with a defective amino acid sequence unable to serve as an enzyme. The defect is probably in a single amino acid residue (perhaps at the active site, or in a key position determining tertiary structure) that can be traced to a single base change of a DNA triplet. Correct reading of the *m*-RNA from this particular DNA gives rise to the defective enzyme, but streptomycin allows significant "mis-reading", so that some of the protein being made is enzymically active. For example, if the enzyme had a serine residue at the active site, this could be coded by the triplet U.C.C. A mutation to C.C.C. would yield a protein with proline in the place of serine, and so would be inert as an enzyme. However, in the presence of streptomycin, C.C.C. is frequently "mis-read" as U.C.C. so the effects of the mutation are partially neutralized. This advantage is balanced by the production of other defective proteins, so that growth is only moderate or poor.

Another class of drug that binds to nucleic acids is the acridine group, which includes proflavin, acriflavin, euflavin and aminacrine. These are popular antiseptics, effective against a wide range of bacteria. A related compound is quinacrine, which has the curious property of binding to only a specific type of DNA, found on one small segment of the Y-chromosome. As the dye binds to this receptor even in metaphase nuclei, resting cells can be "sexed" by the presence or absence of a small fluorescent spot in the nucleus as seen with a UV microscope.

Binding of acridines to DNA is thought to involve some local "untwisting" of the double helix so that acridine molecules stack between the purine–pyrimidine base-pairs without disturbing the hydrogen bonding. An acridine molecule is a flat cation approximately the same size as the base-pairs. At sub-lethal concentrations, many acridines are potent bacterial mutants, inducing genetic changes very similar to spontaneous mutations. Indeed, many *spontaneous mutants* can be reverted by acridine treatment, while *acridine-induced mutants* often spontaneously revert. It is likely that DNA with intercalated acridine molecules is used as a template by the transcriptase enzymes for *m*-RNA biosynthesis, but that the bases in the region of the dye are mis-read. This mis-reading is probably of the "*frame-shift*" type, whereby steric alteration to DNA deletes (or adds) a single base from a triplet, so that the transcription of subsequent triplets becomes out of step. An example of deletion is shown below:

Normal *m*-RNA	—A—G—A—A—A—G—A—G—U—C—C—A—U
Triplets	1 2 3 4
Corresponding peptide	— thr — lys — ser — pro —

m-RNA from acridine treated DNA where the first base of triplet 3 has been deleted.

	—A—C—A—A—A—G—G—U—C—C—A—U—
Triplets	1 2 3 4
Corresponding peptide	— thr — lys — val — his —

Nuclei possess mechanisms for repair of small defects to DNA molecules, caused by local untwisting or breaks in a chain.

Acridine treatment has been found to inhibit these repair mechanisms, so that locally damaged DNA persists for a long time.

Another important class of drugs with nucleic acid receptors is the steroid hormones, both natural and synthetic. There is little doubt that androgens, estrogens, progestogens and corticosteroids are all actively transported on protein carriers through the cytoplasm of their target cells to nuclei, where they bind to specific nucleic acid receptors and so alter the mechanisms of enzyme induction and repression. All steroid hormones are polyfunctional and bring about a series of biochemically unrelated actions, some of which may be mediated by *extra-nuclear* mechanisms. For example, many steroids, especially corticosteroids, have marked effects on the stability of membranes of sub-cellular particles, such as lysosomes, which may be of importance for anti-inflammatory responses.

Steroids are relatively flat, unionized molecules with high lipid-solubility. By studying the effects of chemical alteration to steroid molecules something can be learned of the nature of the interaction with receptors. The problem is complicated, however, by the transport of steroid drugs from the site of administration to the nucleic acid receptors of the target cells. Many steroids bind to specific plasma proteins, from which they exchange onto specific *cytoplasmic* proteins of the target cells. It is very likely that steroid binding to transport proteins is chemically different to binding to nucleic acid receptors. Chemical modifications which change biological activity of a steroid may have altered the groups involved in binding to the transport protein, rather than to the final receptor. The action of a steroid drug depends on its *rate* of metabolism and clearance, which may be significantly altered by chemical modification, and so produce a marked change in *activity*, without influencing the steroid–receptor *interaction*. With these reservations in mind, it is interesting to study the effects of chemical modification of a particular steroid hormone on biological activity. Table 2.19 summarizes effects of modifications to cortisol (Figure 2.15) on its activity in the liver glycogen deposition test.

The interpretation of these findings is complicated, not only for the reasons given previously, but also by the fact that two or more modifications to a corticosteroid may give unexpected

Table 2.19. Glucocorticoid activities

Corticosteroid	Modifications	Relative activities		
		Glycogen deposition	Thymus involution	Anti-granuloma
Cortisol	None	1	1	1
Prednisolone	1-ene	5	4	3
Methylprednisolone	1-ene 6α-methyl	13	3	6
Fluoroprednisolone	1-ene 6α-fluoro	27	4	18
Paramethasone	1-ene 6α-fluoro 16α-methyl	23	45	64
Dexamethasone	1-ene 9α-fluoro 16α-methyl	90	47	104
Betamethasone	1-ene 9α-fluoro 16β-methyl	59	12	36
Fluocinolone	1-ene 6α-fluoro 9α-fluoro 16α-hyroxyl	44	8	20
Fluocinolone acetonide	1-ene 6α-fluoro 9α-fluoro 16α, 17α-acetonide	138	263	446

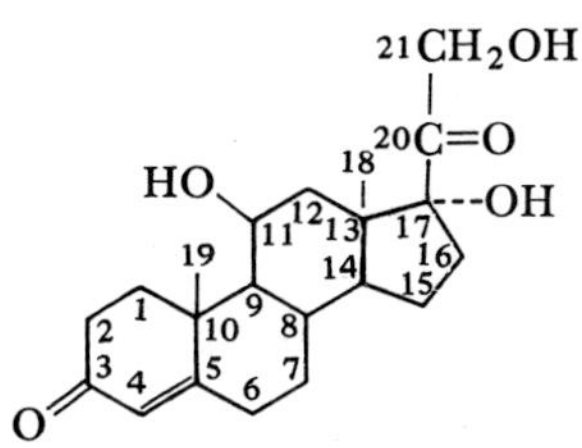

Figure 2.15. Cortisol (see Table 2.19).

activity, or may change one property without altering another. This is illustrated by the following example (Table 2.20).

It appears that an 11β-hydroxyl is essential for activity, though an 11-oxo is acceptable for it can be reduced to β-hydroxyl by the body. Only a fluorine or a chlorine atom in

Table 2.20. Chemical structure and activities of cortisol analogues

Group in cortisol molecule	Modification	Effect on activity
1. 11β-hydroxyl	Deletion	Greatly reduced
	11α-hydroxyl	Greatly reduced
	11β-acetoxy	Greatly reduced
	11β-hydroxy-11α-methyl	Greatly reduced
	11-oxo	Unchanged
	11β-fluoro	Reduced
	11β-chloro	Reduced
	11β-bromo	Much reduced
2. C-19 methyl	Deletion	Greatly reduced
3. C-17 β-sidechain	α-sidechain	Reduced
4. 17α-hydroxyl	Deletion	Small reduction
	Esterification	Unchanged
5. 20-oxo	20-hydroxyl	Greatly reduced
6. 21-hydroxyl	21-methyl	Greatly reduced
	21-al	No change
	21-fluoro	Increased
	21-chloro	Greatly reduced
	21-bromo	Greatly reduced
	Glycine ester	Greatly reduced
	Fatty acid esters	Long-acting
7. 4-ene	Saturation	Greatly reduced
8. C-1 and C-2 saturated	1-ene	Increased
9. Ring A with 3-oxo and 4-ene	Ring A aromatic (with 3-hydroxyl)	Greatly reduced
10. C-2 unsubstituted	2α-methyl	Increased
	2α-ethyl	Greatly reduced
11. C-4 unsubstituted	4-methyl	Greatly reduced
12. 3-oxo	3α-hydroxyl	Reduced
	3β-hydroxyl	Reduced
	3α-acetoxy	Greatly reduced

Table 2.20—continued

Group in cortisol molecule	Modification	Effect on activity
13. C-6 unsubstituted	6-ene	Reduced
	6α-methyl	Increased
	6β-methyl	Reduced
	6α-fluoro	Increased
	6β-fluoro	Reduced
	6α-chloro	Unchanged
	6α-bromo	Reduced
14. C-9 unsubstituted	9α-fluoro	Increased
	9α-chloro	Small increase
	9α-bromo	Reduced
	9α-iodo	Reduced
	9α-hydroxyl	Greatly reduced
	9α-methyl	Greatly reduced
15. C-12 unsubstituted	12α-fluoro	Increased
16. C-16 unsubstituted	16α-methyl	Increased
	16β-methyl	Reduced
	16α-hydroxyl	Reduced
	16α, 17α-acetonide	Increased

the 11β-position can partially replace the hydroxyl, but the resulting compounds are weak. If the 11β-hydroxyl is blocked by the simultaneous presence of an 11α-methyl, activity is lost.

Both the 4-ene and 3-oxo groups are also essential, though 3-hydroxyl compounds are weakly active because they are partly converted *in vivo* to 3-oxo. If the 3-hydroxy is esterified all activity is lost.

The configuration of the pregnane side-chain is important for compounds with the unnatural α-chain at C-17 have only weak activity. The 17α-hydroxyl, however, is not essential. The 21-hydroxyl is also important, but not essential, for replacement by a fluorine, increases activity, as does an aldehyde. The latter group is probably reduced to hydroxyl *in vivo*. Esterification of 21-hydroxyl with glycine decreases activity greatly, but with fatty acids (acetate, caproate, *n*-valerate, etc.) produces active compounds that are long-acting (and so are more suitable for topical administration). It is probable that 21-esters are slowly hydrolyzed to the active parent compounds.

It seems likely, therefore, that corticosteroids are bound to their receptors through the conjugated 3-oxo-4-ene system of ring A, the 11β-hydroxyl of ring C, and the 20-oxo of the pregnane side-chain. The nature of the chemical bonds between steroid hormones and their nucleic acid receptors is uncertain, but probably involves hydrogen bonding at two or three specific sites on the molecule, as is the case with the binding of natural steroid hormones to their transport proteins.

As steroid drugs have flat molecules, it is interesting to know whether they are bound to receptors on the α- or β-sides. Both the 11-hydroxyl and the side-chain carrying the 20-oxo group are β-, so it is likely that this side of the steroid molecule is bound to the receptor. However, α-substitution is not without effect on activity, so that, as was the case for the acridines, steroid drug molecules may fit into a "pocket" of the nucleic acid receptors and that substituents on *either* side of the steroid may alter the binding. Two types of substituent are commonly introduced into steroid drugs: halogens and alkyl groups. Halogens are *electron-attracting*, while alkyl groups are *electro-repelling*, so that electron distribution to functional groups involved in binding to receptor sites may be altered in different ways by the nature of the substituents. Replacement of the 21-hydroxyl by a fluorine will tend to attract electrons from the nearby 20-oxo, which is involved in receptor binding. As this substitution *enhances* activity, this type of electron re-distribution must improve binding, while methyl substitution at the same site, which would have the opposite effect, *decreases* activity. Fluorine atoms are very small, but atomic size increases in halogens from chlorine to bromine to iodine. Electron-attraction decreases with the size of the halogen, so that although a halogen substituent may favour receptor-binding, its *size* may prevent the steroid molecule from coming into effective contact with the receptor site. For substituents at C-21, any bulky substituent (chlorine, bromine, or glycyl) reduces activity, while fluorine increases it.

The differential effects of particular substituents for activity in various tests of corticoid properties suggests that the structural requirements of the steroid binding-site on the different target tissue nuclei may be different. For example, 6α-substituents increase glycogen-deposition and anti-granuloma activity, but have no effect on thymus-involution. A 16α-methyl, however, greatly improves thymus-involution and anti-granuloma activity, but not glycogen-deposition. A methyl at 16β also increases thymus-involution and anti-granuloma activity, but less so that an α-substituent.

It has usually been assumed that steroid drugs react with specific sequences of DNA, but recent work suggests that particular transfer-RNA types may be the receptors.

The anti-malarial, chloroquine, is a further example of a drug with a nucleic acid receptor. It is a 4-aminoquinoline, structurally unrelated to chloroguanide, which was discussed earlier and shown to be metabolized to an inhibitor of dihydro-folate reductase. In contrast, chloroquine binds to DNA in malaria parasites and appears to prevent replication. The nature of the bonding is still uncertain.

Warfarin, the anti-coagulant, may also be an example of a drug with nucleic acid receptors, but the evidence is still questionable. There is no doubt that the drug acts as a vitamin K antagonist, but the precise mode of action of the latter is still unknown. It is likely that vitamin K de-represses the biosynthesis of prothrombin, perhaps by interaction with the appropriate DNA molecules, so that warfarin could compete for the same sites, but be a repressor. Such an action would be similar to, though the reverse of, the de-repression of delta-aminolevulinic acid synthetase in porphyria precipitated by drugs (see p. 147).

It might be thought that drugs with anti-virus activity would necessarily have virus DNA or RNA as their receptors, but this is not necessarily true, for all viruses are nucleo*proteins*, which replicate within cells in a manner very analogous to nucleic acid formation by cell nuclei. A number of chemically unrelated compounds shown anti-virus properties and have found clinical use. For example, β-isatin thiosemicarbazide has been used against smallpox, amanthadine against influenza, and various base-analogues (e.g. 5-iodouracil) against herpes simplex. β-Isatin thiosemicarbazide acts late in the virus growth cycle, even after DNA is formed, while amantadine somehow prevents virus penetration of host cells. We have already examined the receptors for base-analogues and shown that they vary considerably, so that even if an analogue is incorporated into virus nucleic acid, this does not strictly constitute a drug receptor.

Table 2.21 summarizes the interaction of drugs with nucleic acid receptors.

Table 2.21. Drugs with nucleic acid receptors

Drug class	Examples	Receptors
1. Alkylating agents	Nitrogen mustard Melphalan Cyclophosphamide Thio-TEPA Busulphan Mitomycin	RNA, DNA, and proteins
2. Actinomycins	Actinomycin C Actinomycin D Actinomycin E Actinomycin K^1	DNA
3. Streptomycins	Streptomycin Dihydrostreptomycin	Ribosomal-RNA
4. Acridines	Proflavin Acriflavin Euflavin Aminacrine	DNA
5. Steroid hormones	Androgens Estrogens Progestogens Corticosteroids	DNA (?) *t*-RNA (?)
6. Anti-malarials	Chloroquine	DNA
7. Anti-coagulants	Warfarin	DNA (?)

4. Lipids as Drug Receptors

We have seen how many drugs act by chemically combining with proteins, both enzymic and non-enzymic, or with various types of nucleic acid. There remains a large, rather diffuse, group of drugs whose receptors appear to be lipids. Most of these drugs act on biological membranes which are composed of lipid–protein mixtures, though lipids usually predominate. The association between the lipids and the proteins is complex and many different types of chemical bonding are involved. The well-known importance of lipid-solubility for the passage of drugs through biological membranes has already been discussed, but action at the molecular level of drugs whose target receptors are membrane lipids presents considerable problems.

All drugs which act on membranes alter permeability so that

osmotic transfer of ions or other molecular species occurs across the membrane. We have already seen that acetylcholine antagonists act by combining with specific membrane proteins and so prevent increased permeability to Na^+ which normally results from acetylcholine interaction with these receptors. Similarly, many steroid hormones enter the lipid phase of the membranes of lysosomes, and other sub-cellular particles, and either prevent the release of various enzymes (as for corticosteroids), or promote it (as for sex hormones). These actions of steroids, however, are probably a secondary response.

Anaesthetic drugs are probably the most important class with lipid receptors. It has always been puzzling that anaesthesia can be brought about by such a wide range of chemically diverse drugs (e.g. ether, chloroform, nitrous oxide, cyclopropane, carbon disulphide, xenon, etc.). The hypothesis was therefore proposed that anaesthetics physically dissolve in the lipids of nerve membranes, which alters permeability to Na^+. Evidence to support this suggestion came from an investigation of the minimum concentration of various simple aliphatic alcohols required to produce anaesthesia in laboratory animals under standardized conditions. It will be seen from Table 2.22 that anaesthetic potency increases with increasing lipid-solubility, so that the product of minimum anaesthetic potency and the partition coefficient was approximately constant.

It is not certain, however, that this is a satisfactory

Table 2.22. Physical properties affecting anaesthetic action of simple alcohols

Compound	Formula	Minimum anaesthetic potency (M x 10^2) *A*	Lipid–water partition coefficient (x 10^3) *B*	Product *A* x *B*
Methanol	CH_3OH	57	9.7	0.006
Ethanol	CH_3CH_2OH	29	36	0.010
Propanol	$CH_3CH_2CH_2OH$	11	156	0.017
Iso-butanol	$CH_3.CH_2CH(CH_3)OH$	4.5	588	0.026

explanation of anaesthesia, for it is possible to find compounds of similar chemical structure and lipid solubility, which have greatly different anaesthetic properties. A good example is arabinochloralose, where the L-isomer has twice the anaesthetic action as the D-form although the lipid solubilities are identical. However, the lipids of biological membranes are particular steroisomers, so that the interaction with anaesthetic drugs could well be stereo-specific, though this implies that the anaesthetic acts on the lipids in some way other than as a simple solute.

Table 2.23. Relation of solubility to anaesthetic effect

Anaesthetic	Oil–water partition coefficient	Concentration in blood at III_{2-3} mM
Halothane	220	1.0
Chloroform	100	1.8
Vinyl ether	41	2.6
Cyclopropane	34	4.1
Ethylene	14	5.7
Diethyl ether	3.2	20
Nitrous oxide	3.2	29

It has also been proposed that anaesthetics act by the formation of micro-crystalline clathrates in the neurones, but evidence is lacking. The fact remains that volatile anaesthetics are very lipid-soluble and the amount dissolving in adipose tissues has to be taken into account in practical anaesthesia. Table 2.23 lists the oil–water partition coefficients of some common anaesthetics in comparison to their concentration in blood at stage III, plane 2–3 anaesthesia (typical for abdominal surgery).

Other drugs which appear to act on the lipids of various biological membranes are listed below:

Polyene antibiotics (e.g. candicidin, amphotericin B)
Various steroids

IV. SIDE-EFFECTS OF DRUGS

1. General

Most drugs are compounds with potent metabolic effects so that if too much enters the body, toxic reactions are likely. In this section we will consider the biochemical aspects of undesirable side-effects which arise when drugs are given in the correct therapeutic dosage and will not mention toxic effect of over-dosage.

We have already outlined some of the unexpected effects that can arise in particular individuals due to unusual enzymes (e.g. succinylcholine sensitivity, slow metabolism of isoniazid, porphyria, etc.). Here will give details of some of the potentially dangerous drug actions that can arise in *normal* persons.

2. Teratogenicity

Many chemical substances are known with the capacity to affect the cells of a developing embryo and produce congenital abnormalities. Some embryonic defects are incompatible with life, so that abortion results. Many others are carried to term and a viable but abnormal infant is born. In this section we will consider those teratogens affecting only *somatic* cells, so that the congenital defects they induce are not inherited. Other substances are known which can also affect the embryonic *germ* cells and so induce hereditary defects. The problem of mutagenicity is somewhat different and will be dealt with later.

The investigation of *human* teratogenicity is exceedingly difficult and much use has to be made of studies with various laboratory animals. Unfortunately, large species differences have been encountered so that the relevance to humans is often doubtful. Before discussing the biochemistry of teratogenic action, a list is given in Table 2.24 of some drugs believed to be implicated in the induction of foetal abnormalities in humans following administration during pregnancy.

Most laboratory investigations of teratogenicity have been conducted with rats, mice or rabbits. One of the earliest discoveries was foetal sensitivity to deficiency or excess of particular nutrients, or of drugs having their primary action of nutrients. Deficiency of any of the following vitamins during

Table 2.24. Teratogenic effects of drugs

Drug class	Examples	Foetal abnormalities
1. Antibiotics	1. Streptomycin	1. 8th nerve deafness (?)
	2. Tetracyclines	2. Retarded bone age and discoloured teeth
	3. Chloramphenicol	3. Kernicterus
	4. Erythromycin	4. Liver damage (?)
	5. Novobiocin	5. Kernicterus
2. Tranquillizers	1. Phenothiazines	1. Hyperbilirubinaemia
	2. Meprobamate	2. Retarded development (?)
	3. Reserpine	3. Respiratory obstruction
3. Steroid hormones	1. Corticoids	1. Cleft palate (?)
	2. Progestogens	2. Masculinization
	3. Androgens	3. Masculinization
	4. Estrogens	4. Advanced bone age
4. Thyroid drugs	1. Propylthiouracil	1 and 2. Goitre and mental retardation
	2. Methimazole	
5. Anti-coagulants	1. Warfarin	1, 2 and 3. All can induce haemorrhages and death
	2. Dicumarol	
	3. Ethylbicoumacetate	
6. Sedatives, etc.	1. Thalidomide	1. Phocomelia
	2. Barbiturates	2. Neonatal bleeding
	3. Morphine	3. Neonatal bleeding
7. Diabetic drugs	1. Sulphonylureas	1. Various anomalies (?)
	2. Phenformin	2. Lactic acidosis (?)
8. Cancer drugs	1. Aminopterin	1, 2 and 3. All can induce a variety of anomalies which may lead to death of the foetus
	2. Methotrexate	
	3. Chlorambucil	
9. Anti-malarials	1. Quinine	1. Thrombocytopenia
	2. Chloroquine	2. Retinal damage (?)
10. Anti-histamines	1. Meclozine	1 and 2. Various anomalies (?)
	2. Cyclizine	
11. Diuretics	1. Thiazides	1. Thrombocytopenia
12. Miscellaneous	1. Sulphonamides	1. Kernicterus
	2. Hexamethonium	2. Neonatal ileus
	3. Nitrofurantoin	3. Hemolysis
	4. Smallpox vaccine	4. Foetal vaccinia.

pregnancy was associated with foetal malformations: ascorbic acid, riboflavin, thiamin, niacin, folic acid, pantothenic acid, vitamins A, D and E. Consequently malformations can be induced in pregnant animals receiving adequate vitamin intake by vitamin antagonists such as anti-folate drugs, 6-amino-nicotinamide, or 3-acetylpyridine. It was also found that maternal hypervitaminosis A is strongly teratogenic, while administration of large doses of nicotinic acid also can induce malformations.

The effects of hypervitaminosis A are a good illustration of how the nature of the foetal abnormality is determined by the stage of pregnancy, and consequently of embryonic development, at the time of exposure to the teratogen. The data in Table 2.25 illustrates the effects of 60 000 IU vitamin A given

Table 2.25. Time-dependency for teratogenic action of vitamin A

Foetal abnormality	Days of pregnancy Vitamin A given	Incidence of abnormality in surviving foetuses
Anencephaly	5, 6, 7	Few
	8, 9, 10	Many
	11, 12, 13	None
	14, 15, 16	None
Eye defects	5, 6, 7	Few
	8, 9, 10	Most
	11, 12, 13	None
	14, 15, 16	None
Cleft palate	5, 6, 7	Few
	8, 9, 10	Some
	11, 12, 13	Most
	14, 15, 16	Many
	18, 19, 20	None
Spina bifida	5, 6, 7	None
	8, 9, 10	Few
	11, 12, 13	None
	14, 15, 16	None
Cataract	5, 6, 7	None
	8, 9, 10	None
	11, 12, 13	None
	14, 15, 16	Few
	18, 19, 20	Few

by mouth for three consecutive days to pregnant rats: (average gestation period 23–25 days).

Other nutrients involved in teratogenicity are amino acids and sugars, but effects are usually seen only following the administration of a synthetic analogue. For example, the amino acid analogues diazo-oxo-norleucine or azaserine are both teratogens, as is the sugar analogue, 2-deoxyglucose. Large doses of D-galactose are also teratogenic in laboratory animals, perhaps by competing with glucose.

The foetus is a rapidly growing system, so that any substance which interferes with metabolism is likely to induce a malformation. Anything reducing the supply of oxygen is a potent teratogen and malformations are found after relatively brief exposure of pregnant animals to carbon monoxide. As would be expected, all drugs acting on nucleic acid biosynthesis are teratogens (e.g. alkylating agents, base-analogues, acridines, colchicine, etc.). So are such general enzyme inhibitors as lead, thalium, mercury, strontium and selenium salts.

Hormonal relationships between foetus and mother are complex and, as hormones are the primary regulators of many types of metabolism, teratogenic effects are seen following hormone imbalance during pregnancy. Increased incidences of malformations occur in deficiencies of pituitary, thyroid or pancreatic hormones, whether these are induced by damage to the glands, or following the administration of agents which block their secretions (e.g. progestogens, thiouracils, alloxan). As with the vitamins, excess of some hormones is teratogenic and malformations are seen following the administration of high doses of any of the following: thyroxine, corticosteroids, insulin, vasopressin, sex hormones, adrenaline.

There are complex, little understood interactions between hormones and other teratogens. Doses of corticosteroids that are *not themselves teratogenic* will increase the incidence of malformations following excess vitamin A, but a non-teratogenic dose of insulin has the opposite effect. Thyroxine will also reduce the teratogenicity of vitamin A, but thiouracil compounds increase it. In a particular strain of laboratory mice there is a high incidence (about 10%) of cleft lip of spontaneous origin. If thyroxine is given to pregnant mice of this strain on days 11 and 12, the foetal abnormality is prevented, but administration on any other day is totally without effect.

Azo-dye	Structure	Teratogenicity
Trypan blue	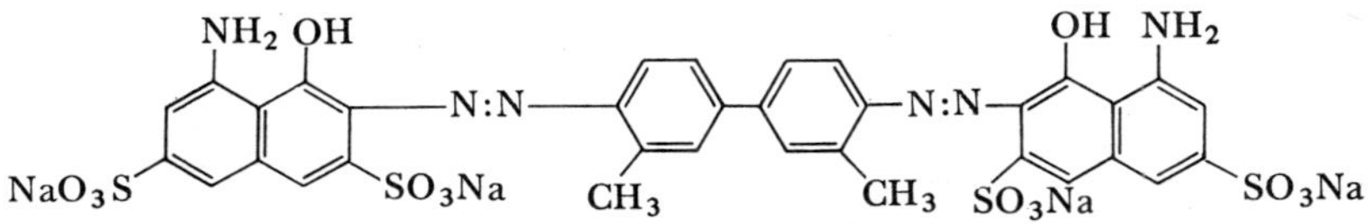	Potent
Evans blue	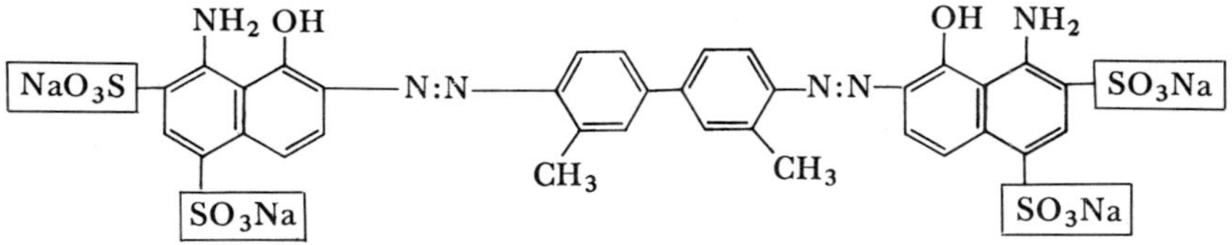	Weak
Niagara blue	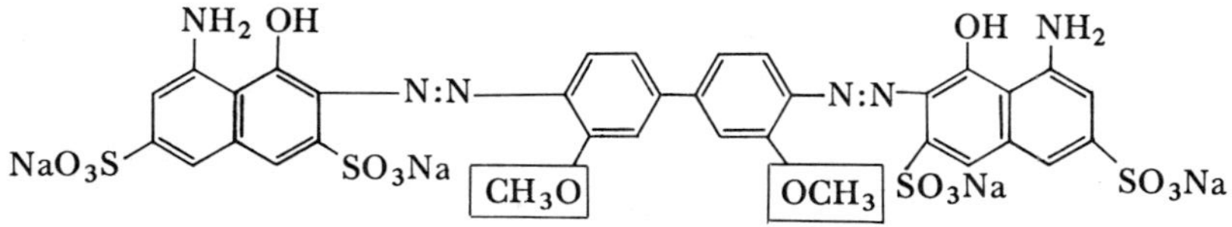	Very Weak
Azo blue	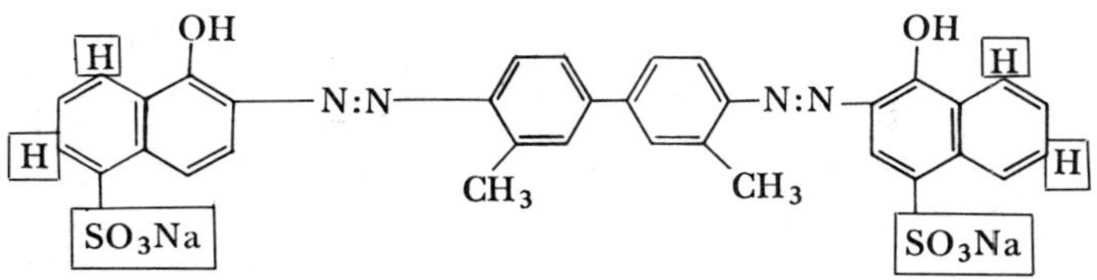	None

Structural changes are indicated

Figure 2.16. Azo-dye teratogenicity.

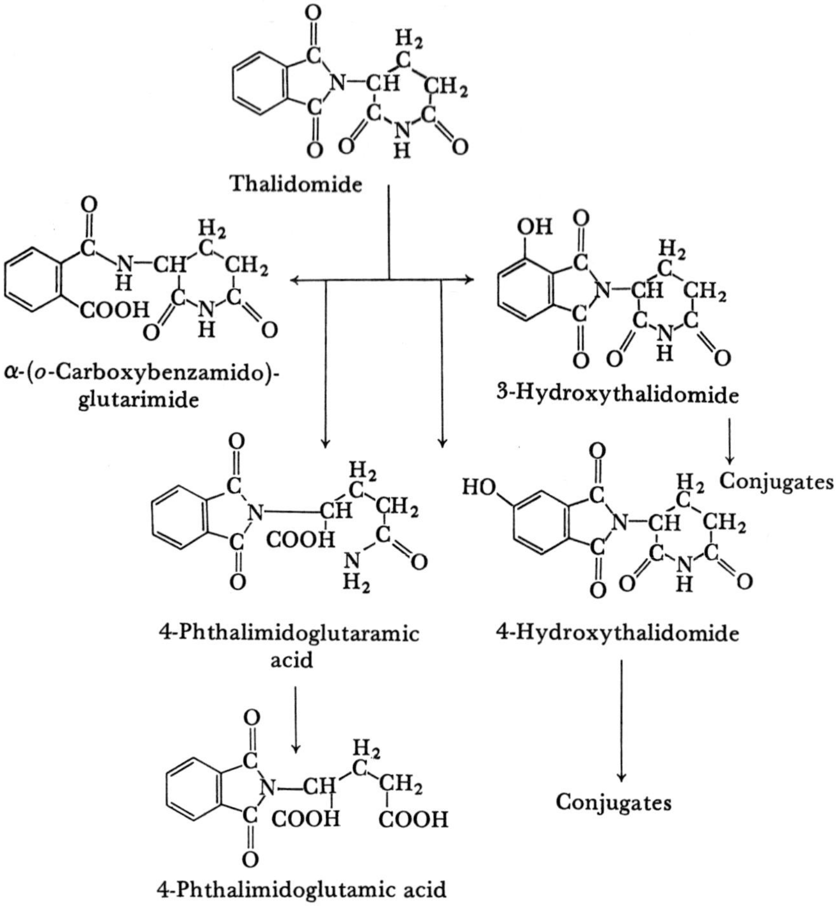

Figure 2.17. Metabolites of thalidomide.

The relationships of chemical structure to teratogenic activity have received some attention, but no very clear-cut relationships have been forthcoming. Figure 2.16 shows the structures of a series of azo-dyes, some of which are teratogens in the rat.

It will be seen that the substitution of methoxy for methyl, or altering the position of sulphonate groups, has a dramatic effect on teratogenicity.

One of the best investigated human teratogenic drugs is thalidomide, which was introduced in Europe and elsewhere in the late 1950s as a hypnotic and tranquillizer. Prescribed as an anti-nauseant in early pregnancy it produced a catastrophic increase in phocomelia. It has been estimated that about 10 000 infants throughout the world were affected. Thalidomide is metabolized in humans to a wide range of products and it is suspected that one or more of the metabolites, rather than the parent compound, is the teratogen. This view is supported by the fact that thalidomide shows only weak teratogenicity in rats, mice, rabbits and dogs, and these species form different metabolites to humans. Moreover, the penetration of particular metabolites into the foetus differs markedly with the species. Figure 2.17 summarizes the major metabolic pathways of thalidomide.

In humans, the major urinary metabolite of thalidomide is 4-phthalimidoglutaramic acid, with small amounts of its deaminated derivative, but in dogs and rats the major metabolite is carboxybenzamido-glutarimide. The 3- and 4-hydroxylated compounds appear in the urine of rabbits. The mode of action of thalidomide, or its metabolites, in producing foetal malformations is still uncertain, but may involve acylation of nucleic acids.

3. Allergic Reactions

Exposure to certain drugs may be associated with reactions having an immunological basis. All allergic responses to drugs require two exposures; a *sensitizing* contact, followed about 7 to 10 days later by the *eliciting* contact. During the interval between the contacts, drug-specific antibodies have been produced by the body, so that drug–antibody interaction occurs following the second exposure and is accompanied by the usual manifestations of allergy. These can range from minor skin lesions to fatal anaphylactic shock.

Allergic responses to a drug given at therapeutic doses should be clearly distinguished from toxic reactions caused by overdoses, and from idiosyncratic responses due to metabolic peculiarities (e.g. porphyria, primaquine-sensitivity, etc.).

Two distinct types of drug allergy can be recognized, depending upon the chemical composition of the drug. If proteins are present, as in anti-sera or vaccines, the proteins *themselves* act as foreign antigens. However, most drug allergies arise from the use of numerous different types of non-protein compounds, usually of low molecular weight. These drugs act as *haptens* and combine with body proteins, so that it is the drug–protein conjugate which serves as the antigenic stimulus to antibody formation. In many cases, the body protein forming conjugates with the drug is one of the plasma proteins, but hapten–erythrocyte interactions can occur, while for topically applied drugs one of the skin proteins can be involved. In each case, the drug hapten is combined with the protein by *covalent* linkage; drugs bound to proteins by hydrogen bonds or ionic linkage do not appear to serve as antigens.

Clinically, there are two important biochemical problems of drug allergy. The first is the cross-reaction phenomenon, whereby the sensitizing contact is with one drug, while the eliciting contact is with a chemically different compound. Secondly, there is the closely related problem of whether the antigenic stimulus depends on a metabolite of the drug or the parent compound. We will consider these problems by taking penicillin allergy as an example.

Allergic reactions to penicillin have been reported in between 1 and 10% of patients, according to different surveys. It is, therefore, one of the commonest drug allergies. Penicillins are metabolized *in vivo* to a number of different products as shown in Figure 2.18.

It will be seen that there is an initial rearrangement of the penicillin molecule to yield penicillenic acid, which can then be transformed to either free penicilloic acid or to penicilloyl derivatives of proteins by the formation of peptide-like bonds with free amino groups, such as those of lysyl residues. It would seem very likely, therefore, that penicillin allergy is produced by penicilloyl-proteins acting as antigens, rather than by penicillin itself.

Experimental investigation of antigen–antibody reactions is easily undertaken in the laboratory and, for low molecular weight drugs, the technique of hapten-inhibition of haemagglutination is particularly suitable. Antibodies to penicillin are produced by repeated injection into rabbits. The drug is then absorbed onto erythrocytes, which are incubated with the antiserum, and agglutination results. The degree of inhibition of agglutination of this system may then be determined by adding penicillin metabolites, or related compounds, separately to the antiserum before incubation with the erythrocytes. In this way the structural specificity of the antibody binding-sites can be determined. The results are shown in Table 2.26.

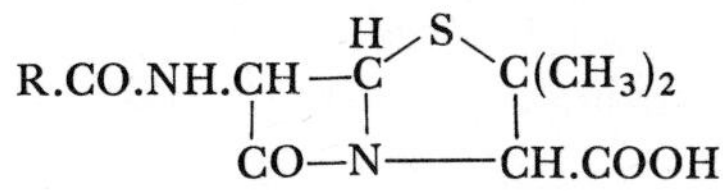

Penicillins (R = benzyl in G, etc.)

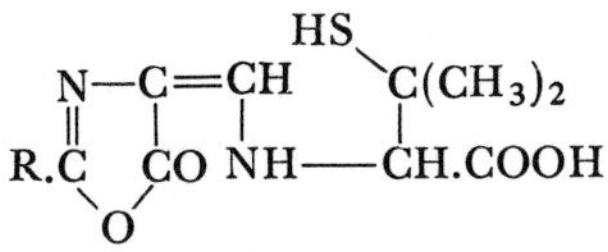

Penicillenic acid

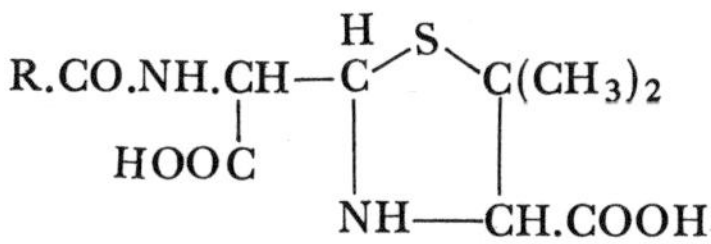

Penicilloic Acid

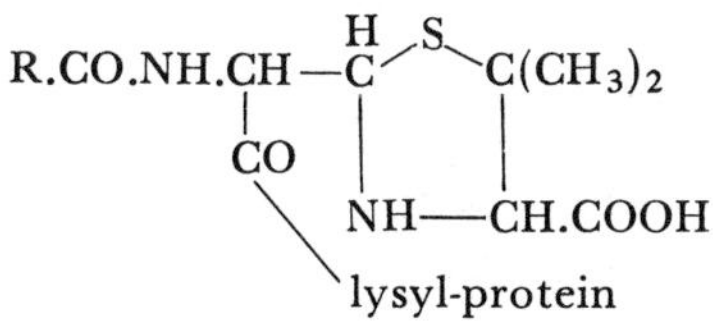

Penicilloyl-protein

Figure 2.18. Penicillin derivatives.

These results are strong evidence that antibodies formed following penicillin administration are to penicilloyl- derivatives rather than to the drug itself, or its two metabolites in the free state. The four structurally related compounds all failed to respond in this test, even at high concentrations, but the various penicillins in therapeutic use, which differ only in the ester group, all respond about the same as penicillin G. It is clear, therefore, that cross-reaction may be expected between structurally related compounds, though not if the modifications are

Table 2.26. Activity of penicillin analogues

Compound	Relative inhibition
Penicillin G	1
Penicillenic acid	1
Penicilloic acid	1
Penicilloyl-lysine	*c.* 100
Penicilloyl-cysteine	*c.* 100
Penicilloyl-aminocaproate	*c.* 100
Penillic acid	0
Penilloic acid	0
Penicillamine	0
Penamaldylbenzylamine	0

extensive, and that metabolites of drugs may be of major importance in determining allergic reactions.

4. Mutagens and Carcinogens

We have already discussed in detail drugs which act on nucleic acid metabolism, as well as the mode of action of the various drugs used in cancer chemotherapy. We will mention only one or two aspects of drug-induced mutations.

A mutation involves genetic alteration, and may be in somatic or germ cells. Cancer is widely believed to be a somatic mutation, but by no means all somatic mutations are harmful. Human albinos often develop freckles due to a somatic mutation restoring melanin-synthesizing ability to one or two melanocytes in particular regions of skin. Germ cell mutation, however, means that the genetic change will be inherited.

Very few drugs are known to be mutagenic or carcinogenic. Among the exceptions are thorium dioxide, a mildly radioactive compound once used as radiological contrast medium, and synthetic estrogens, which have been implicated in breast cancer in men. Acridine dyes, caffeine and most of the drugs used in cancer chemotherapy are potential mutagens, as are ethanol and nicotine, but there is little evidence of significant effects in humans. By extrapolating data from other species it can be shown that a man who is an habitual coffee-drinker can expect to have caffeine-induced mutations in chromosomes of 3% of his sperms, if human sensitivity is similar to that of rodents.

V. SELECTED REFERENCES

1. Binns T. B. and Dodds C. (editors) (1964) *Absorption and Distribution of drugs.* Williams and Wilkins.
2. Busch H. and Lane M. (1967) *Chemotherapy—An Introductory Text* Year Book, Medical Publishers Inc.
3. Goldstein A., Aronow L. and Kalman S. M. (1968) *Principles of Drug Action.* Hoeber Medical Division, Harper and Row.
4. Sollmann T. (1957) *A Manual of Pharmacology.* W. B. Saunders Company, eighth edition.
5. Blacow N. W. and Wade A. (editors) (1972) *Extra Pharmacopoeia—Martindale* The Pharmaceutical Press, twenty-sixth edition.
6. Wolstenholme G. E. W. and O'Connor C. M. (editors) (1960) *Congenital Malformations.* Little, Brown and Co. Inc.

Chapter 3

NATURAL ANTI-METABOLITES*

I. INTRODUCTION

In the previous chapters we have seen that humans are exposed to a wide range of foreign substances, many of which are toxic. Drugs for human use are often synthetic analogues of naturally occurring compounds, although considerable use is made of anti-metabolites derived from plant and microbial sources (e.g. atropine, penicillin, etc.). In this chapter we look in detail at the range of anti-metabolites produced by animals, plants and micro-organisms, and the way in which these act on human metabolism, or the metabolism of human pathogens.

People are exposed to anti-metabolites from plant and animal foods, from purified (or partly purified) preparations used as medicines, and from venomous animals, such as snakes, spiders and insects. There is also a small group of "venomous" plants, such as poison ivy, stinging nettle, etc., which release toxins on contact with human skin.

In the discussion on the mode of action of drugs, we introduced the concept of a "receptor", and showed that the majority of drugs act by combining with specific macromolecules to bring about a biological change. Only a few drugs do not require specific receptors. An analogous situation exists for most toxins. Almost all require specific receptors to produce their actions, and this applies equally to toxins producing human diseases, as for toxins used to control parasitic organisms within the body.

The basic principles of anti-metabolic effects should be borne in mind. There is only a limited number of ways in which a toxic compound can produce its actions. The following list covers the major possibilities, although items are *not* mutually exclusive for multiple action is occasionally possible:

* In this chapter, "anti-metabolites" is used in its widest (and literal) meaning to include all substances interfering with metabolism.

Mode of action of natural toxins:

(i) Inhibit the action of *specific* enzymes by:
  (a) combining with the active site
  (b) combining elsewhere, but by so doing, alter the access of substrate molecules to the active site
  (c) repressing formation of enzyme protein
  (d) removing essential inorganic ion activators
  (e) altering the biosynthesis or binding of organic cofactors.

(ii) *Increase* the activity of a specific enzyme; usually by de-repressing its formation

(iii) *Increase or decrease* activity of *several* enzymes, usually by altering the concentration of cofactors required by several enzymes.

(iv) Combine with non-enzymic protein-receptors of active body constituents, such as neuro-transmitters.

(v) Combine with constituents of biological membranes and alter their permeability.

(vi) Inhibit the function of nucleic acids by:
  (a) combining with DNA molecules and preventing replication
  (b) inhibiting enzymes concerned in nucleic acid biosynthesis
  (c) inhibiting enzymes concerned in the biosynthesis of nucleic and precursors
  (d) being incorporated into nucleic acids and disrupting their information content.

We will discuss first the biochemistry of microbial toxins, particularly of pathogenic bacteria and of those micro-organisms likely to contaminate human foods. This will be followed by an account of microbial toxins acting on organisms causing human diseases, or which find other application in *treatment* of pathological conditions, such as cancer. We will then turn to plant toxins, which we will consider in two groups; those present in plants used as human foods, and those present in medicinal plants. Finally, we will look at toxins present in animals, particularly of venomous species, or present in animal tissues used as food. Throughout we will stress what is known of the *chemistry* of these natural anti-metabolites, and of their *mode of action*.

II. MICROBIAL TOXINS

1. General

We can distinguish two types of clinically important microbial toxins: those of pathogenic organisms and those used as antibiotics.

Toxins of microbial origin affect human metabolism either by being secreted from organisms present within the body tissues, or by entering with foods containing the toxin, although live micro-organisms may not be present.

Infections are the commonest type of human disease. An understanding of each particular condition requires a knowledge of the effects of its toxins on human metabolism, while successful treatment requires the selection of a suitable agent to disrupt the metabolism of the particular pathogenic micro-organism. While great advances have been made in the biochemistry of microbial toxins in recent years, much still remains to be learned. Human diseases are caused by a very wide range of micro-organisms, but bacteria and (to a lesser extent) fungi are the most important types, acting via toxic anti-metabolites. Viruses, which also cause many types of disease, do not produce toxins, but alter nucleic acid metabolism of human cells so that viral rather than normal nucleoproteins are formed. They will not be considered further in this section, but some of the toxins produced by pathogenic algae will be mentioned.

It has been known since 1888 that symptoms of many bacterial diseases can be produced in laboratory animals by administering cell-free filtrates of cultures from pathogenic bacteria. This was shown first for diphtheria, then for tetanus, and finally for very many bacterial diseases. It was found that, in some cases, bacterial toxins are excreted in urine of infected patients and that injection of sterile preparations of the urine produces symptoms of the disease. The first relatively pure preparations of bacterial toxins were obtained as early as 1890, and since then, many have been isolated and chemically identified.

Bacterial toxins are macro-molecules of various types and two sorts of toxin are usually recognized: *exotoxins*, which are *released* by bacteria, and *endotoxins*, which are present *within* the bacteria, but may be prepared by extracting the cells, and are probably released *in vivo* if cell walls are disrupted. While this classification is open to some criticism, it has pragmatic value and extends into chemistry (exotoxins are simple proteins, while endotoxins are lipoglycoproteins) and morphology (exotoxins are produced by many *gram-positive* bacteria, but only by very few gram-negative species, e.g. *Shigella* and *Vibrio* sp.; while endotoxins are *all* produced by *gram-negative* bacteria).

Exotoxins are continuously secreted by actively growing bacteria, while endotoxins, which form part of cell-walls, are released only by autolysis. Exotoxins are usually more toxic than endotoxins, but are readily inactivated by heat; endotoxins are heat-stable. Homologous antibody will neutralize exotoxins, but not endotoxins, for it is directed against the protein and carbohydrate portions of the molecules, whereas, as we shall see later, toxic effects are mediated by lipid residues.

It is important to appreciate the *extreme* toxicity of bacterial exotoxins. Table 3.1 lists relative toxicities of bacterial exo- and

Table 3.1. Relative toxicities

Poison	Source	Relative animal toxicity
Strychnine	Alkaloid of strychnos seeds ($C_{21}H_{22}O_2N_2$)	1.0
Fluoroacetic acid	Synthetic compound ($FCH_2.COOH$)	10
Crotactin (crystalline)	Protein component of rattle-snake venom	10
Ricin (crystalline)	Protein of castor beans	20
Endotoxins (purified)	Gram-negative bacteria, *irrespective* of species	*c.* 0.1
Exotoxins (purified)	Gram-positive bacteria, *dependent* on species	300 to 3 000 000
Examples:		
(i) ε-Toxin	*Clostridium welchii*	300
(ii) α-Toxin	*Staphylococcus* sp.	350
(iii) Diphtheria	*Corynebacterium diptheriae*	2 000
(iv) Tetanus	*Cl. tetani*	1 000 000
(v) Botulinum B	*Cl. botulinum*	3 000 000

endotoxins in comparison with some other protein and non-protein poisons.

While these figures are only approximate, and apply for parenteral administration, but take no account of species, weight, etc., they nevertheless illustrate the enormous toxicity of bacterial exotoxins, which have a lethal capacity many orders of magnitude higher than any other poisonous substance. It has often been said about 100 g of botulinum B is sufficient to kill the entire human population of the Earth. In contrast, bacterial endotoxins are much less potent.

2. Bacterial Exotoxins

Most species secrete more than one with different biological actions.

Table 3.2 is a summary of exotoxins produced by common

Table 3.2. Some bacterial exotoxins causing human disease

Species of bacteria	Human disease	Exotoxins	Major effects
Clostridium botulinum (the botulins formed depends on type)	Botulism	Botulinum A	Neurotoxic
		Botulinum B	Neurotoxic
		Botulinum C	Neurotoxic
		Botulinum D	Neurotoxic
		Botulinum E	Neurotoxic
		Botulinum F	Neurotoxic
		Hemagglutinin	Blood coagulation
Cl. oedematiens	Gas-gangrene	α-Toxin	Dermonecrotic
		β-Toxin	Dermonecrotic and haemolytic
		γ-Toxin	Haemolytic and dermonecrotic
		δ-Toxin	Haemolytic
		ε-Toxin	Haemolytic
Cl. septicum	Gas-gangrene	α-Toxin	Haemolytic
		β-Toxin	DNase
Cl. sordellii	Gas-gangrene	Hemorrhagin	Oedema and haemolysis
Cl. tetani	Tetanus	Tetanospasmin	Neurotoxic
		Tetanolysin	Haemolytic and cardiotoxic

Table 3.2—continued

Species of bacteria	Human disease	Exotoxins	Major effects
Cl. welchii	Enteritis necroticans and gas-gangrene	α-Toxin	Haemolytic and dermonecrotic
		β-Toxin	Lethal
		γ-Toxin	Lethal
		δ-Toxin	Lethal
		ε-Toxin	Dermonecrotic
		η-Toxin	Lethal
		ι-Toxin	Dermonecrotic
		θ-Toxin	Haemolytic and cardiotoxic
		κ-Toxin	Proteolytic
		μ-Toxin	Spreading factor
Corynebacterium diphtheriae	Diphtheria	Shick-toxin	Lethal and dermonecrotic
Shigella dysenteriae	Dysentry	Hemorrhagin	Neurotoxic
Staphylococcus aureus	Pyogenic infections	α-Toxin	Dermonecrotic and haemolytic
		β-Toxin	Haemolytic
		γ-Toxin	Haemolytic
		δ-Toxin	Haemolytic
		ε-Toxin	Haemolytic
		Hyaluronidase	Spreading factor
		Coagulase	Blood coagulation
		Staphylokinase	Fibrinolytic
		Enterotoxin	Emetic
		Leukocidin	Kills W.B.C.
Streptococcus pyogenes	Pyogenic infections Tonsillitis	Dick-toxin	Erythrogenic
		Streptolysin-O	Cardiotoxic and haemolytic
		Streptolysin-S	Haemolytic
		Hyaluronidase	Spreading factor
		Streptokinase	Fibrinolytic
		Streptodonase	DNase
Vibrio cholerae	Cholera	Enterotoxin	Diarrheagenic
		Haemolysin	Haemolysis

human pathogenic bacteria, together with their principal actions. The *nomenclature* of the toxins is the one in most common use, but some alternatives will be mentioned later.

The study of bacterial exotoxins involves their purification, which can be achieved only by culturing the bacteria on a large scale in the laboratory. There is every reason to believe, however, that toxins obtained in this manner are identical to those released by infective organisms within the human body. Some curious features of exotoxin production in cultures mentioned below may be relevant to the human situation:

(i) The quantities of toxins released by *Clostridium welchii* depend on the carbohydrate supplied; glucose suppresses both α- and σ-toxins, while dextrins increase α-toxin release.
(ii) At concentrations of iron above about 100 μg/100 ml, toxin production ceases for many bacteria.
(iii) Production of staphylococcal hemolysins require Mg^{++} whereas Ca^{++} is inhibitory.

It seems possible, therefore, that a particular strain of bacteria may produce different toxins in different parts of the body, depending on supply of carbohydrates or metal ions.

In culture media, many exotoxins are formed and released as *protoxins*, which are larger molecules than exotoxins and require metabolic activation in a manner very similar to that for proteolytic enzymes of the pancreas (e.g. trypsinogen and chymotrypsinogen). Activation of protoxins can result from either proteolytic enzymes of the host, or similar bacterial enzymes (not themselves toxins). It must also be noted that proteolytic *destruction* of exotoxins is catalyzed by both mammalian and bacterial enzymes.

The molecular weight of bacterial exotoxins ranges from about 25 000 to over one million. As for many other proteins, molecular association and dissociation occurs in solution. The dissociation of a high molecular weight toxin into smaller components has often revealed the presence of "sub-units" with different toxicities. For example, most preparations of botulinum A have a M.W. about one million, but can be made to dissociate into smaller units, each about 70 000, but differing in electrophoretic mobility. Only some fractions are toxic, and these are more toxic per unit weight than the original preparation. Botulinum A under certain conditions can be shown to give rise to a sub-unit of molecular weight of only about 12 000, which shows clearly that exotoxin molecules are composed of a variety of sub-units, only some of which have toxic activity.

Purification of crude tetanus toxin leads to a spontaneous loss of activity, due to the formation of a *dimer* of the toxic *monomer,* which has a molecular weight of about 67 000. It seems likely that the two sub-units join together so that the parts of the molecules responsible for toxicity are masked. The dimer can be dissociated by reduction with cysteine, with restoration of toxicity, which suggests that the dimer is formed by a disulphide bridge between the two smaller units. Similar findings have been reported for diphtheria toxin.

We have seen that some toxins are secreted as inactive protoxins, while many associate and dissociate in solution. Association can lead to loss of activity, if the toxic sites of the protein are masked in the process. Dissociation usually provokes an increase in toxicity, sometimes only for one or two fractions, but occasionally there is complete loss of toxicity. The leukocidin of *Staphylococcus* sp. dissociates into two components (known as F and S), neither of which has any toxic properties, but which act synergistically.

All exotoxins so far purified have proved to be simple proteins lacking any prosthetic group (unlike the endotoxins which possess both lipid and carbohydrate residues). Amino acid compositions are known for several, and are shown in Table 3.3.

Before considering the more important exotoxins individually, it is of value to consider their mode of action in general. Three types of biochemical response will be seen in the later examples as the basis for all the biological properties:

(i) The toxin may be an *enzyme* which catalyses a chemical change in body constituents.
(ii) The toxin may *not* be an enzyme, but chemically combines with important body constituents.
(iii) The toxin may be a foreign protein which acts as an *antigen* and produces allergic responses.

All exotoxins act partly through mechanism (iii) but, for most, this is secondary to a primary action mediated by one of the other two possibilities.

It should be noted that the *antigenic sites* of an exotoxin are different to its *toxic sites*. An exotoxin preparation occasionally loses its toxicity spontaneously, or can be induced to do so by chemical reagents (e.g. nitrous acid, ascorbic acid, iodine). An

Table 3.3. Amino acid composition of some bacterial exotoxins

Amino acid	Botulinum type A	Tetanus	*Clostridium welchii* ϵ-toxin	Staphylococcus α-toxin	Staphylococcus enterotoxin	Diphtheria	Streptococcus erythrogenic
Alanine	3.9	3.4	3.0	2.5	1.3	5.1	1.8
Arginine	4.6	3.8	2.6	3.9	2.5	4.1	2.1
Aspartic acid	20.3	15.6	17.3	13.2	19.1	12.6	15.3
½ Cystine	0.8	0.9	0.6	0	—	0.8	2.2
Glutamic acid	15.6	10.7	10.0	7.2	9.3	13.2	15.7
Glycine	1.4	3.2	3.1	3.9	2.8	4.7	3.6
Histidine	1.0	1.2	1.3	1.4	3.0	3.9	3.5
Isoleucine	11.9	8.8	3.7	3.8	4.0	6.7	5.9
Leucine	10.3	8.7	5.8	4.3	7.3	7.7	9.5
Lysine	7.7	9.8	14.2	7.7	17.1	9.1	12.2
Methionine	1.1	2.1	1.9	1.9	2.5	1.9	1.7
Phenylalanine	1.2	5.1	1.7	3.6	5.7	4.4	5.3
Proline	2.6	3.9	3.8	1.9	1.7	3.6	4.6
Serine	4.4	5.7	6.1	5.2	6.1	8.5	5.7
Threonine	8.5	5.2	10.0	6.2	5.0	6.1	6.1
Tryptophan	1.9	0.9	0.5	?	0.8	1.0	0.8
Tyrosine	13.5	7.0	6.0	3.8	10.7	4.8	11.3
Valine	5.3	4.6	7.0	3.2	5.8	7.4	5.8
Total	116.0	100.6	98.6	73.7	104.7	105.6	113.1

exotoxin preparation which has lost its toxicity is called a "*toxoid*", yet antigenic properties remain and antibodies produced in response to the toxoid will combine with *both* the toxoid *and* the active exotoxin. Combination of antibodies with exotoxins *always* leads to a loss of activity *in vivo*, although not necessarily *in vitro*. For example, diphtheria toxin will inhibit protein biosynthesis in intact HeLa cells, as well as in cell-free extracts, but whereas the diphtheria toxin–antibody complex continues to inhibit the biosynthesis in the cell-free extracts, it no longer has any action on the intact cells.

3. Tetanus Toxin

It will not be possible to discuss all the important exotoxins of bacteria, so a few have been selected to illustrate the various biochemical actions involved.

The toxin of *Clostridium tetani* was among the first to be isolated and investigated towards the end of the nineteenth century. Tetanus is an important human disease arising from contaminated wounds. Even with modern treatment, the mortality may be as high as 40%, depending upon the general health and age of affected persons. All signs and symptoms seen in human patients can be imitated in animals by injection of the purified toxin.

The exotoxin of this bacterium is sometimes called "*tetanospasmin*", for a second exotoxin, known as "*tetanolysin*" has been discovered. The latter is of little clinical importance as all the major effects are caused by the former. The following discussion is confined to tetanospasmin, which will be referred to as "tetanus toxin".

Highly purified toxin preparations are available and prove to be of a simple protein containing 1.04% sulphur and a trace of phosphorus. For the dimeric form (see above), the M.W. is about 140 000 and contains 10 cysteine residues, of which four occur as two disulphide bridges. The molecule is a single polypeptide chain, but lacks a free *N*-terminal amino group.

The toxin is one of the most poisonous substances known, producing hypertonic muscular paralysis followed by convulsions and death in all species of laboratory animals. It is most active when injected parenterally, and is only weakly active by mouth. The LD_{50} for the mouse is about 20 pg per adult animal by the i.m. route.

From its actions in both human and animals, it appears that the toxin acts primarily on the CNS and the convulsions and spasticity are due to a suppression of inhibition within the spinal cord. There are several similarities between the effects of tetanus toxin and strychnine.

Investigations with the toxin have failed to show any particular biochemical actions on isolated cells or tissues, and cholinesterase and choline acetylase are unaffected. The toxin is firmly bound by intact brain cells, or even by a cell-free brain homogenate. Non-neural tissues, such as liver, spleen, bone marrow, etc., do not bind the toxin. Interestingly, tetanus *toxoid* (see above) is *not* bound by the brain. Of other bacterial exotoxins so far examined, only tetanus toxin is absorbed by neural tissues, so that the interaction is clearly of considerable significance to its mode of action.

Further knowledge of tetanus-toxin receptors was obtained in two ways. Firstly, by studying binding to different sub-cellular fractions of neural tissues, and secondly by studying binding to chemically characterized fractions of brain lipids. The toxin does not bind to mitochondria, microsomes or soluble fractions, but it combines avidly with synaptosomes (nerve endings). Synaptic vesicles do not bind to toxin; it appears to be specifically linked to fractions rich in synaptic membranes. Only one fraction of brain lipid has the ability to combine with tetanus toxin *in vitro: gangliosides*. These are water-soluble, chloroform-soluble mucolipids. Despite their water-solubility, gangliosides cannot be extracted from brain with water, for they occur complexed to cerebrosides. Pure cerebrosides do not combine with tetanus toxin, but a complex of ganglioside and cerebroside (often called "*protagon*") binds the toxin as well as free gangliosides.

The binding of tetanus toxin to gangliosides is very specific. Other exotoxins, with the exception of strychnine, are not bound. No other known substance is capable of binding with "protagan", except tetanus toxin and, to a much lesser extent, its toxoid. It does not necessarily follow that the mode of action of the toxin is determined by this binding to mucolipids of the synaptic membranes in the CNS, but the part played by lipids in membrane permeability makes it a strong possibility.

As gangliosides are relatively small molecules, they can be chemically modified with comparative ease and interaction of tetanus toxin with modified derivatives then measured *in vitro*.

Table 3.4 summarizes the toxin-binding capacity of various ganglioside derivatives:

Combined with the previously mentioned fact that cerebro-

Table 3.4. Binding ot tetanus toxin to various gangliosides

Ganglioside	Structure (carbohydrate moiety only)	Tetanus-toxin binding (g toxin/g ganglioside)
G IV	—Glu—Gal—GaN—Gal—Sial Sial (on Gal) Sial	19.0
G III	—Glu—Gal—GaN—Gal Sial (on Gal) Sial	19.3
G II	—Glu—Gal—GaN—Gal—Sial Sial (on Gal)	3.6
G I	—Glu—Gal—GaN—Gal Sial (on Gal)	2.6
G 0	—Glu—Gal—GaN Sial (on Gal)	0
G'lact	—Glu—Gal Sial (on Gal)	0

Where: Glu = glucose, Gal = galactose, GaN = *N*-acetylgalactosamine, Sial = Sialic acid.

sides do not bind the toxin, this information shows the degree of specificity required for ganglioside–toxin interaction. For a significant binding, gangliosides require:

(ii) *Two* sialic acid residues *linked to galactose*
(ii) A *second* galactose residue beyond the acetylgalactos-amine residue.

A terminal sialic acid residue is *not* essential. A possible explanation of these facts is that tetanus toxin had a two point contact with the ganglioside—to the second galactose and second sialic acid residues. On the basis of the known molecular weights, it can be calculated that each tetanus toxin monomer combines with two ganglioside molecules (G IV or G III), which implies that the toxin monomer contains *two* toxic binding-sites. It is not known if both these sites are used for *in vivo* binding to synaptic membranes, but their existence suggests that the toxin could act by binding with two ganglioside residues and so distort the lipoidal structure at this point.

Linkage of the protein toxin to the galactose and sialic acid residues of the receptor may involve hydrogen bonding, ionic bonding, or both.

It is an interesting fact that tetanus toxin does not bind to ganglioside G O, which is the defective ganglioside of Tay-Sachs disease, so that patients with this condition may perhaps be immune to this toxin.

Injection of a ganglioside solution will protect laboratory animals against tetanus toxin, but only if given *before* the toxin. Once the toxin is bound to the CNS, there is no known way of removing it. Presumably, patients who recover from tetanus lose the toxin from their synaptic membranes by gradual repair of the membranes and degradation of the involved molecules.

Precisely how combination of tetanus toxin with synaptic membranes of the CNS leads to a suppression of inhibition is not clear. A number of "inhibitory-transmitters" of the CNS are known and each may have a particular anatomical region of action. In the spinal cord, glycine is strongly favoured as the inhibitory transmitter and it has now been shown that, while tetanus toxin blocks inhibitory impulses in the cord, it does not interfere with the action of glycine applied through a micro-pipette, although strychnine does. This suggests that tetanus toxin acts *pre-synaptically* to prevent the release of glycine, while strychnine acts *post-synaptically* to block the glycine receptors.

There is also evidence to suggest that tetanus toxin blocks neuromuscular transmission and interferes with muscular relaxation mechanisms, but little is known of the biochemistry of these effects.

Unlike any other exotoxin, tetanus toxin is transported in

the body partly in the blood and partly in the tissue spaces between individual fibres of nerve trunks.

4. Clostridium Perfringens α-Toxin

This bacterium is one of the group producing gas-gangrene in humans. It produces five or more different exotoxins, but we will consider only α-toxin. Early studies of this material were complicated by cross-contamination of preparations by one or more of the other exotoxins, but it is now clear that α-toxin is a potent lecithinase (phosphatidase or phospholipase).

Lecithins are hydrolysed by four specific classes of enzyme: Lecithinase A hydrolyses the α-fatty acyl residue, lecithinase B the β-fatty acyl residue, lecithinase C (a phosphatidase) hydrolyses the α′-phosphorylcholine residue. The fourth group catalyse the removal of the choline residue. The α-toxin is a member of the C group and the products formed by its action on lecithins are an α,β-diglyceride and phosphorylcholine (Figure 3.1).

Some lecithinases are activated by the physical properties of lecithins or the products of hydrolysis (e.g. the lecithinase of snake venoms, discussed on p. 343), but this is not true of α-toxin which has a *direct* haemolytic action. Very similar lecithinases to α-toxin are produced by many of the gas-gangrene clostridia.

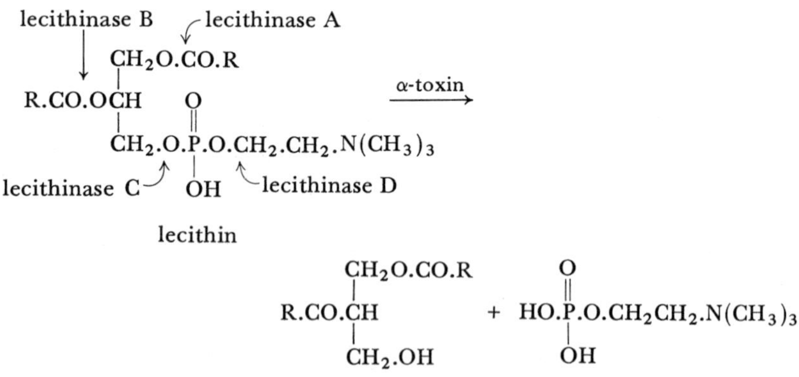

α,β-Diglyceride + phosphorylcholine

Figure 3.1. Lecithinase classification.

The α-toxin has an optimum pH *in vitro* between 7.0 and 7.6, depending on the buffer system used, so that it is most active in human blood plasma (pH 7.4). The enzyme is activated by Ca^{++} up to 0.1 M, though higher concentrations are inhibitory. A number of other ions are also activators (Mg^{++}, Zn^{++}, Co^{++}). Interestingly, α-toxin inhibited *in vitro* by EDTA-chelation of Ca^{++}, cannot be fully reactivated by calcium ions, although it can by zinc or cobalt; which suggests that one or other of these metals is an essential cofactor and that Ca^{++} is only a secondary activator.

The enzyme is relatively heat-stable, but is denatured by saline. This can be prevented by serum albumin, which appears to strongly protect the toxin from inactivation. The α-toxin protein has a molecular weight of about 106 000, but appears to contain two similar sized components, both with enzymic activity, which may be separated chromatographically, although antigenicity is identical.

Many attempts have been made to investigate the active site of α-toxin, but results have been negative. The enzyme is not inhibited by sulphhydryl-blocking agents, or by denaturation with urea. The only known inhibitors are compounds which remove the metal ions required for activation (e.g. EDTA, H_2S, *O*-phenanthroline, etc.), or which interfere with the interaction of the substrate (e.g. detergents such as sodium dodecyl sulphate). As mentioned above, zinc will fully re-activate α-toxin inhibited by EDTA, and analyses of the purified toxin have revealed the presence of bound zinc, but not of other metals. As zinc stimulates α-toxin production by the micro-organism in culture, it is likely that the active site of α-toxin contains this metal ion.

The substrate specificity of α-toxin is high, for only lecithins are acted upon to any significant extent. Of the other phospholipids, there is some evidence for a mild action on phosphatidyl ethanolamine.

The haemolytic action of α-toxin is certainly related to its lecithinase activity. The toxin acts on lecithins in the membrane of erythrocytes, which results in their disruption. A significant increase in serum concentration of one product of the enzyme (phosphorylcholine) has been demonstrated *in vitro* for blood treated with α-toxin *before* haemolysis was seen. This toxin has a haemolytic action on erythrocytes of all laboratory animals so far tested, with the single exception of the horse, the reason for which is unknown. α-Toxin is occasionally called the "hot-cold" haemolysin for in low concentration it will not induce haemolysis at 37°C, but if treated blood is cooled, haemolysis occurs. However, the purified toxin is potently haemolytic *in vivo* at body temperature.

A number of biological effects other than haemolysis are seen in animals treated with purified α-toxin. It is uncertain to what extent these are mediated by lecithinase activity. Long-lasting contractions develop in smooth muscle after exposure to α-toxin, which correlate with a loss of microsomal ATPase activity. There is also increased ionic permeability in membranes of striated muscle, accompanied by degeneration of mitochondria. All these changes could be due to partial removal of lecithins from various important membranes.

5. Staphylococcal Leukocidin

Many species of staphylococcus produce a toxin which kills polymorphonuclear leukocytes and macrophages, but has no action on any other type of cells. This group of bacteria produce many other toxins (see Table 3.2) and leukocidin is merely one factor in their pathogenicity.

Leukocidin has been purified and shown to consist of two distinct fractions (named "F" and "S"), neither of which is toxic by itself. The two components give rise to *separate* antibodies, which do not cross react. Leukocidin F is a simple protein of M.W. about 32 000 with very low solubility at pH 7.0, though readily soluble in acids. Leukocidin S is a somewhat larger simple protein of M.W. about 38 000, much more soluble than the F component at most pHs. A mixture of the two gives good separation on electrophoresis at pH 5, but at pH 9 they run as a single band. Similarly, in the ultra-centrifuge a solution of the mixture at low ionic strength gives a major fraction containing both components, but at higher ionic strength the two sediment separately.

It is clear that the toxic action of leukocidin is mediated by the two components, each acting separately. While leukocidin can be shown to interact with leukocytes, only a small amount is absorbed by cells; the rest is inactivated and remains in solution. The first biochemical change seen in the leukocytes after treatment with the toxin is a marked increase in permeability to K^+. The cell membrane becomes permanently depolarized, Ca^{++} is also accumulated, while some proteins are lost. There is increased ATPase activity, which results in decreased ATP concentrations.

The mode of action of leukocidin is obviously complex and it is difficult to arrive at the sequence of events in its reaction with receptors. It seems likely that synergistic interaction occurs between the two components of the toxin and some component of the leukocyte membrane, resulting in inactivation of leukocidin and alteration of membrane structure and function. The toxin has been shown to be inactivated by homogenates of leukocytes, as well as by preparations of leukocyte cell membranes *in vitro*. Investigations have been made to determine whether any of the known components of these membranes interact with or inactivate leukocidin.

By testing the effects of various phospholipids on the two components *separately* it became clear that leukocidin S is not affected but leukocydin F is polymerized by the following:

phosphatidylserine
diphosphoinositide
triphosphoinositide
monophosphoinositide
phosphatidylcholine

but *not* by sphingomyelins, lecithins, cerebrosides, gangliosides, or triglycerides. The polymerization induced by triphosphoinositide is not blocked by triphosphinositol, which is inactive. A likely explanation of these results is that leukocidin F interacts with fatty acid side-chains of specific charged phospholipids.

While leukocidin S is not polymerized by phospholipids, if it is present in solution with the F form when a suitable phospholipid is added, *both* leukocidins polymerize. This suggests a conformational change in leukocidin F, induced by the phospholipid, to "expose" a reactive site capable of polymerization, or of reaction with the S form.

While leukocidin *S* is not polymerized by phospholipids, leukocidin *F* is polymerized by *several*. One phospholipid (triphosphoinositide) *inactivates* both components. The toxins do not chemically alter this phospholipid, which can be recovered unchanged from inactivated toxin solutions. This particular phospholipid produces conformational changes in *both* leukocidins, with loss of toxicity.

Extrapolation of the *in vitro* findings to the toxic actions of leukocidin in humans has produced the following hypothesis:

(i) Leukocidin F interacts with triphosphoinositide molecules of leukocyte membranes, so that two fatty acyl residues of the phospholipid penetrate hydrophobic regions of the protein toxin and a change in tertiary structure results.

(ii) Adsorption of leukocidin S onto exposed reactive groups of leukocidin F follows, so that the S molecule is in close proximity to the triphosphoinositide.

(iii) Interaction of leukocidin S with the inositol ring of the phospholipid changes the *conformation* of the ring.

(iv) This change in conformation is accompanied by separation of the two leukocidins, now joined, from the membrane.

This hypothesis is supported by evidence of a change in conformation of the inositol ring *in vitro* and *in vivo* following treatment with leukocidin. Alteration of the special relationships of the highly charged phosphate groups joined to the inositol ring is assumed to create a defect in the cell membrane, which allows penetration of ions formally unable to pass.

The only known *specific* inhibitor of the toxin actions of leukocidin is tetraethylammonium ion, which is positively charged and may well block interaction of toxin with negatively charged triphosphoinositide.

6. Other Exotoxins

We have looked in detail at the biochemical actions of three important bacterial exotoxins and seen very different primary interactions. Tetanus toxin binds avidly and specifically to gangliosides of synaptic membranes within the CNS; *Cl. perfringens* α-toxin is a highly active enzyme catalysing the hydrolysis of lecithins of erythrocyte membranes; while staphylococcus leukocidin binds specifically to phosphoinositide molecules of leukocyte cell walls. We will attempt to briefly summarize what is known of the mode of actions of other exotoxins in Table 3.5.

Table 3.5. Enzymic and non-enzymic exotoxins

Type of toxin	Organism and name	Target molecules
Enzymes:		
1. Phospholipase C (lecithinase C)	*Clostridium welchii α-toxin*	Membrane lecithins
	Clostridium oedematiens β-toxin	
	Clostridium oedematiens γ-toxin	
	Clostridium haemolyticum μ-toxin	
	Bacillus anthracis toxins	
2. Phospholipase B	El tor vibrio toxins	Membrane lecithins
	Staphylococcus sp. toxins	
3. Sphingomyelinase	*Staphylococcus* sp. β-toxin	Membrane sphingomyelins
4. Hyaluronidase	*S. aureus* spreading factor	Cell hyaluronic acids
	S. pyogenes spreading factor	
5. Collagenase	*Cl. welchii* κ-toxin	Collagen on connective tissue
6. DNase	*S. pyogenes* streptodornase DNA	
	Cl. septicum β-toxin	
7. Thrombin-like	*S. aureus* staphylokinase	Plasminogen
	S. pyogenes streptokinase	
Non-enzymic proteins:		
1. Neurotoxins	*Cl. botulinum* toxins	Bind to presynaptic cholinergic and neuro-muscular fibres to prevent ACh release
	B. anthracis	Centre of brain
2. Haemolysis (non-enzymic)	*Streptococcus* streptolysin 0	Cholesterol esters of erythrocyte membranes
	B. cereus cerolysin	Ditto (?)
	Streptococcus streptolysis S	Phospholipids of erythrocyte membranes
	Staphyloccocus α-toxin?	
3. Enzyme inhibitors	*Pasteurella pestis murine* toxin	Mitochondrial oxidations
	Diphtheria toxin	Oxidative phosphorylation

7. Bacterial Endotoxins

We have already discussed the problems of definition for toxic products of bacteria and shown that exotoxins are simple proteins secreted by gram-positive organisms; although they may be formed *within* bacterial cells, or even *stored* in granules. In contrast, the term "endotoxin" was derived historically to describe toxic extracts of gram-negative bacteria. Such extracts contain complex proteins, with lipid and carbohydrate inclusions, and cannot be converted to toxoids, as can exotoxins.

There is now no doubt that endotoxins are concentrated at the surface of bacteria and are, in fact, constituents of cell walls or membranes. When endotoxins are administered to laboratory animals, several clear-cut responses can be recognized:

(i) Death—Only parenterally administered endotoxin causes death, but toxicity is much less than for exotoxins; resistance increases with age, except in germ-free animals.

(ii) Fever—In most species, but not in rats or mice, endotoxins are pyrogenic, and at high doses, a double response is observed to a single dose.

(iii) Dermal reaction—If endotoxin is infiltrated into the skin of a rabbit, then a second dose given i.v. about 24 hours later, a striking haemorrhagic necrosis develops at the site of the infiltration (dermal "Shwartzman reaction").

(iv) Generalized reaction—A generalized "Shwartzman reaction", involving haemorrhagic lesions to the mesentery and kidneys, is caused by the second administration of endotoxin.

(v) Miscellaneous—Among other responses to endotoxins are a necrotic action on malignant tumors, a non-specific increased resistance to infections, a stimulation of interferon biosynthesis, hyperglycemia followed by hypoglycemia, hypoferremia, antibody formation, and several more.

With such wide ranging biological actions by a single preparation from particular bacteria, it is relevant to know whether the effects are produced by one or more components of endotoxin. The data in Table 3.6 show the effects of acid hydrolysis on a particular endotoxin preparation for its action on several previously mentioned parameters. It will be seen that these do not change together, as would be expected it all were mediated by a single component gradually decomposed by acid.

There is no doubt that similar biological responses are seen with *live* bacteria, *dead* bacteria, and bacterial *extracts*. The problem of determining the chemical composition of endotoxins, however, is complex. Unlike exotoxins, which can be purified from cell-free culture fluids, endotoxins must first be extracted from bacterial cells. The chemical composition of extracts varies not only with the method of extraction and purification, but also with bacterial species and its method of

Table 3.6. Effect of acid treatment of exotoxins on toxic potencies by different tests

Treatment with O. N acetic acid (min)	Relative potencies (mouse)				
	Pyrogenic	Inflammatory	Lethal	Anti-tumor	Anti-infective
0	100	100	100	100	100
5	27	41	49	50	141
15	39	41	21	12	83
30	18	18	14	3	27
45	14	1	1	0	1
60	1	0	0	0	1

cultivation. Most extracts contain proteins (or polypeptides), lipids, carbohydrates, and various metal ions, but the nature of the distribution of these within the various macromolecular fractions is uncertain. There is evidence for lipoglycoproteins, lipoproteins, glycoproteins, glycolipids, and phospholiposaccharides. Which are present in bacterial cell walls, and which are artefacts of extraction, is not known. Further extraction and chemical treatment of extracts produces solutions containing low molecular weight compounds with endotoxic properties. The smallest M.W. fractions lack protein, but contain a mixture of phospholiposaccharides, which are presumably the functional groups of the native endotoxin, now broken away from macromolecular carriers on which it occurs in living bacteria. The model of bacterial endotoxins is now widely accepted (Figure 3.2). Region I consists of repeating oligosaccharide units, containing specific sugars in specific

O-antigen specific chains	Basal core	Lipid A	Cell wall proteins
Region I	Region II	Region III	

Figure 3.2. Generalized molecular model for bacterial endotoxins.

linkages, which determine the immunological specificity of the toxin. This is linked to the Basal Core, which is probably the same (or very similar) in bacteria of a particular genus. Region III, unlike Regions I and II, contains lipids, together with carbohydrate residues. Linkage of the toxin to proteins of the bacterial cell wall occurs through Lipid A.

The best investigated endotoxins are those of *Salmonella* species and Table 3.7 gives the structures of repeating oligosaccharide units found in Region I of endotoxins from various species.

Table 3.7. Chemical structures of region I for bacterial andoxins

Species	Region I oligosaccharides
Salmonella typhi	2α-Man-1,4-Rha-1,3α-Gal-1- | 1,3 | 1,4 α-Tyv α-Glc and 2-α-Man-1,4α-Rha-1,3α-Gal-1- | 1,3 | 1,4 Tyv Ac-0-2α-Glc
S. typhimurium	2α-Man-1,4-Rha (?)α-Gal-1 | (?) | 1,6 (or 1,4) α-Abe α-Glc and 2α-Man-1,4β-Rha-1,3α-Gal-1 | 1,3 | 1,4 Ac-0-2α-Abe α-Glc
S. bredeney	4α-Man-1,4-Rha-1,3α-Gal-1 | 1,3 | 1,6 (or 1,4) α-Abe α-Glc and 2α-Man-1,4β-Rha-1,3α-Gal-1 | 1,3 | 1,6 α-Abe α-Glc

Table 3.7 continued

Species	Region I oligosaccharides
S. paratyphi A	2α-Man-1,4α-Rha-1,3α-Gal-1 1,3 | 1,4 α-Par Ac α-Glc
S. anatum	6β-Man-1,4-Rha-1,3α-Gal-1 | Ac
S. newington	6β-Man-1,4-Rha-1,3β-Gal-1 and 6β-Man-1,4-1Rha-1,3β-Gal- | 1,4 α-Glc
Escherichia coli 0111.84	α-Col-1,6β-GlcNAc-1,2-Glc (?) Gal | 1,4 α-Col
E. coli 086.87	α-Gal-1,3β-Gal-1,3(or 4)-GalNAc | ? α-Fuc
Shigella flexneri	6-GlcNAc-1,2-Rha-1,4-Rha-1- | 1,4 α-Glc

Abbreviations: Ac = Acetyl; NAc = *N*-Acetyl; Man = Mannose; Rha = Rhamnose; Par = Paratose; Fuc = Fucose; Gal = Galactose; Glc = Glucose; Tyv = Tyvelose; Abe = Abequose; Col = Colitose

It will be seen that the basal unit of Region I oligosaccharides is a three sugar-fragment in almost every case; although one or two side-chains may be present. For salmonelli, the three sugars are mannose, rhamnose and galactose in every case so far investigated, but side-chain sugars vary considerably. It follows that linkage to the basal core must be from a terminal galactose residue of Region I oligosaccharide. The linkage appears to be fairly constant and has the structure shown in Figure 3.3.

The basal core itself is a polysaccharide containing the following five sugars:

D-glucose (Glc)
D-galactose (Gla)

(Region I) oligosaccharide-Gal-1, ----------- 4-Glc-basal core
terminal | 1,2 (Region II)
GlcNAc

Figure 3.3. Linkage of O-antigen specific chains to basal core.

N-acetyl-D-glucosamine (Glc*N*Ac)
L-glycero-D-mannoheptose (Hep)
2-keto-3-deoxy-D-mannooctonic acid (KDO)

We have seen that the antigenic Region I is joined through its terminal glactose residue to a glucose residue of the basal core, and that this glucose carries a Glc*N*Ac side-chain. It has proved possible to obtain a pentasaccharide from this end of the basal core of salmonella endotoxins by chemical degradation and it has the structure shown in Figure 3.4. The remainder of the

Region I-Glc-1,2-Gal-1,3-Glc-remainder of core
| 1,2 (?) | 1,6
Glc*N*Ac Gal

Figure 3.4. Pentasaccharide of salmonella basal core.

basal core has proved more difficult to investigate, but is a polymer of *only* Hep and KDO, though some of the residues are phosphorylated, while others are condensed with phosphoryl ethanolamine residues.

The overall basal core structure of the endotoxins of bacteria other than salmonella is the same as that described above, except that the initial pentasaccharide varies with the genus.

In *Shigella* it is shown in Figure 3.5.

Glc*N*Ac
| 1,4
-Gal-1,3-Glc-
| 1,3 | 1,4
Glc Glc

Figure 3.5. Pentasaccharide of shigella basal core.

Several heptose and ketodeoxymannooctonic acid containing polysaccharides have been obtained from the basal core of endotoxins of *Salmonella* species and *E. coli* and the structures of a few are shown in Figure 3.6.

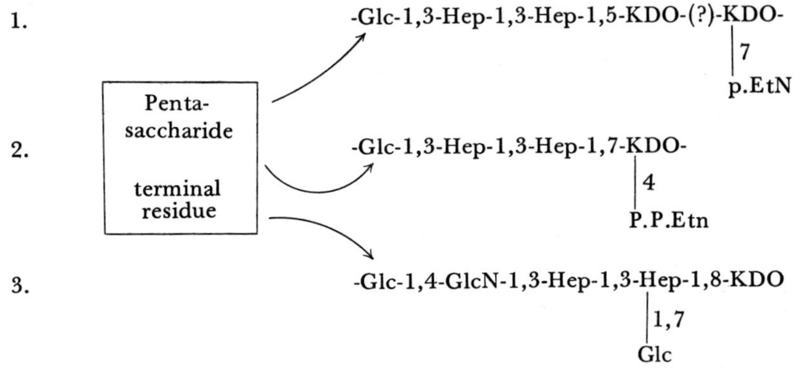

Figure 3.6. Three different saccharides from bacterial endotoxin basal cores.

There remains Region III of the endotoxin molecule, which is known as "lipid A". The basic chemical units are:

D-glucosamine
phosphate
fatty acids (mainly lauric acid, myristic acid, palmitic acid, D-β-hydroxymyristic acid)

Enzymic analysis has shown that the phosphate is largely esterified to position-4 of the glucosamine residues, while hydroxymyristic acid forms a peptide-like bond to the amino group on C-2. Other fatty acids are esterified to the hydroxyl of C-6. The basic monomeric unit of lipid A is probably as drawn in Figure 3.7. Several alternative structures are compatible with

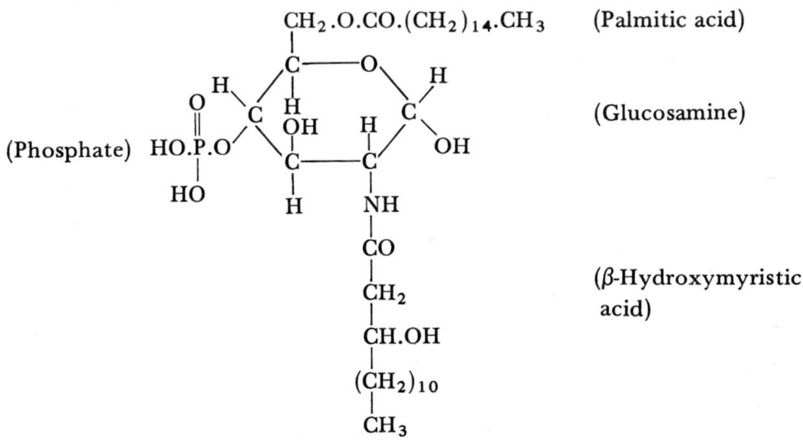

Figure 3.7. Lipid A monomer

the evidence so that various monomers may be included. The major linkage is probably β-1,6 between the substituted glucosamines, but 1,4 crosslinking by phosphate is possible.

Lipid A polymer appears to be linked 1,3 to a terminal KDO residue of the basal core, but nature of the linkage to proteins of bacterial cell walls is uncertain.

The overall structure of bacterial endotoxins may therefore be represented by the Figure 3.8.

Region I — *O*-antigen specific chain,
e.g. (-6β-Man-1,4-Rha-1,3-β-Gal-)c.8
linked through -Gal-1,4- to:

Region II — initial pentasaccharide of basal core,
e.g. -Glc-1,2-Gal-1,3-Glc-
(first Glc: 1,2 branch to GlcNAc; last Glc: 1,6 branch to Gal)

which links to remainder of core through
-Glc-1,3-:
e.g. -Hep-1,3-Hep-1,5-KDO-2,8-KDO-
(last KDO: 7 branch to P.EtN)

which links through -KDO-2,3- to:

Region III — lipid A,
e.g. poly-fatty acylglucosamine phosphate
which links to:

Proteins of bacterial cell wall

Figure 3.8. Generalized chemical structures of endotoxin regions.

We have seen that the antigenic portion of endotoxin molecules is Region I, but the toxins produce a wide range of effects, including many that are not of an immunological basis. The search for primary receptors may be illustrated by the following example:

(i) Administration of endotoxin leads to a rapid development of hyperglycaemia.
(ii) Hepatic activation of phosphorylase precedes the hyperglycaemia.
(iii) These observations suggest that endotoxins act on phosphorylase, but a simultaneous increase in circulat-

ing adrenaline occurs, which is a known activator of hepatic phosphorylase.

(iv) Even after adrenalectomy, endotoxin stimulates glycogen breakdown in the liver, but this is prevented by cortisol, which also increases survival.

(v) If protein biosynthesis is inhibited by actinomycin D (see p. 235), cortisol no longer protects against the actions of endotoxins.

(vi) Very low doses of endotoxin, which produce little or no stress, produce a *fall* in blood sugar, while hypoglycaemia follows several hours after *non-lethal* high doses.

(vii) Endotoxin also lowers the activity of hepatic tryptophan pyrrolase and decreases concentration of total nicotinamide coenzymes.

(viii) The effect of endotoxin on tryptophan pyrrolase and coenzymes is antagonized in normal animals by cortisol, and is not seen at all in animals resistant to endotoxin.

All these observations suggest some hormone-like action of endotoxins or carbohydrate metabolism in the liver, but the receptor and initiator of these changes at the subcellular level is not known, nor is it clear which part of the endotoxin molecule brings about these non-antigenic actions.

8. Toxins of Fungi

We may distinguish three clinically important types of fungal toxins:

(i) Toxins of fungi used for food (e.g. mushrooms, fermented products, etc.)

(ii) Toxin of fungi accidentally present in foods (e.g. aflatoxins, yellowed-rice, ergotism, etc.)

(iii) Toxins released by fungi pathogenic to humans (e.g. ringworm, histoplasmosis, actinomycosis, etc.)

As many derivatives of poisonous fungi are used medicinally, we will reserve a discussion of these toxins until a later section (p. 332); as will drugs derived from ergot and related fungi.

There is a relationship of dietary thiamine intake to beri-beri, but the *acute* cardiac form of this disease has always presented a number of mysterious features not readily explained by a vitamin deficiency. This condition appeared in Japan in the late seventeenth century and no earlier accounts can be found. It was originally known as "Edo-sickness", for it was prevalant in the city of Edo (which is now known as Tokyo), a city with over one million inhabitants three centuries ago. There can be no doubt that the disease appeared about the time polished rice began to be used in the place of husked grains, but it was the rich gourmets who first used the new rice who developed the disease. It was not until much later that acute cardiac beri-beri began to be seen among the poor as the use of polished rice spread. A vitamin deficiency in rich gourmets, who clearly were eating many other foods seems very unlikly as is a condition often fatal within 3 days from the onset of symptoms.

The key to understanding the condition came with appreciation that mouldy rice appears normal after it has been polished, so that careless (or unscrupulous) grain-merchants could add contaminated rice to their bulk-stores without detection. This was realized in Japan by about 1898 and official rice-inspectors were appointed for limited areas in 1901. By 1922, more than 60% of Japan was regularly inspected and the incidence of acute cardiac beri-beri had begun to fall. It was in this year that thiamine was discovered and shown to be a successful treatment for dry beri-beri in other countries. After a few years, thiamine was introduced into Japan as a treatment for acute cardiac beri-beri, but was only mildly successful. The historical evidence, therefore, points strongly to a fungal toxin as *the* aetiological agent in this condition although thiamine-deficiency could well be a complicating factor in poor people.

A telling experiment was conducted in 1922, when liver obtained post-mortem from acute cardiac beri-beri patients was fed to rats with artifically induced thiamine-deficiency. It rapidly cured the animals. Despite these facts, acute cardiac beri-beri is *still* widely considered to be thiamine avitaminosis, but as the condition is now exceedingly rare, this is of little practical importance. Nevertheless, the elucidation of the cause of a disease responsible for the deaths of millions of people in Japan, and other rice-eating countries, is not without interest.

Although most moulds found on rice belong to the genus

Penicillium, other are occasionally present. Mouldy rice with toxicity to laboratory animals often contains *P. citreo-viride* (sometimes called *P. toxicarium*). Extracts of this fungus cultured *in vitro* are toxic to animals by any route of administration, the cause of death being cardio-vascular and respiratory failure preceded by a progressive ascending paralysis with transient convulsions. The extracts fluoresce strongly in UV and a similar fluorescence is seen in blood, urine, milk, and internal organs of treated animals. The active constituent is thought to be *citreoviridin*, which has the structure shown in Figure 3.9. The mode of action is not known, but its structure suggests a similarity to ubiquinones, whose functions it may disrupt.

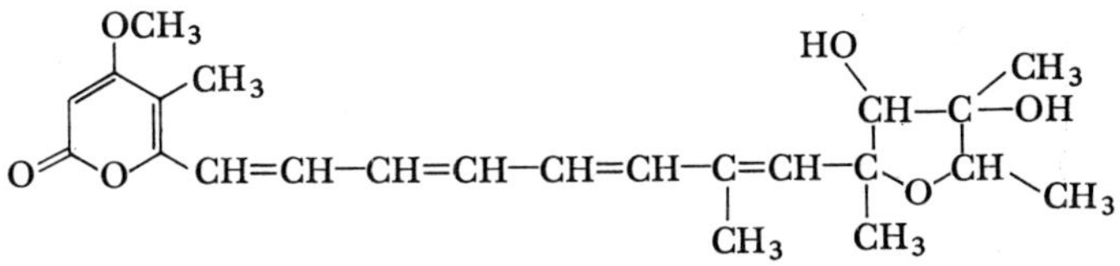

Figure 3.9. Citreoviridin.

Several other toxic substances have been detected from fungal contaminants of rice, but their biological actions are unlike those of citreoviridin, and so are unlikely to contribute to acute cardiac beri-beri, although they may well be related to a variety of other conditions seen in Asia and Africa. Table 3.8 summarizes some data on these toxins.

Some of these toxins have been studied in detail, so we will briefly consider their biological effects.

Rice contaminated by fungi is often yellow, due to the toxic metabolites, many of which are coloured. "*Yellow Rice Research Teams*" have been established in several countries, particularly in Japan, to investigate the biochemistry and importance of these fungal toxins. *P. islandicum* was found as a common contaminant of rice from many tropical countries and to be associated with human liver cirrhosis in Egypt, S.E. Asia, and elsewhere. In culture, this fungus produces three hepatoxic agents: luteoskyrin, cyclochlorotidine and islanditoxin. Structures are shown in Figure 3.10. Luteoskyrin is a bis-dihydroanthraquinone, but the others are chlorine-containing cyclic peptides. These two compounds have very similar effects, so we

Table 3.8. Some fungal exotoxins

Toxic compound	Fungi	Biological actions
Citrinin	*Penicillium citrinum* *P. implicatum* *P. urticae* *Aspergillus candidus*	Nephrotoxin
Patulin	*P. expansum* *P. urticae* *P. claviforme* *A. clavatus* *A. ciganteus* *A. candidus*	Neurotoxin
Rugulosin	*P. rugulosum* *P. tardum* *P. wortmanni* *P. brunneum*	Hepatotoxin
Luteoskyrin	*P. islandicum*	Hepatotoxin
Cyclochlorotidine	*P. islandicum*	Hepatotoxin
Islanditoxin	*P. islandicum*	Hepatotoxin
Rubratoxin	*P. rubrum*	Hepatotoxin and Nephrotoxin
Sterigmatocystin	*A. versicolor*	
Ochratoxin	*A. ochratoxin*	Hepatotoxin
Aflatoxins	*A. flavus*	Hepatotoxin

will consider only cyclochlorotidine and compare its actions with luteoskyrin (Table 3.9).

These results show that luteoskyrin is a fairly specific liver toxin, but cyclochlorotidine has actions on a variety of tissues,

Table 3.9. Fast and slow hepatotoxins of fungi

Toxin	Toxicity	Major actions
Luteoskyrin	LD_{50} mouse 150 mg/kg (s.c.) 220 mg/kg (p.o.)	Slow poison (2–4 days), prostration, hypothermia, and acute liver injury.
Cyclochlorotidine	LD_{50} mouse 0.5 mg/kg (s.c.) 6.5 mg/kg (p.o.)	Fast poison ($<$ 1 day), respiratory and circulatory collapse, convulsions, hepatic coma only in long survivors.

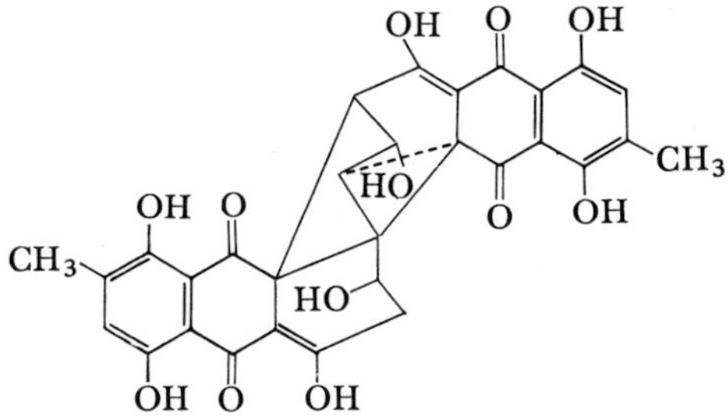

Luteoskyrin

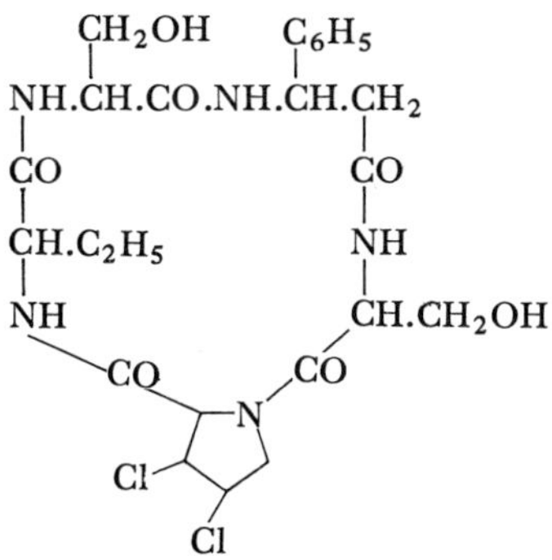

Cyclochlorotidine

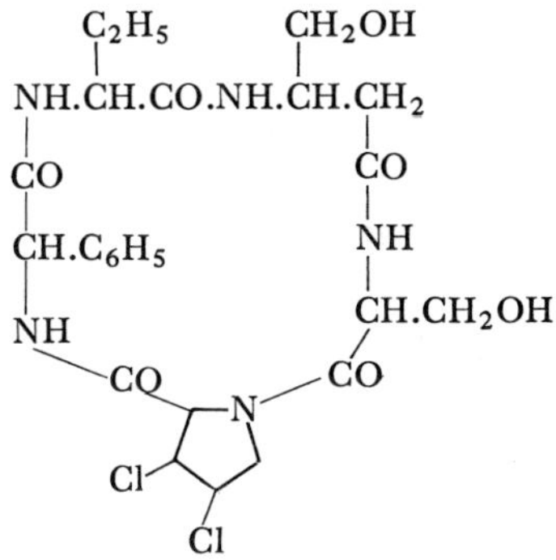

Islanditoxin

Figure 3.10. Chemical structures of some mould toxins.

including the liver. If either compound is labelled with ^3H, then given i.v., 60 to 80% of the dose accumulates in the liver within about 30 minutes. Whereas cyclochlorotidine is excreted principally in urine as the unchanged compound, luteoskyrin passes into bile and faeces, probably as a mixture of metabolites.

Luteoskyrin acts on the liver to inhibit *and* uncouple oxidative phosphorylation in the mitochondria. As with many other substances, receptors for the toxin are molecules of the mitochondrial membranes. A second action is shown by *in vitro* luteoskyrin binding to DNA in a Mg^{++} dependent reaction, although it does not become intercalated between base-pairs (contrast with actinomycin D, p. 174). As the toxin readily binds to synthetic pyrimidine-base polymers, but not to purine-base polymers, it is likely that the binding is to pyrimidine bases of DNA. The DNA–luteoskyrin complex cannot serve as template for transcriptase enzymes and 10^{-6} M concentrations of toxin inhibit *m*-RNA biosynthesis *in vivo*.

In contrast, the primary action of cyclochlorotidine is on carbohydrate metabolism and a rapid depletion of liver glycogen follows its administration. In normal animals there is a transient hyperglycaemia, followed by a drop in blood glucose to very low values. The hyperglycaemic response is not seen in adrenalectomized, nor alloxan-diabetic rats following the toxin, but hypoglycaemia develops rapidly. At the enzyme level, a major increase in glucose-6-phosphate dehydrogenase activity is seen, but other enzymes of carbohydrate metabolism are unaltered, except for glycosyl-transferase, which is reduced. Increased G6PD activity raises $NADPH_2$ levels to favour biosynthesis of fat, but it is difficult to account for the fast toxic and lethal effect of cyclochlorotidine in this way. Cause of death is presumably directly related to the hypoglycaemia. Chlorine in the molecule of this toxin is essential to toxicity.

Both compounds are long-term hepatic carcinogens when given at sub-lethal doses for periods of 200 days or more.

Other hepatotoxins are metabolites of *A. Flavus*, a mould occasionally found on rice, but more frequently on groundnuts. In 1960 about 100 000 turkeys died in Britain of liver necrosis induced by aflatoxins in contaminated feeds over a three-month period. This "turkey-X disease", as it was originally called, was followed in 1961 by the death of 14 000 ducklings in Britain from a similar cause, which was eventually traced to Brazilian groundnuts. At least eight chemically related alfatoxins have now been identified from cultures of the fungus and all are potent hepatotoxins and carcinogens. The structures of the two parent compounds are shown in Figure 3.11, together with the substituents in the various metabolites (Table 3.10). All aflatoxins fluoresce in UV and may be separated by thin layer chromatography in various solvents. Toxicities range from 36 μg/kg for *B 1* to 345 μg/kg for *G 2* (LD_{50} for ducklings).

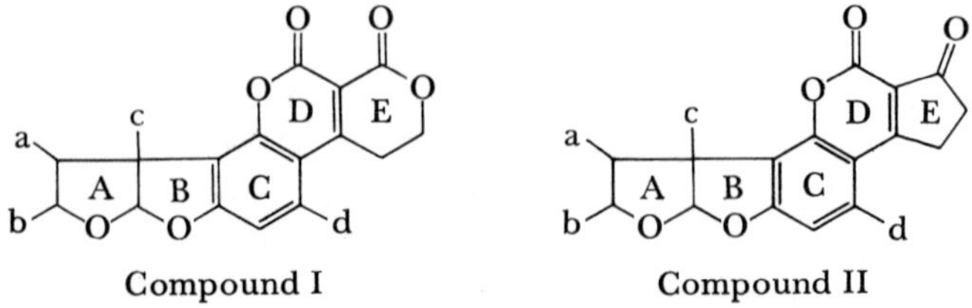

Figure 3.11. Chemical structures of aflatoxins (see Table 3.10).

Table 3.10. Nomenclature of aflatoxins

Aflatoxin	Parent compound	Substituents			
		a	b	c	d
B 1	II	H	H	H	OCH_3
B 2	II	H_2	H_2	H	OCH_3
B 2a	II	H_2	H(OH)	H	OCH_3
M 1	II	H	H	OH	OCH_3
M 2	II	H_2	H_2	OH	OCH_3
G 1	I	H	H	H	H
G 2	I	H_2	H_2	H	H
G 2a	I	H_2	H(OH)	H	OCH_3

Chemical structures are shown in Figure 3.11.

Structure-activity investigations suggest that rings A and E are involved in binding to receptors and that unsaturation of ring A and a cyclopentanone structure for E are most favourable for interaction.

Aflatoxin receptors have not been precisely identified, but protein biosynthesis is markedly reduced. There is evidence for inhibition of *s*RNA-amino acid ligases (amino acid activating enzymes), but both DNA and RNA biosynthesis are also reduced. Several aflatoxins bind to DNA *in vitro* and block its use as a template for transcriptases, as well as inhibiting DNA replication. Whether these latter effects are concerned in the acute toxicity, or only in long-term carcinogenic actions, is not clear.

Another interesting series of mycotoxins are *red moulds* of wheat, barley, oats and related crops. While these are more frequently associated with poisoning of farm animals, human toxicity has occurred. Red moulds of wheat (sometimes referred to as "wheat scab" or "wheat blight") belong to the genus *Fusaria* and produce toxins which induce vomiting, nausea, drowsiness, dizziness and headache. Continued ingestion leads to depression, loss of weight, hypothermia and eventually, death. At autopsy, haemorrhages of many fatty tissues are seen, together with infiltration and congestion of lungs and liver.

Table 3.11 summarizes toxins of known chemical composition obtained from fungi. Chemical structures are shown in Figure 3.12. Many others are thought to be present and signs

Table 3.11. Effects of some fungal toxins

Toxin	Fungi	Biological effects
Fusariogen	*Fusariam sporotrichiodes*	Alimentary toxic aleukia
Zearalenone	*F. trichinctum*	Estrogenic
Butelonides	*F. nivale*	Oedema, necrosis
Fagicladosporic acid	*C. epiphyllum*	Oedema, haemorrhage
Diacetoxyscirpenol	*F. equiseti*	Necrosis, phytotoxic
Scirpenol	*F. equiseti*	Necrosis, phytotoxic
Lycomarasmine	*F. oxysporum*	Phytotoxic
Sporidesmins	*Pithomyces chartarum*	Eczema, liver damage
Psoralens	*Sclerotinia sclerotiorum*	Skin photosensitizer
Oosponol	*Oospora* sp.	Muscular spasms
Maltoryzine	*Aspergillus oryzae*	Paralysis

Chemical structures are shown in Figure 3.12.

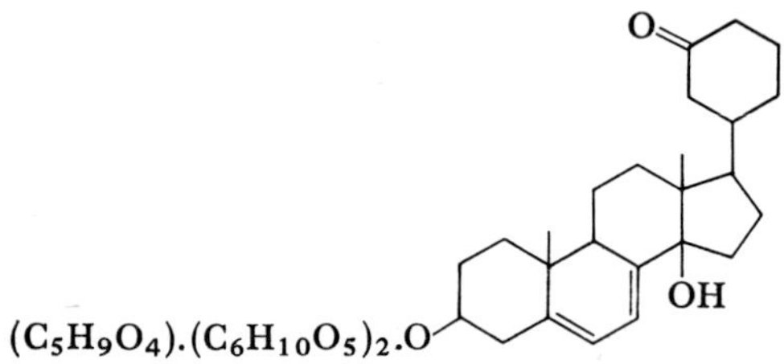

Fusariogenin

$$CH_3.(CH_2)_9.CH{=}CH.(CH_2)_9.CO.SH$$

Fagicladosporic acid

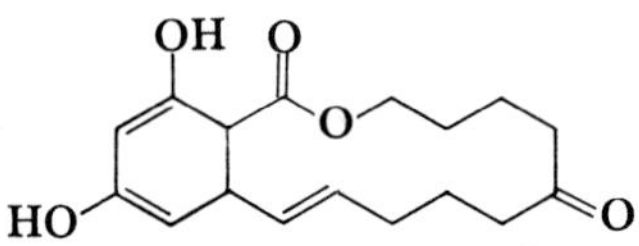

Zearalenone

$$CH_3.(CH_2)_9.CH{=}CH.(CH_2)_{13}.CO.SH$$

Epicladosporic acid

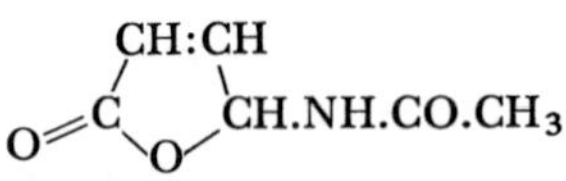

Acetyl-γ-amino-α,β-butelonide

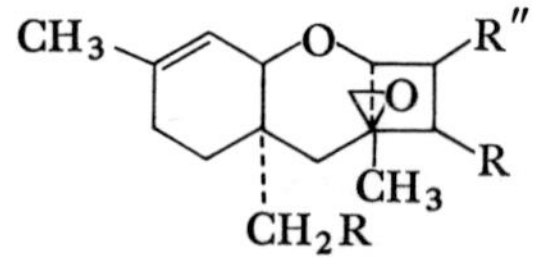

Diacetoxyscirpenol

(R = OAc, R′ = OAc, R″ = OH)

Scirpenol

(R = R′ = R″ = OH)

Figure 3.12. Some other fungal toxins (see Table 3.11).

and symptoms of fusarium poisoning cannot be exactly reproduced by any of the purified toxins given alone, or in combination.

Very little is known of the mode of action or receptors for these toxins. Some, such as zearalenone, are weak estrogens and bind to the estrogen receptors of tissues such as the uterus. Several others inhibit protein biosynthesis *in vitro*.

Toxins of fungi, not previously mentioned, are included in Table 3.11. Little is known of their mode of action.

Many fungi produce endotoxic substances, analogous to pathogenic bacteria. Fungal exotoxins are antigenic and pyrogenic, but little is known of their chemistry or mode of action. In several cases, serious reactions to glucose solutions have been traced to fungal endotoxins, but are much less frequent than those of bacteria.

III. ANTIBIOTICS

1. General

Antibiotics are toxins of micro-organisms (usually fungi, but occasionally bacteria or algae) which act on other micro-organisms to inhibit their growth, reproduction or viability. These toxins are widely used in human medicine for the treatment of infections, while some have value in cancer chemotherapy. Some antibiotics are chemically modified before use in pharmaceutical products, but these will not be considered further. This section is devoted to the biochemistry and mode of action of *naturally occurring* toxins.

Several thousand different antibiotic substances are now known, so we will select some widely used compounds for a detailed discussion. Table 3.12 briefly summarizes the names, source and general use of some antibiotics.

We have already encountered antibiotics in previous chapters. Chemical structures will be found on p. 67–72, while we have seen that antibiotics used in cancer chemotherapy (e.g. actinomycin D, mitomycin C), bind to guanine residues of DNA molecules (p. 235). In contrast, streptomycin binds to the 30 S component of ribosomes to allow serious "mis-reading" of *m*-RNA (p. 175). In this section we will consider the different ways in which antibiotics act against pathogenic micro-organisms.

Table 3.12. Fungal toxins as antibiotics

Antibiotic	Source	Use
Albomycin	*Actinomyces subtropicus*	Most gram +ve bacteria
Bacitracin	*Bacillus licheniformis*	
Benzylpenicillin	*Penicillium notatum*	
Cephalothin	*Cephalosporium sp.*	
Erythromycin	*Streptomyces erythreus*	
Gramicidin	*B. brevis*	
Lincomycin	*S. lincolnensis*	
Novobiocin	*S. niveus*	
Oleandomycin	*S. antibioticus*	
Pristinamycin	*S. pristina spiralis*	
Ristocetin A & B	*Nocardia lurida*	
Sodium fusidate	*Fusidium coccineum*	
Spiramycin	*S. ambofaciens*	
Tyrothrinin	*B. brevis*	
Vancomycin	*S. orientalis*	
Colistin	*B. polymyxa*	Most gram –ve bacteria
Polymyxin B	*B. polymyxa*	
Chloramphenicol	*S. venezuelae*	Broad spectrum
Demethylchlor-tetracycline	*S. aureofaciens*	
Framycetin	*S. decaris*	
Fusafungine	*Fusarium lateritium*	
Gentamycin	*Micromonospora purpurae*	
Neomycin	*S. fradiae*	
Oxytetracycline	*S. rimosus*	
Spectinomycin	*S. spectablis*	
Xanthocillin X & Y	*P. notatum*	
Capreomycin	*S. capreolus*	Anti-tubercular
Cycloserine	*S. orchidaceus*	
Rifamycins	*S. mediterranei*	
Streptomycin	*S. griseus*	Anti-fungal
Viomycin	*S. griseus*	
Amphotericin B	*S. nodosus*	
Griseofulvin	*P. griseofulvum*	
Natamycin	*S. natalensis*	
Nystatin	*S. noursei*	
Pecilocin	*Paecilomyces varioti banier*	
Trichomycin	*S. hachijoensis*	
Aminosidin	*S. chrestomyceticus*	Anti-amoebal
Fumagillin	*Aspergillus fumigatus*	
Paromomycin	*S. rimosus*	
Puromycin	*S. albo-niger*	
Actinomycins	*S. antibioticus*	Anti-cancer
Mitomycin C	*S. caespitosus*	
Porfiromycin	*S. ardus*	

2. DNA Anti-metabolites

We have just mentioned that cancer chemotherapy antibiotics bind to DNA to prevent their replication and use as templates. Another antibiotic with DNA as its target receptor is *griseofulvin*, an extremely useful agent for fungal diseases. It inhibits the growth of mycelial fungi, but has no action on yeasts or bacteria. Only fungi with chitinous cell walls are affected (e.g. *Trichophyton* sp.); those with cellulose walls are not inhibited (e.g. *Neurospora* sp.). After treatment with griseofulvin, sensitive fungi continue to grow for a short time, but produce grossly abnormal spores, hyphae and germ-tubes. Mitosis is abnormal and huge polyploid nuclei are formed.

Griseofulvin has no effect on cell wall formation, respiration, or protein biosynthesis. It will complex with nucleic acids *in vitro*, but the inhibition of growth is partially reversed by *mixtures* of purine-containing nucleotides, e.g. AMP, GMP, etc., though individual compounds have little effect. Further evidence for an effect on nucleic acid biosynthesis comes from studies of griseofulvin in higher organisms. In rats, mitosis of bone marrow cells is arrested in metaphase, with chromosomes scattered throughout the cytoplasm, rather than in an organized array. Bean shoots react to this toxin by incomplete spindle formation at metaphase, with delayed mitosis.

The uptake of radioactive precursors of proteins and nucleic acids by griseofulvin-sensitive fungi, before and after exposure to the antibiotic reveals some interesting effects (Table 3.13).

Biosynthesis of DNA appears to be *stimulated* by griseofulvin, and this could explain the large nuclei seen in treated fungi. As RNA and protein biosynthesis are both reduced, it is likely that there is something "wrong" with the extra DNA formed. The mode of action at the molecular level remains

Table 3.13. Metabolic effects of griseofulvin

Radioactive precursor	Incorporated into	Relative incorporation at 12 hours with griseofulvin	without griseofulvin
Uridine	RNA	57	100
Thymidine	DNA	125	100
Leucine	Proteins	67	100

unknown, but it seems likely that the toxin initiates DNA replication inappropriately.

Humans and laboratory animals extensively metabolize griseofulvin to the 6-demethyl compound, while fungi and other micro-organisms produce both 6-demethyl compound and 4-demethylgriseofulvins. It would be of considerable interest to know whether these methyl transfers are involved in the mode of action (Figure 3.13).

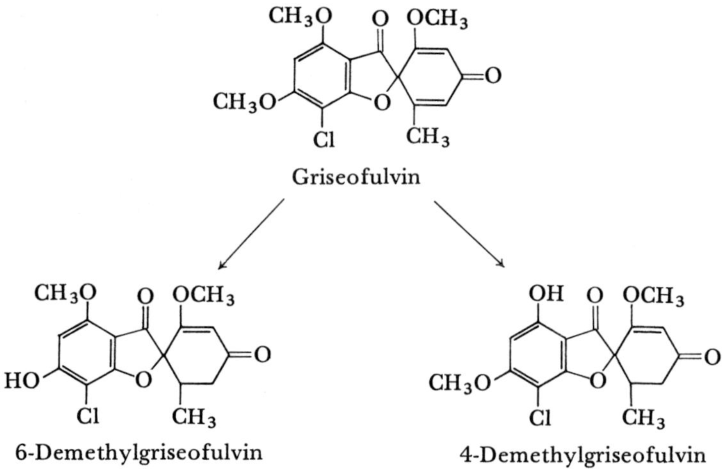

Figure 3.13. Metabolism of griseofulvin.

Other antibiotics acting via the DNA of sensitive species are *novobiocin* and the *porfiromycins,* though the action is different in the two cases. Treatment of *E. coli* cultures with novobiocin produces an inhibition of DNA biosynthesis to 20%, and of RNA to 50%, of untreated controls. The antibiotic prevents the use of DNA as a template by transcriptase enzymes, but the precise molecular interaction is uncertain, for novobiocin does not bind to the nucleic acids. This antibiotic will inhibit the replication of T_2 and T_3 phages which contain DNA, but it also blocks replication of MS-2 phage, which contains RNA. It is likely that novobiocin acts primarily on DNA replication, but also disrupts transcription. The result is decreased formation of many cell enzymes. Also, the antibiotic is a potent inhibitor of electron transport; indeed, on a molar basis, it is more active than either cyanide or azide. A proposed molecular mechanism for the action is removal of Mg^{++} ions by chelation, for there is an absolute requirement for Mg^{++} in DNA and RNA biosynthesis, amino acid activation, and electron transport. The formation of stable magnesium chelates of novobiocin has been shown *in vitro*.

Porfiromycin is the *N*-methyl derivative of *mitomycin C*, while hydroxyporfiromycin is a 7-hydroxy analogue in which NH_2- is replaced by OH-. We have seen that mitomycin C acts as an alkylating agent for guanine residues on DNA and porfiromycins act the same way. The minimum inhibitory concentrations of porfiromycin for sensitive bacteria are approximately the same as for mitomycin C, but those for hydroxyporfiromycin are about ten times higher. For this reason, the hydroxy compound has found little clinical application.

Table 3.14. Antibiotics acting on DNA

Antibiotic	Actions
Griseofulvin	Stimulates increased formation of faulty (?) DNA
Novobiocin	Blocks DNA replication and transcription (by chelating Mg^{++})
Porfiromycin	Alkylates guanine residues of DNA
Hydroxyporfiromycin	Alkylates guanine residues of DNA
Mitomycin C	Alkylates guanine residues of DNA
Actinomycins	Alkylates guanine residues of DNA
Cycloheximide	Blocks DNA replication, without binding
Bruneomycin	Promotes DNA degradation
Daunomycins	Binds to DNA

Table 3.14 summarizes the actions of antibiotics discussed above, together with several others, which affect the biosynthesis of DNA.

3. RNA Anti-metabolites

Clearly, any antibiotic which prevents replication of DNA, or inhibits use of DNA as a template for RNA biosynthesis will disrupt the formation and metabolism of the various types of cellular RNA. There are several antibiotics, however, which have little effect on DNA metabolism, but considerably alter utilization of RNA in quite different ways.

The inter-relations of DNA, RNA and proteins are complex but in order to understand the different modes of action of antibiotics we will briefly review the salient features:

(i) Proteins are formed as polypeptide chains on sub-cellular particles known as "ribosomes".
(ii) An intact bacterial ribosome has a sedimentation value of about 70 S, but can be shown to have two components (30 S and 50 S), *both* of which are required for it to function, but which appear to be formed separately, then united.
(iii) Messenger-RNA is formed on DNA templates by a "transcriptase" enzyme and has a base-sequence determined by its template.
(iv) A triplet of bases of an *m*-RNA molecule is bound to an "active site" on the 70 S ribosome and is matched by a triplet "anti-codon" of an aminoacyl-*t*-RNA molecule.
(v) A second *m*-RNA triplet is matched to a second aminoacyl-*t*-RNA anti-codon and the aminoacyl group of the first *t*-RNA molecules transferred to the second to form a dipeptide-*t*-RNA.
(vi) The process continues until all the *m*-RNA triplets have passed over the ribosome active site and matched to *t*-RNA anti-codons: the final *m*-RNA triplet being left with a polypeptide-*t*-RNA, which detaches from the ribosome.
(vii) One molecule of *m*-RNA can be "read" by several ribosomes simultaneously ("polysomes").
(viii) Aminoacyl-*t*-RNA molecules are formed by ATP-dependent "activating enzymes".

Anti-metabolites can disrupt this complex process at several points, all of which give the same end result: decreased protein biosynthesis. The following are the major points of disruption by various antibiotics:

(a) Inhibition of ribosome formation
(b) Blocking *m*-RNA uptake by ribosomes
(c) Preventing matching of *m*-RNA codons by *t*-RNA anti-codons
(d) Promoting mis-reading of *m*-RNA codons

The following is a list of some important antibiotics which inhibit protein biosynthesis in susceptible micro-organisms and which also chemically combine with ribosomes:

erythromycin
lincomycin
spiramycin
chloramphenicol
tetracyclines
framycetin
gentamicin
rifamycin
streptomycin
viomycin
paramomycin
puromycin

It is important to distinguish between antibiotics which completely block polypeptide formation and those which partially block, although the peptide formed has the "wrong" amino acid sequence, due to "mis-reading" of *m*-RNA codons. In the previous chapter (p. 175) we discussed in detail the action of *streptomycin*, which binds to bacterial ribosomes and promotes this type of mis-reading. Other antibiotics which appear to have a similar action are *paromomycin*, *gentamicin*, *neomycin*, *framycetin* and many structurally related compounds.

Investigation of structure–activity relationships among compounds promoting *m*-RNA mis-reading has revealed some curious results. A consideration of the complete chemical structure of *streptomycin* (p. 68) shows the presence of diaminohexanol residue, known as streptamine, carrying two guanido groups. This residue is also present in *bluensomycin*, while 2-deoxystreptamine residues are found in *neomycin*, *kanamycin* and *paromomycin*. The diaminohexanol residue of *spectinomycin* and *kasugamycin*, however, is actinamine. The structure of these diaminohexanols and their relationship to *m*-RNA mis-reading is shown in Table 3.15.

All these antibiotics bind to the 30 S component of bacterial ribosomes, but only those with a streptamine or 2-deoxystreptamine residue cause *m*-RNA mis-reading. In streptamine, the hydroxyl at C-2 is α-, while it is β- in its stereoisomer,

Table 3.15. Relationship of antibiotic structure to m-RNA mis-reading

Diaminohexanol residue	Antibiotics	Mis-reading of *m*-RNA
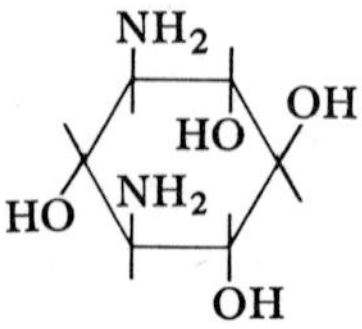 Streptamine	Streptomycin Bluensomycin Gentamicin (?)	Yes
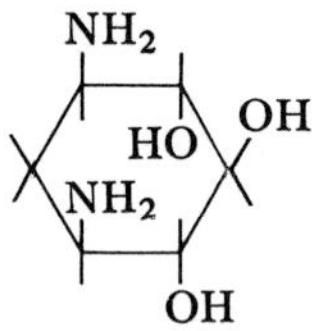 2-Deoxystreptamine	Neomycin B Kanamycin Paromomycin	Yes
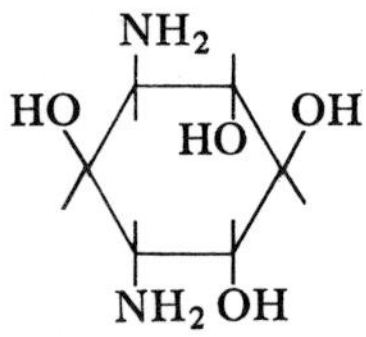 Actinamine	Spectinomycin Kasugamycin	No

actinamine. It could be a reasonable hypothesis that these antibiotics bind to ribosomes via the C-2 hydroxyl of their diaminohexanol residues, and that α-configuration gives a spatial arrangement allowing *m*-RNA mis-reading, while a β-group produces some other effect, were it not for the absence of a C-2 hydroxyl in the neomycin group. However, it is possible that these antibiotics must first undergo a 2α-hydroxylation before they act. Some support for this idea comes from studies of cross-resistance: streptamine and 2-deoxystreptamine antibiotics show mutual cross-resistance, but actinamine compounds do not.

The antibiotics which promote *m*-RNA mis-reading all bind to the 30 S component, but antibiotics which block protein biosynthesis without including mis-reading usually bind to the 50 S component. Some of these prevent the combination of *m*-RNA with the ribosome, while others allow this combination but prevent the matching of *m*-RNA codons by blocking uptake of aminoacyl-*s*-RNA. A number of interesting antibiotic interactions have been discovered at the ribosomal level and are of possible clinical importance where combined antibiotic therapy is being considered. For example, *puromycin*, which from its structure (Fig. 3.14) might be expected to disrupt utilization of phenylalanyl-*t*-RNA or tyrosyl-*t*-RNA, has been found to be a potent synergist for *streptomycin*. The latter binds to a specific ribosomal site and *m*-RNA mis-reading occurs only while the *streptomycin* is bound to this site. *Puromycin* acts by accepting growing peptide chains from any *t*-RNA molecule.

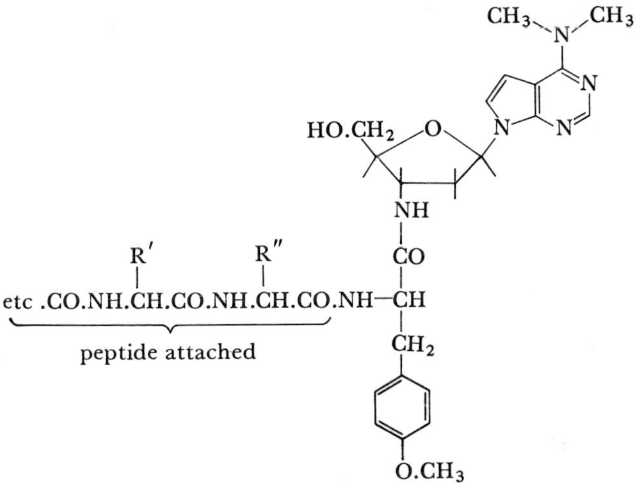

Figure 3.14. Puromycin peptides.

Once the peptide is transferred to puromycin, further reading of *m*-RNA stops and ribosomes dissociate from the *m*-RNA molecule. It is likely that *streptomycin* cannot bind to a ribosome while it is combined to *m*-RNA, so that after puromycin treatment there are more free ribosomes available to interact with streptomycin. The structure of puromycin peptides formed by bacteria from this antibiotic is shown in Figure 3.14

The binding of *puromycin* to ribosomes is rather a special case, in that its structure is analogous to aminoacyl-*t*-RNA, so that it is taken up as though it were a normal cell constituent, without influencing uptake by ribosomes of aminoacyl-*t*-RNA molecules. Nor does it interfere with *m*-RNA uptake, even though it promotes dissociation once peptide bond formation has stopped.

In contrast, *chloramphenicol* binds to the 50 S ribosome component and blocks protein biosynthesis by preventing peptide bond formation. As *chloramphenicol* has a relatively simple molecule (see p. 71), there has been considerable interest to identify the group(s) involved in binding to its ribosomal site. The side-chain is a substituted triose sugar residue, with two asymmetric centres. The natural antibiotic is a D(—)-threo compound, but the L(+)-threo, L(+)-erythro, and D(—)-erythro compounds have all been prepared. None has any significant antibiotic or ribosome binding activity; which suggests that binding is stereospecific and involves the substituted-triose side-chain. As *chloramphenicol* is one of the few known naturally occurring nitro-compounds, it might be thought that this group and its location on the benzene ring are critical, but this is not so. Any strongly electronegative group can substitute for nitro, while the latter need not be *para-* to the side-chain. Such changes to the aromatic ring all reduce biological activity, but not completely unless they are major. An aromatic ring, however, *is* essential, the ring and its electronegative substituent probably activate the hydroxyl at C-1 of the side-chain. The dichloroacetamido group linked to C-2 is also critical, for removal leads to almost complete loss of activity. Analogues with monochlor-, or acylamides higher than acetyl also have much reduced activity. A likely linkage of the antibiotic to ribosomes is therefore through C-1 hydroxyl and either ". NH ." or ". CO ." of the dichloroacetamido group.

m-RNA structure influences inhibition by *chloramphenicol*, for this increases with the proportion of cytidylic acid residues, though not of other bases. We have mentioned that *puromycin* can act as a peptide-acceptor from *t*-RNA, so that puromycin-containing peptides are found in bacteria treated with this antibiotic. Chloramphenicol inhibits the formation of puromycin peptides, so that, unlike streptomycin, no synergism exists between puromycin and chloramphenicol. Indeed, puromycin cannot act in the presence of chloramphenicol.

Another interaction is between spiramycin and chloramphenicol. The binding of chloramphenicol to 50 S ribosomes is readily reversed, but spiramycin binds to the same component very strongly and will displace any chloramphenicol present. It is likely that the two antibiotics compete for the same ribosomal site. Naturally occurring spiramycin is a mixture of three compounds with closely related structures; all bind strongly to ribosomes by a process requiring the presence of NH_{4^+} ions. Little is known of this binding, but the terminal mycarose residue is not involved for it can be removed without loss of activity. Like chloramphenicol, spiramycin blocks puramycin-peptide formation and so probably has a similar mode of action.

Two other macrolide antibiotics, structurally related to spiramycin, are erythromycin and oleandomycin, both possessing large lactose rings linked to desosamine and a second sugar residue: cladinose in erythromycin and oleandrose in oleandomycin (see p. 72 for full structure). As might be expected, both bind to 50 S ribosome components at the same site as chloramphenicol which they will displace, though less markedly than spiramycin. However, once bound to ribosomes they cannot be displaced by chloramphenicol. Both antibiotics inhibit peptide formation, though not by affecting ribosomal uptake of *m*-RNA or aminoacyl-RNA. As for chloramphenicol, the degree of inhibition is proportional to the cytidylic acid content of *m*-RNA.

Precise details of the action of this group of antibiotics are still unknown. Polysome formation occurs in treated micro-organisms, so that *m*-RNA uptake is normal. Aminoacyl-*s*-RNA is incorporated by ribosomes possessing an antibiotic molecule bound to the 50 S component, but peptide-bond formation is very much reduced. It is uncertain whether transfer of a peptide from one *t*-RNA molecule to an adjacent aminoacyl-*t*-RNA occurs on the ribosome *spontaneously*, or requires *catalysis* by an enzyme incorporated into the ribosomal structure. The evidence favours an enzyme, for peptide-chain lengthening can be blocked by -SH group inhibitors, while K^+ ions appear to be activators. The hypothetical enzyme has been named "transfer factor II" and could be the target receptor for these antibiotics, though they do not inhibit other -SH dependent enzymes.

Of the remaining antibiotics which bind to ribosomes, *lincomycin* and *tetracyclines* do so to inhibit protein biosynthesis by blocking uptake of aminoacyl-*t*-RNA, though leaving *m*-RNA uptake unaffected. *In vitro* lincomycin binds reversibly to the 50 S component, but not to the 30 S component, or to intact 70 S ribosomes. Anti-bacterial properties of the compound must therefore be due to binding to a component of ribosomes during their *de nova* formation, rather than to existing ribosomes. The site to which lincomycin binds on ribosomes must be the same, or very close, to that of other 50 S-binding antibiotics, for lincomycin inhibits chloramphenicol binding; while lincomycin is displaced by erythromycin. Again ammonium ions are involved in the binding. Despite this common ribosomal site, these two antibiotics have different actions on protein biosynthesis. Lincomycin strongly inhibits uptake of aminoacyl-*s*-RNA by ribosomes, but the inhibition is reversed by erythromycin, which displaces lincomycin, even though protein biosynthesis remains blocked.

In contrast, tetracyclines such as *chlortetracycline* and *oxytetracycline* bind to the 30 S ribosome component, but also inhibit uptake of aminoacyl-*t*-RNA. As the binding sites are on different parts of the ribosome, tetracyclines do not interfere with antibiotics binding to the 50 S component. Indeed, the effects of lincomycin and tetracyclines on aminoacyl-*t*-RNA uptake by ribosomes are additive. Unlike lincomycin, tetracyclines also bind to intact 70 S ribosomes. Protein biosynthesis is blocked in bacteria treated with tetracyclines, but RNA formation is often increased, perhaps by a compensatory process to overcome inhibition of ribosomes.

Streptomycin and several related antibiotics also bind to 30 S ribosome components of susceptible bacteria, but there is no antagonism between these compounds and tetracyclines; so that two different sites are probably involved. It is likely that the sites are close together, for *kasugamycin*, though not *spectinomycin*, (two members of the streptomycin series that do *not* induce mis-reading of *m*-RNA, see p. 238) also inhibits uptake of aminoacyl-*s*-RNA.

Large numbers of tetracyclines have been tested for antibacterial properties and ribosome-binding so that some information on the tetracycline–ribosome linkage is available. The 11-oxo-12-ene chromophore is essential for activity; as is the 2-carboxamido group (structure, p. 68). Substituents along the C-4 to C-7 "upper edge" of the molecule have little effect. The exception is the 4-dimethylamino group, for activity is greatly reduced when it is removed, or in the *epi*- position. *In vitro*, tetracyclines chelate divalent metal ions and Mg^{++} is essential for binding aminoacyl-*t*-RNA to ribosomes. It is possible that tetracyclines link to a site on the 30 S component through their 2-carboxamido group and chelate magnesium ions required to link *t*-RNA molecules. Addition of other chelating agents, such as EDTA, to ribosomes containing tetracycline, releases about 20% of the antibiotic, while addition of excess Mg^{++} partially reverses some inhibitory effects *in vivo*.

Two antibiotics inhibiting protein biosynthesis by yet another ribosomal effect are *rifamycin* and *viomycin*. Both inhibit uptake of natural and synthetic *m*-RNA by ribosomes, so that protein biosynthesis cannot begin. There is evidence that rifamycin inhibits the combination of 30 S and 50 S components of ribosomes, though it does not disrupt pre-existing particles. Very little is known of details of these interactions.

There remains a group of antibiotics which inhibit protein biosynthesis at the ribosomal level by other mechanisms. We have mentioned the actinamide-containing member of the streptomycin series, *spectinomycin*, which binds to 30 S ribosome components, but does not cause mis-reading of *m*-RNA; nor does it block ribosome uptake of either *m*-RNA or aminoacyl-*t*-RNA. The steroidal antibiotic, sodium fusidate (fusidic acid), also markedly inhibits protein biosynthesis in cell-free extracts, as well as in intact bacteria. It does not bind to ribosomes to any appreciable extent, nor inhibit uptake of *m*-RNA or aminoacyl-*t*-RNA. Unlike puromycin, it can be recovered unchanged from inhibited ribosomes. Polysome formation is unaffected, as is synthesis of aminoacyl-*t*-RNA. The inhibitory action of either chloramphenicol or streptomycin in combination with sodium fusidate is additive, even when the other antibiotics are present at concentrations producing their maximum effects; so that the action of fusidate is at a different site. It may act by inhibiting GTP-dependent movement of *m*-RNA on ribosomes, but conclusive evidence is lacking.

4. Cell Wall Actions

Micro-organisms are surrounded by cell walls which protect their cytoplasm from undesirable influences in the environment. The walls are complex biological membranes and are targets for several important antibiotics. Any anti-metabolite capable of upsetting protein biosynthesis will eventually have a non-specific effect on cell walls, so that antibiotics acting on DNA, RNA, or protein biosynthesis will have some actions. We will be concerned in this section with compounds with a mode of action *principally* on cell walls. Two distinct classes can be recognized:

(i) Antibiotics which disrupt *existing* cell walls
(ii) Antibiotics which inhibit the *biosynthesis* of cell wall components.

Microbial cell walls are broken down by many chemical and physical agents, including strong acids and alkalies, osmotic shock, detergents, sonic vibration, freezing and thawing, etc. Antibiotics which disrupt existing cell walls do so by chemically combining with substances present in membranes and so changing their physical properties.

The first example is the peptide antibiotics represented by *polymyxin B* and *colistin*. These closely related compounds have 10 amino acid residues linked to 6-methyloctanoic acid. Seven amino acids are linked in a chain and are unusual in containing both L- and D- residues.

At neutral pH, these antibiotics are positively charged and, as with cationic detergents, their anti-bacterial activity is decreased by anionic compounds, such as soaps, phospholipids, etc. Both polymyxin and colistin have hydrophilic and hydrophobic portions in their molecules (amino acid chain and methyl-octanoic residue respectively) and so are well suited to interact with cellular membranes. Treatment of bacterial cultures with either antibiotic induces inhibition of cell division, accompanied by loss of cytoplasmic material into the medium. Several dyes are known which fluoresce in UV only when combined with peptides (e.g. *N*-tolyl-α-naphthylamine-8-sulphonic acid). Bacteria exposed to such dyes do not fluoresce as they possess no protein in their outermost layers, but polymyxin-treated bacteria exposed to the dye show a strong fluorescence, indicating a loss of proteinaceous material induced by the antibiotic.

Damage to cell walls is confirmed in electron micrographs of polymyxin-treated bacteria.

A likely explanation for the action of these compounds on bacteria is that antibiotic molecules combine with cell membranes and become interspersed between lipid and protein layers, so that membranes cease to function as osmotic barriers and death of the cell follows.

As would be expected, acetylation of all free amino groups results in loss of anti-bacterial activity, but little else is known of structural features necessary for successful interaction with bacterial cell walls (for only gram –ve organisms are effected).

Similar compounds are the polyene antibiotics, such as amphotericin B, nystatin, and trichomycin. The detailed chemical structures of these remains uncertain, but they possess large lactone rings with highly conjugated polyene systems, linked to amino sugars and/or *p*-amino-phenylacetone. The three compounds mentioned are amphoteric due to possession of carboxyl and amino groups and *N*-acetylation converts them to acids which form water-soluble salts.

Polyenes also interact with membranes, both of cell walls and sub-cellular particles. They disrupt erythrocyte membranes as well as artifical lipid bilayers. It is likely that the mode of action is similar to polymyxins, and "pits" are found in membranes treated with polyenes. A comparison of effects on artificial membranes composed of lecithin alone, or of lecithin–cholesterol mixtures, reveal a rapid induction of randomly sited circular holes in the latter, but not the former. This confirms earlier findings that many sterols (e.g. cholesterol, ergosterol, sitosterol, etc.) antagonize the antibiotic action of polyenes, and suggests that they form complexes with neutral steroids of cell membranes. As bacterial cell walls do not contain sterols, but those of fungi do, this interaction explains the potent anti-fungal, yet absence of anti-bacterial, action of polyene antibiotics.

No apparent correlation is found between anti-fungal activity and the number of conjugated bonds, presence or absence of ionizable groups, or net charge. However, polyenes with 33 to 37 C atoms (e.g. filipin, etruscomycin, etc.) are much more damaging than those with 46 to 63 C atoms (e.g. nystatin, candidin, etc.); for metabolic effects of the latter group can be at least partially reversed by K^+ and $NH_4{}^+$ salts, while those of the former cannot.

Polyene antibiotics have certain features similar to those of surface active detergents, but their penetration and disruption of cell walls probably lies in chemical interaction with membrane components, rather than a simple physical solubilization. Saponins of various plants also interact with steroids of cell membranes, probably by a similar mechanism.

While polymyxins and polyene antibiotics disrupt existing cell walls of gram-negative bacteria and fungi respectively, the following compounds act mainly to block the biosynthesis of cell walls of gram-positive bacteria:

penicillins
cephalosporins
ristocetin
vancomycin
cycloserine
bacitracin.

To appreciate the biochemical actions of these antibiotics it is important to understand the metabolic pathways involved in the formation of cell wall polymers. In a previous section (p. 216) we have discussed in detail the chemical structures of lipoglycoproteins found in cell walls of *gram-negative* bacteria; for these are antigenic endotoxins. In contrast, pathogenic *gram-positive* bactera tend to release protein exotoxins (p. 201), and their cell walls play little role in disease processes. Electron microscopy reveals a distinct bacterial wall with several layers that may be up to 25% of the dry weight of the organism. The chemical composition of cell walls is summarized in Table 3.16.

Generally speaking, cell walls of gram-positive bacteria and mycoplasma tend to contain more hexosamines, protein and lipid than those of gram-negative bacteria, though there is far more variation in this group. Gram-negative endotoxins have a complex structure involving many different monosaccharide residues, substituted and unsubstituted, together with a variety

Table 3.16. Composition of microbial cell walls

	Amount in cell walls (range of %)		
Component	Gram-negative bacteria	Gram-positive bacteria	Mycoplasma
Proteins	38 to 63	50 to 75	50 to 60
Lipids	11 to 30	20 to 30	37 to 42
Carbohydrate	8 to 46	10 to 20	2 to 4

of fatty acids; all linked to proteins. The structure of the cell walls from gram-positive bacteria differs in significant respects and Figure 3.15 gives the basic structure.

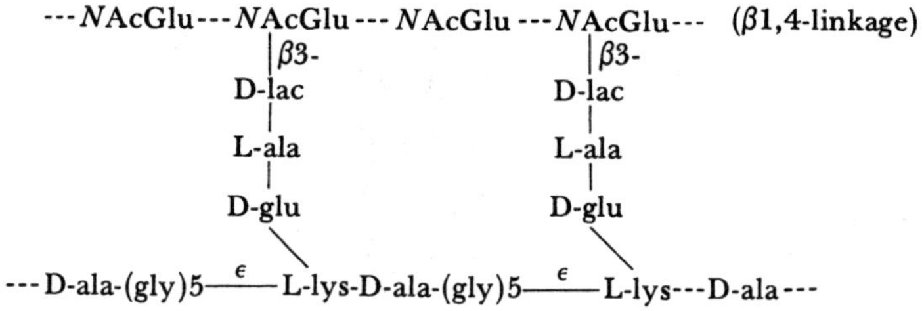

Key = *N*AcGlu = *N*-acetyl-D-glucosamine
D-lac = D-lactic acid
L-ala = L-alanine
D-ala = D-alanine
D-glu = D-glutamine
L-lys = L-lysine
gly = glycine

Figure 3.15. Gram-positive bacteria cell wall mucopeptides.

It will be seen that there are two polymer chains; the first is a β-1,4 linked chain of *N*-acetylglucosamine residues, while the second is a branched polypeptide containing D- and L-alanine, L-lysine, and D-glutamine. The branches are linked to *N*-acetylglucosamine residues through D-lactic acid residues which join L-alanyl residues of the polypeptide branches in the manner shown in Figure 3.16.

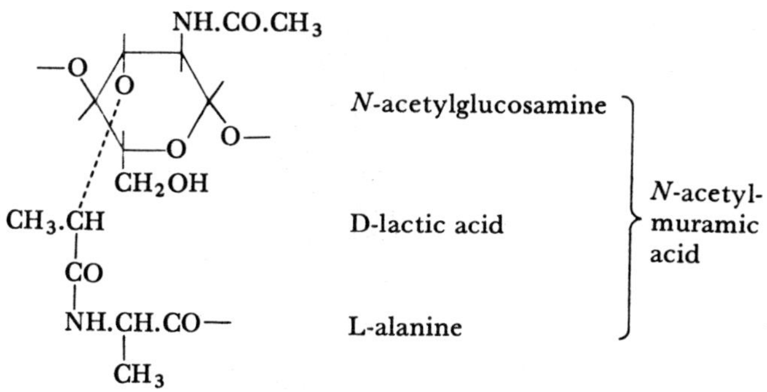

Figure 3.16. Linkage of N*-acetylglucosamine chain to polypeptide.*

The combination of *N*-acetylglucosamine linked to a D-lactic acid residue through 3β-hydroxyl is known as "*N*-acetylmuramic acid", so that the polysaccharide chain could be regarded as alternating residues of *N*-acetylglucosamine and *N*-acetylmuramic acid.

Several peptide bonds of the polypeptide chain are unusual and rarely found elsewhere. The L-lysyl residue is linked to glycine through its ε-amino group, while its α-amino is linked to the γ-carboxyl of D-glutamine.

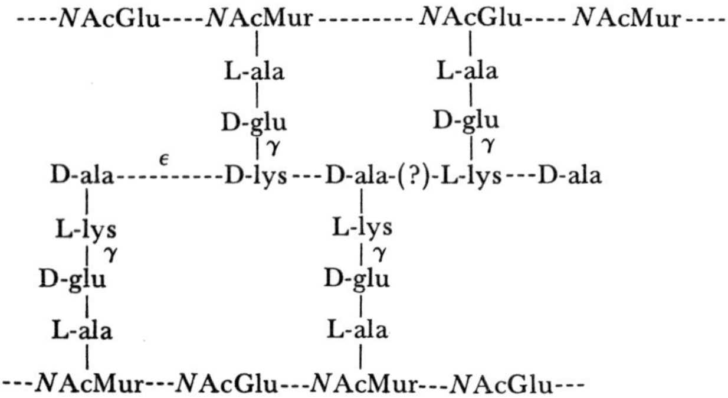

Figure 3.17. "Sandwich" structure for cell wall mucopeptides.

The structure given in Figure 3.15 for cell wall mucopeptide should be regarded as only a first approximation, though it probably closely resembles those of *S. aureus*. Other species show variations and the generalized structure has been proposed (Figure 3.17), in which it is assumed that polypeptides are "sandwiched" between polysaccharide chains.

In this structure, the polypeptide is shown as alternating residues of *N*-acetylglucosamine and *N*-acetylmuramic acid (*N*AcMur) linked to peptides (through the D-lactyl residue). The D-glutamine may be replaced in some species by D-glutamic acid. Similarly, L-lysine may be replaced by any other di-amino acid, such as ornithine, α,ε-diaminopimelic acid, α,γ-diaminobutyric acid, etc. In some species the D-alanyl residue is linked directly to the ε-amino of L-lysine, while in others there are one or more glycyl residues interspersed. Occasionally, other amino acids are included, such as aspartic acid in lactobacilli. Up to 7% of hydroxyl groups on the polysaccharide chains are acetylated.

Despite the problems of arriving at a generalized chemical

model for gram-positive cell walls, it will be seen that they differ appreciably from those of gram-negative organisms. A further difference is the presence of teichoic acids almost exclusively in gram-positive organisms. These polymers occur at the interface between the relatively "stiff" cell wall and the inner cytoplasmic membrane. These compounds occur in bacteria both as polymers of ribitol phosphate carrying single *N*-acetylglucosamine side-chains, and as polymers of glycerol phosphate, also with *N*-acetylglucosamine side-chains (Figure 3.18). Teichoic acids link to mucopeptides of cell walls through

Ribitol teichoic acid

Glycerol teichoic acid

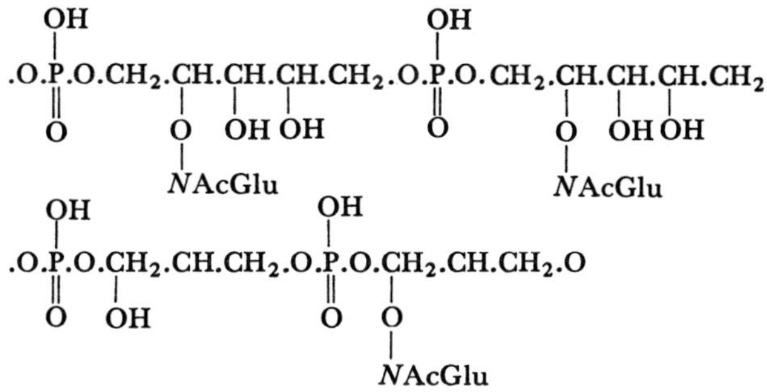

Figure 3.18. Teichoic acids.

D-alanyl residues esterified to hydroxyls of either ribitol or glycerol.

The biosynthesis is via the metabolic sequence shown in Figure 3.19. Partial structure of the membrane-bound phospholipid complex of the *N*-acetylglucosamino-*N*-acetylmuramic acid pentapeptide after addition of pentaglycine, as in *S. aureus*, is shown in Figure 3.20.

The following stages can be recognized:

(1) Condensation of UTP with *N*-acetyl-D-glucosamine-1β-monophosphate to give UDP-*N*-acetylglucosamine, with the elimination of inorganic pyrophosphate.
(2) Condensation of UDP-*N*-acetylglucosamine with phospho*enol* pyruvic acid to form an ether at the 3β-position.
(3) Reduction of UDP-*N*-acetylglucosamine-*enol*pyruvic ether to the corresponding D-lactyl derivative (which is UDP-*N*-acetylmuramic acid).

UTP + *N*AcGlu-1-phosphate

↓ 1

UDP-*N*AcGlu + pyrophosphate

↓ 2 ← phosphoenolpyruvate

UDP-*N*AcGlu-enolpyruvate ether

↓ 3

UDP-*N*AcMUR (= UDP-*N*AcGlu-lactate ether)

↓ 4 ← L-alanine

UDP-*N*AcMur-L-ala

↓ 5 ← D-glutamic acid

UDP-*N*AcMur-L-ala-D-glu

↓ 6 ← L-lysine

UDP-*N*AcMur-L-ala-D-glu-L-lys

↓ 9 ← D-ala-D-ala ←(8)— D-alanine ←(7)— L-alanine

UDP-*N*AcMur-L-ala-D-glu-L-lys-D-ala-D-ala

↓ 10 ← phospholipid

UMP + phospholipid.P.P --------(1β-)-------- *N*AcMur-L-ala-D-glu-L-lys-D-ala-D-ala

↓ 11 ← UDP-*N*AcGlu

UDP + phospholipid.P.P---- *N*AcMur-L-ala-D-glu-L-lys-D-ala-D-ala
| (4α-)
*N*AcGlu

↓ 12

The next step (12) depends upon the species. For *S. aureus* the alpha-carboxyl of the D-glutamyl residue is condensed with a glycylpentapeptide, while in *M. lysodeikticus* a single glycyl is added. The modified peptide chain (*mod. p.c.*) then undergoes the final reactions of the sequence:

phospholid.P.P.------*N*AcMur-*mod.p.c.*
|
*N*AcGlu

↓ 13 ← Acceptor

acceptor-(*N*AcGlu-*N*AcMur-*mod. p.c.*) + phospholipid.P.P.

↓ 14

phospholipid.P + P_i

Figure 3.19. Mucopeptide biosynthesis.

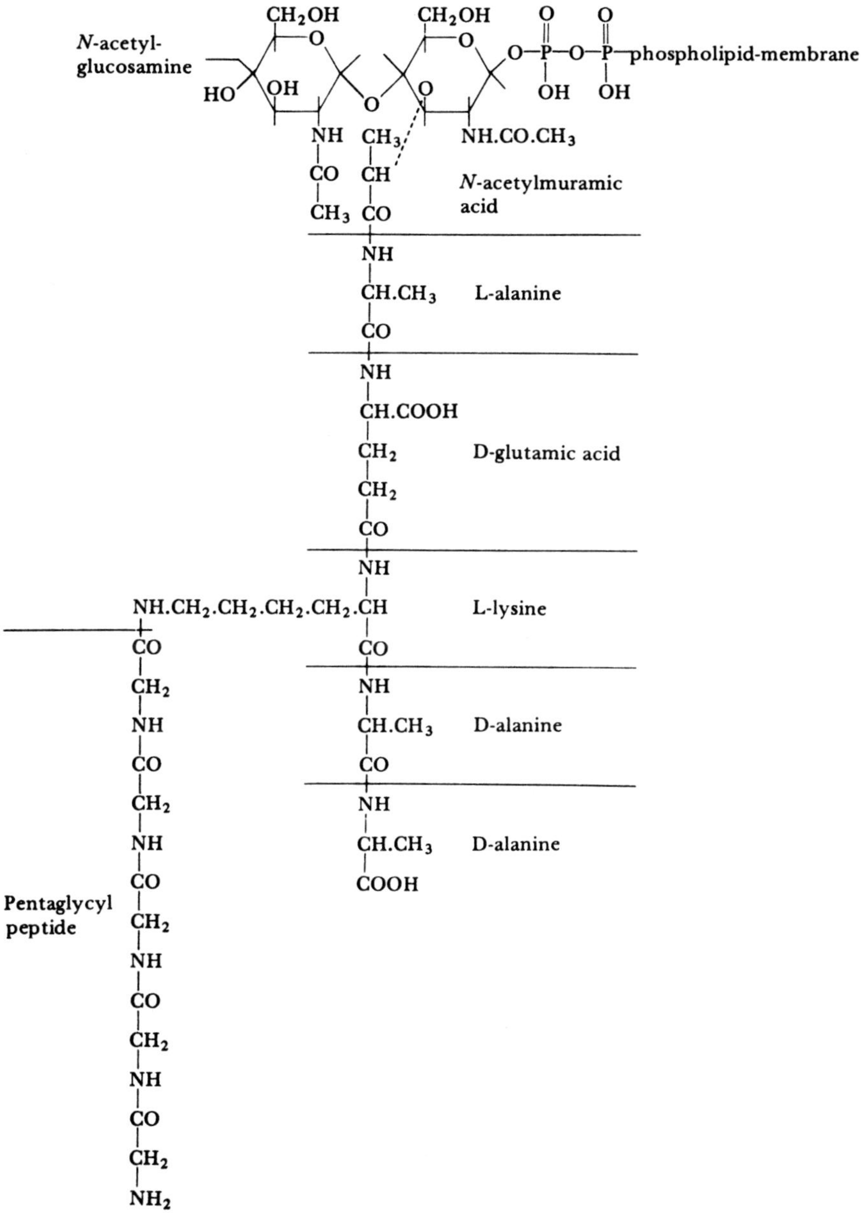

Figure 3.20. Partial structure of membrane bound phospholipid complex.

(4) Condensation of carboxyl of the lactyl residue of UDP-*N*-acetylmuramic acid with the amine group of L-alanine.
(5) Addition of D-glutamic acid by condensation of its α-amine group to the carboxyl of L-alanyl residue.
(6) Addition to L-lysine by condensation of its α-amine group to the γ-carboxyl of D-glutamyl residue (in some species the α-amine of other di-amino acids replaces lysine).
(7) Conversion of L-alanine to D-alanine by a specific racemase.
(8) Condensation of two D-alanine molecules into a dipeptide by an ADP-dependent synthetase (or ligase).
(9) Condensation of the α-carboxyl of D-alanyl-D-alanine with ϵ-amino of the L-lysyl residue (or other di-amino acid residue).
(10) Interaction of UDP-*N*-acetylmuramic acid-pentapeptide with a phospholipid molecule bound to cell membrane, so that UMP is split off and *N*-acetylmuramic acid-pentapeptide is linked to the phospholipid by a pyrophosphate bridge in the 1β-position.
(11) Condensation of this phospholipid complex with UDP-*N*-acetyl glucosamine to form a β1,4-linked disaccharide-type structure.
(12) The previously mentioned species-specific modification of the pentapeptide chain.
(13) Transfer of *N*-acetylglucosamyl-*N*-acetylmuramic acid-pentapeptide (now modified) to a suitable acceptor (another mucopeptide) of the growing cell wall, to leave membrane-bound phospholipid with a pyrophosphate group.
(14) Dephosphorylation of the phospholipid to leave only one phosphate group, with elimination of inorganic orthophosphate.

Of these 14 stages, inhibition of a specific enzyme catalysing the reaction at *any* point is sufficient to block cell wall formation. Antibiotics known to inhibit cell wall biosynthesis are listed in Table 3.17, with the particular stage blocked by each.

Gentian violet is not strictly an antibiotic, but is included as a medicinal substance with a specific action on cell wall biosynthesis. It blocks formation of cell wall mucopeptides by preventing incorporation of the first amino acid of pentapeptides (L-alanine). In contrast, cycloserine blocks synthesis at the tripeptide stage by preventing *both* racemization of L-alanine *and* formation of D-alanyl-D-alanine. The other three antibiotics listed in the table all allow pentapeptide formation, but prevent its use. Bacitracin specifically blocks dephosphorylation of membrane-bound phospholipid-pyrophosphate, so that the acceptor for UDP-*N*-acetylmuramic acid-pentapeptide is lost. Ristocetin and vancomycin act at the same stage to block utilization of phospholipid-bound-*N*-acetylmuramic acid-pentapeptide.

Table 3.17. Inhibitory effects of antibiotics

Antibiotic	Stage inhibited
Bacitracin	14
Ristocetin	13
Vancomycin	13
Cycloserine	7 and 8
Gentian violet	4

Two most important groups of antibiotics affecting cell wall biosynthesis (the penicillins and cephalosporins) have not yet been mentioned, because they act at the terminal stage of cell wall formation, and this has still to be discussed. It is cross-linking of peptide chains at the outer cell surface; a reaction catalysed by *transpeptidase*, an enzyme inhibited by both penicillins and cephalosporins.

Peptide chains incorporated into cell wall mucoproteins vary with the species, so two examples of cross-linking between chains are given in Figure 3.21.

In both cases, transpeptidase catalyses removal of a terminal D-alanyl residue from a peptide-chain by cross-linking it to its neighbour for *E. coli*, occasional terminal D-alanyl residues are removed by a second enzyme (*D-alanine carboxypeptidase*); only some bacteria possess this, but it is also inhibited by penicillins and cephalosporins.

As the cell wall inhibitors so far mentioned (including Gentian Violet) act at *six* different stages in the biosynthetic pathway, it

Peptide cross-linking at outer cell walls

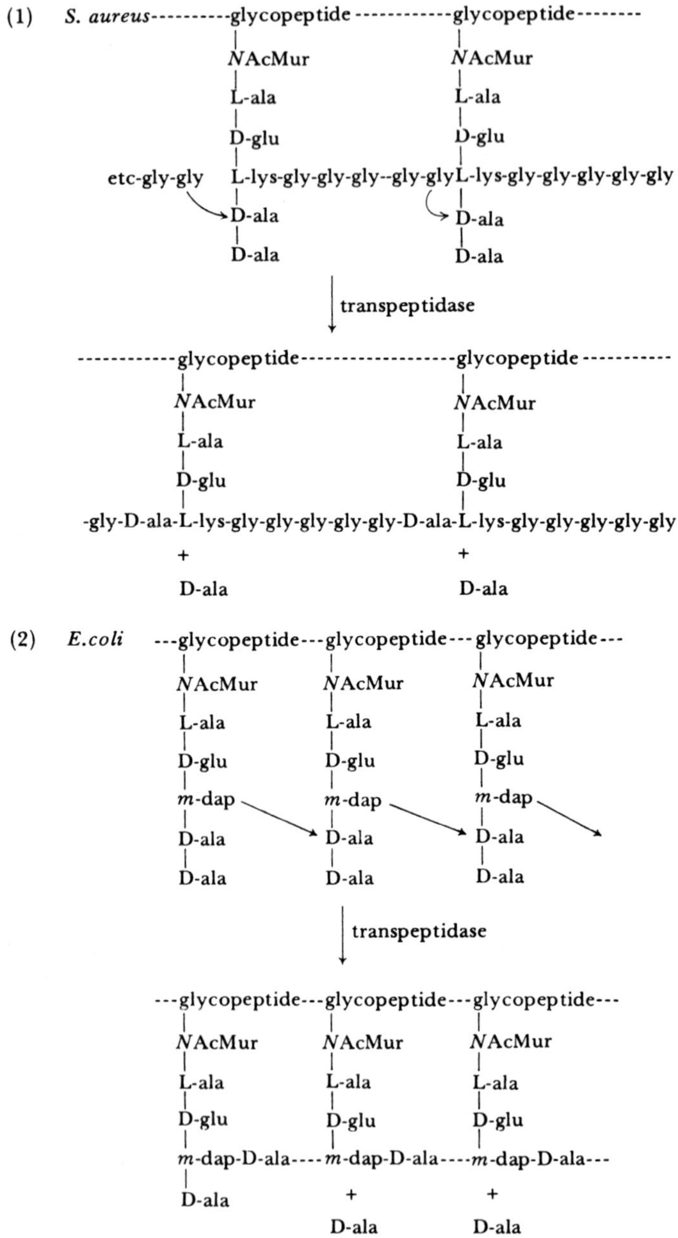

Figure 3.21. Examples of mucoprotein cross-linkages.

is not surprising they show little chemical similarity. The action of most of these anti-metabolites at molecular level is now understood and is outlined below.

(i) Transpeptidase Inhibitors

Penicillins and cephalosporins inhibit transpeptidase, which cross-links chains of cell wall mucopeptides (*note*—despite its name, *cephalosporin P* is chemically unrelated to other members of this group and resembles fusidic acid in structure and action).

Transpeptidase catalyses hydrolysis of peptide linkage between terminal D-alanyl residues of the mucopeptide residue remaining to amino of a residue (usually glycyl) on a second nearby peptide. The enzyme is thought to possess a hydroxyl group at its active site, perhaps from a seryl residue, which splits the D-alanyl-D-alanine linkage and leaves the enzyme esterified to peptide. By an internal molecular re-arrangement, the ester splits and a new cross-linking peptide bond is formed. Details are shown in Figure 3.22.

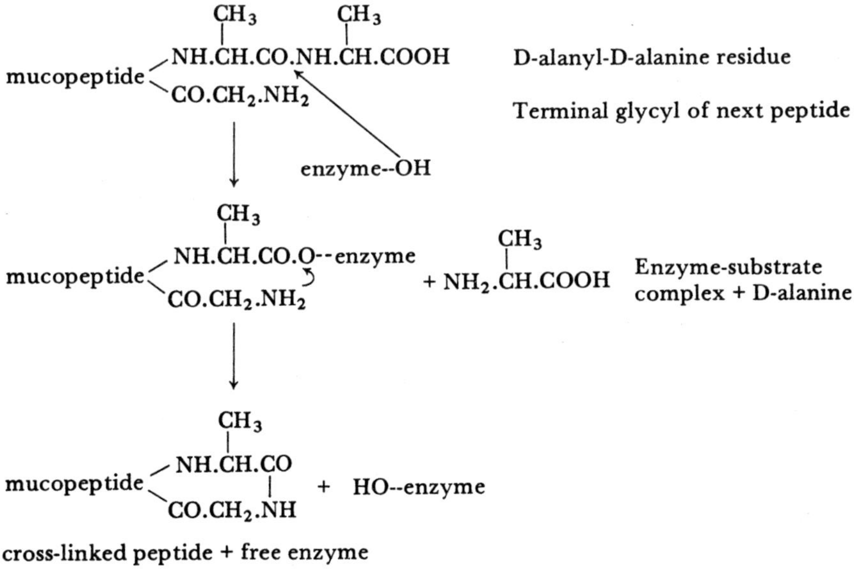

Figure 3.22. Mechanism of transpeptidase.

The critical molecular structure determining activity of this enzyme is D-alanyl-D-alanine, which forms the C-terminal end of mucopeptides. Examination of molecular structures of penicillins and cephalosporins reveals a size and shape very

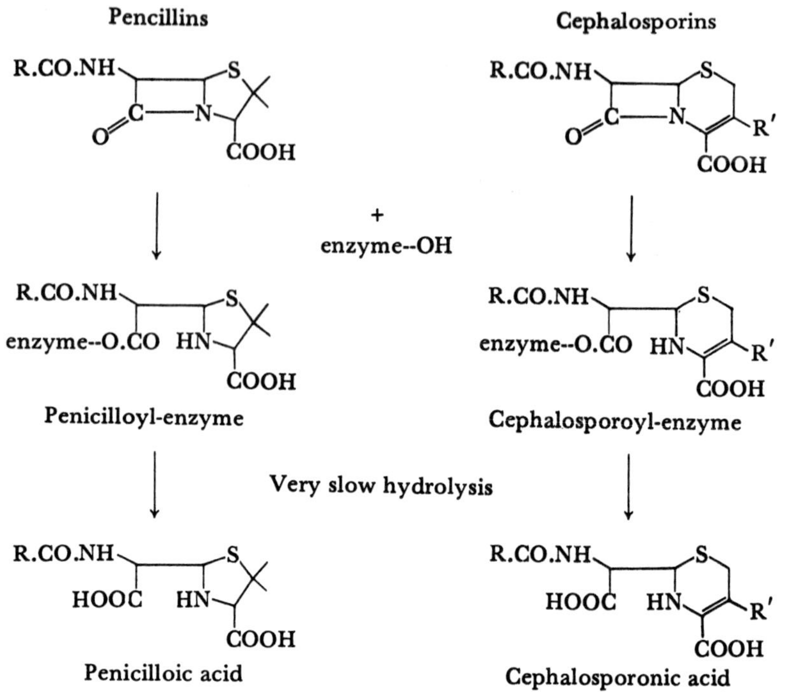

Figure 3.23. Inhibition of transpeptidase by penicillins and cephalosporins.

similar to that of the D-alanyl-D-alanine. Both types of antibiotic will irreversibly inhibit transpeptidase preparations *in vitro* and chemically combine with the active site, from which they are only slowly released by hydrolysis. During the inhibition, the highly reactive –CO–N– bond of the β-lactam ring of the antibiotics is broken in a manner analogous to the breaking of the –CO . NH– bond of D-alanyl-D-alanine. The mechanism of inhibition is explained by the model drawn in Figure 3.23.

Confirmation of this explanation is provided by the antibiotic activity of structural analogues of penicillins and cephalosporins. If the β-lactam ring is opened, as in penicilloic acid, all activity is lost. Similarly, reduction of the carbonyl group of this ring to hydroxyl also eliminates activity. However, substituents elsewhere either increase or decrease antibiotic properties, presumably by changing reactivity of the β-lactam ring. Numerous secondary metabolic defects can be detected in bacteria exposed to transpeptidase inhibitors, due to alterations in cell wall permeability.

(ii) Inhibitors of Lipid-bound Mucopeptide Utilization

Ristocetin and vancomycin both allow formation of *N*-acetylglucosamino-*N*-acetylmuramic acid-pentapeptide bound to phospholipid, but prevent its separation and transfer to acceptors. They presumably inhibit an enzyme catalysing the

Table 3.18. Properties of ristocetin and vancomycin

Antibiotic	Properties
Ristocetin	1. Separates into two components (A and B) on electrophoresis, both with antibiotic properties (B more active).
	2. Approximate proportions of A to B are 9 to 1.
	3. Molecular weights are about 5000.
	4. Both amphoteric and contain amine and phenolic groupings.
	5. Ristocetin A contains 4 moles of D-mannose and 2 each of D-glucose, D-arabinose and L-rhamnose.
	6. Ristocetin B contains 2 moles each of D-mannose, D-glucose and L-rhamnose and 1 mole of D-arabinose.
	7. Sugars and an amine are removed from A and B by acid hydrolysis; with a considerable increase in antibiotic activity.
Vancomycin	1. Complex mucopeptide of M.W. about 3300.
	2. Amphoteric and contains amine and phenolic groupings.
	3. Contains 2 moles each of D-glucose and L-aspartic acid, 1 mole each of *N*-methyl-L-leucine, *o*-cresol, and 3-methyl-4-ketohexanoic acid, and 4 moles of 3-chloro-4-hydroxy-toluene.
	4. Acylation of *N*-methyl-leucine markedly decreases activity.
	5. Hydrolysis of a single glucose residue reduces activity by 25%.
	6. Vancomycin binds irreversibly to bacterial cell walls.

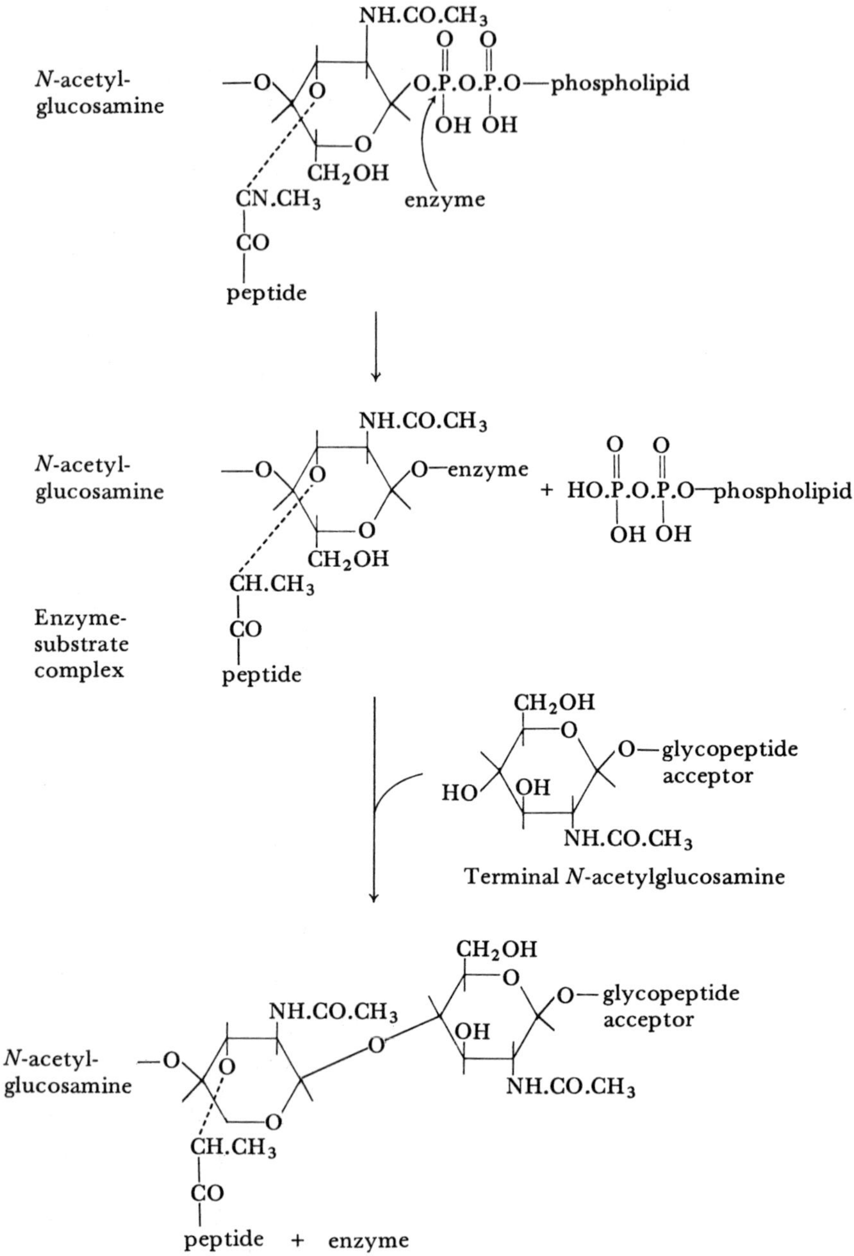

Figure 3.24. Mechanism of N-acetylmuramic acid phosphatase.

splitting of phosphate linked to *N*-acetylmuramic acid residue at the 1β-position, as shown in Figure 3.24.

Little is known of *mucopeptide-transferase* enzyme, but catalyses transfer of *N*-acetylmuramic acid residue of mucopeptide from phospholipid to a glycopeptide acceptor, by joining the residue to a terminal *N*-acetyl glucosamine residue of the acceptor through β1,4-linkage.

The action of ristocetin and vancomycin on this enzyme are difficult to investigate as both antibiotics are high molecular weight compounds, possibly with several components. A brief summary of properties is given in Table 3.18.

It will be seen that both antibiotics are glycopeptides containing unusual sugars and amino acids, not found in bacterial cell walls. It is likely that they act as "false" acceptors for mucopeptide transferase.

(iii) Cycloserine

This antibiotic acts early in the biosynthesis of mucopeptides by inhibiting formation of D-alanine from L-alanine, whilst simultaneously blocking synthesis of D-alanyl-D-alanine.

Cycloserine is a simple molecule (structure, p. 261) which is found as the D-isomer. Both L- and DL- compounds have been prepared and tested for antibiotic actions. For most systems, D-cycloserine is more effective than its L-isomer, but the DL-racemate is sometimes more active than either. High concentrations of D-alanine overcome the inhibitory effects of D-cycloserine.

Alanine racemase is a pyridoxal phosphate-requiring enzyme and its action involves formation of Schiff bases, as shown in Figure 3.25. As cycloserine also forms stable Schiff bases with pyridoxal, even *in vitro*, this is an explanation of its action on alanine racemase. Cycloserine inhibits *in vitro* and *in vivo* at high doses *any* pyridoxal phosphate-dependent enzyme, including transaminases, decarboxylase, etc.

Schiff bases formed by reaction of cycloserine and pyridoxal phosphate are more complex than expected, and stable complexes containing two molecules of cycloserine and two of pyridoxal have been obtained. The interactions occur as shown in Figure 3.26

Even more complex reactions between cycloserine and

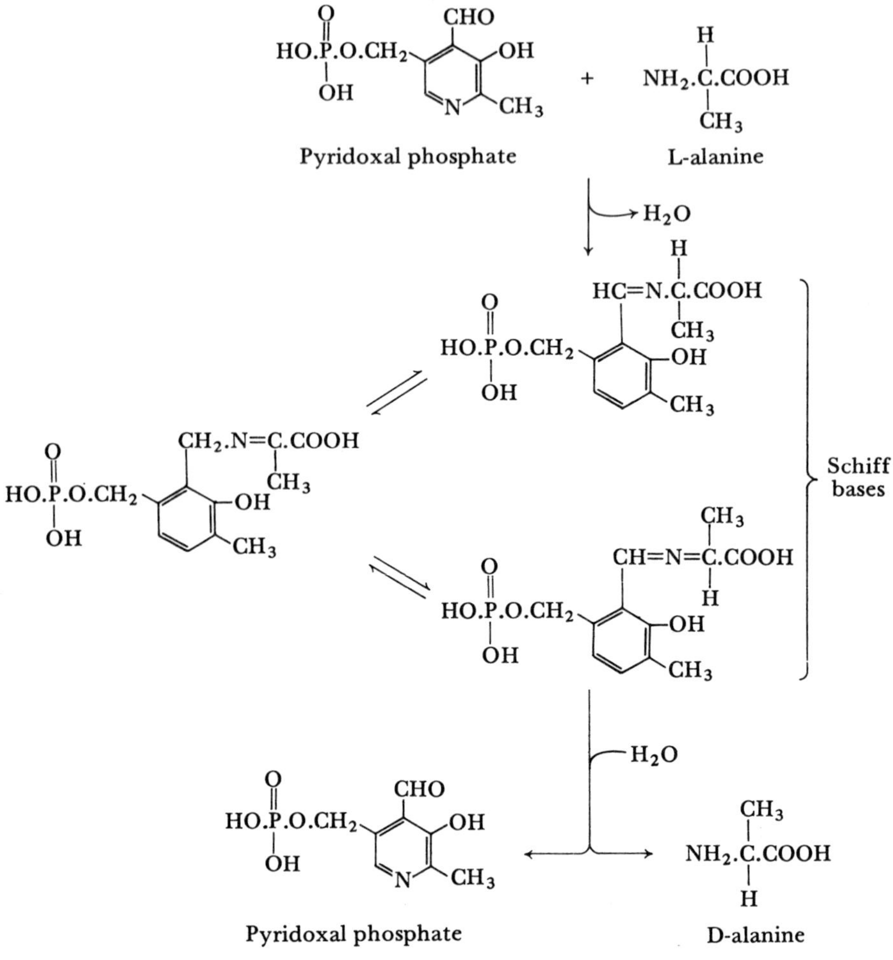

Figure 3.25. Role of pyridoxal phosphate in alanine racemase.

pyridoxal phosphate occur even *in vitro*, for at least 9 different products can be separated by chromatography.

Further evidence that Schiff bases with pyridoxal are responsible for the antibiotic properties of cycloserine comes from study of activity of analogues. If the amino group is alkylated, or substituted by hydroxyl or methoxy groups (so that Schiff base formation cannot occur), all antibiotic activity is lost. Substitution in the ring, or ring cleavage, also decreases activity.

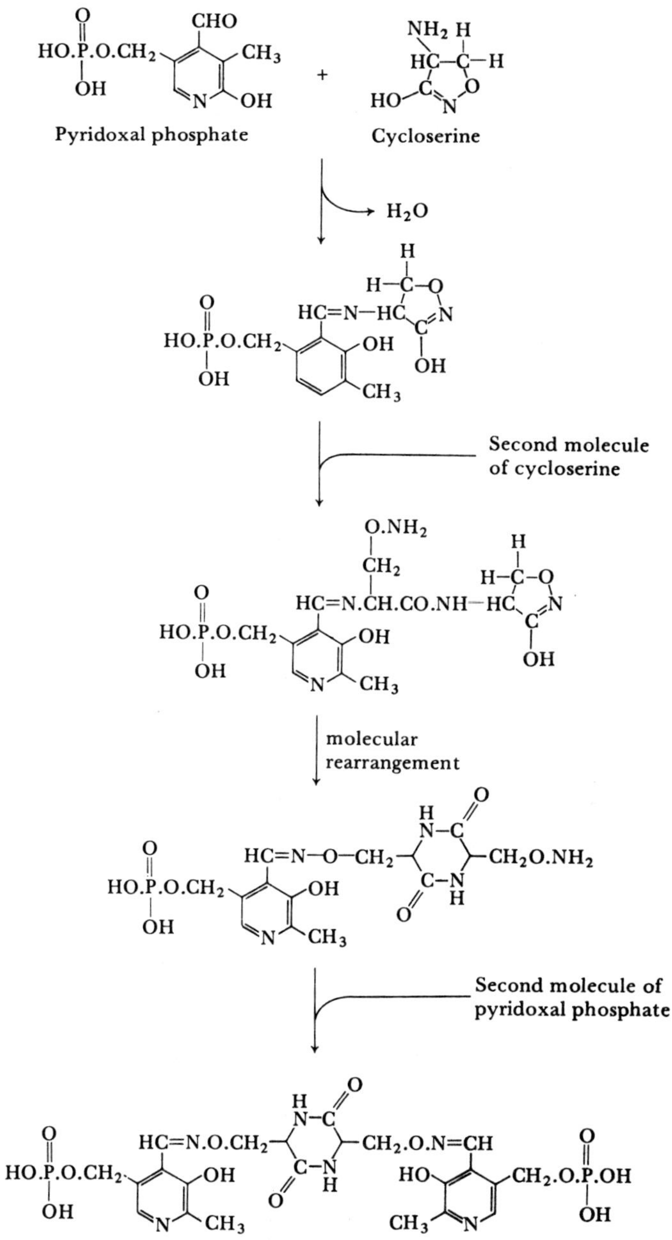

Figure 3.26. Schiff bases of cycloserine and pyridoxal phosphate.

Alanine racemase contains protein-bound pyridoxal phosphate, with which D-cycloserine, but not the L-isomer, combines. The other cell wall enzyme inhibited by cycloserine is D-alanyl-D-alanine synthetase, which does *not* require pyridoxal phosphate, so some other explanation of inhibition is required. Again only the D-isomer is inhibitory *in vitro*.

The synthetase is not specific for D-alanine and produces mixed di-peptides if incubated with a mixture of, say, D-alanine and D-threonine; though L-amino acids are *not* used. In contrast to the racemase, inhibition of the synthetase by cycloserine is completely reversible.

Synthetase enzyme-protein has two closely associated binding-sites for D-alanine, which differ slightly in Michaelis constants and specificity. One binds D-alanine only, but the other binds D-alanine, D-threonine, or D-*allo*threonine; which is

$$\begin{array}{l} CH_3.\underset{\displaystyle NH_2}{\underset{|}{C}}H.CO.NH.\underset{\displaystyle CH_2.O.CO.NH_2}{\underset{|}{C}}H.COOH \end{array}$$

Figure 3.27. D-alanyl-O-carbamyl-D-serine.

why the enzyme catalyses formation of mixed dipeptides, though not of D-threonyl-D-threonine, for example. D-cycloserine binds to *both* sites, but the "donor site" which reacts with D-alanine molecules (that supplies the N-terminal residue of di-peptide) will bind the inhibitor better than the "acceptor site" (that supplies an alanine molecule providing the C-terminal residue).

Inhibitory action of cycloserine on these enzymes is quite different. The racemase is blocked by formation of stable Schiff-bases between its pyridoxal phosphate coenzyme and cycloserine, while the synthetase is blocked by competition between D-alanine and D-cycloserine, (acting as a substrate analogue) for active sites.

The structurally related antibiotic, *O-carbamyl-D-serine*, has similar properties to D-cycloserine and acts at the same sites. It forms stable Schiff bases with pyridoxal phosphate and blocks alanine racemase, but surprisingly it is used as a substrate by the synthetase, with formation of D-alanyl-*O*-carbamyl-D-serine (Figure 3.27) which is incorporated into cell wall mucopeptides the same as D-alanyl-D-alanine, but probably disturbs their functions.

(iv) Bacitracin

This antibiotic is a mixture of about 10 closely related polypeptides, of which the major component is bacitracin A. It inhibits cell wall formation by preventing mono-dephosphorylation of phospholipid-pyrophosphate used as acceptor during mucopeptide biosynthesis. This inhibition is highly specific and only bacitracin is known to act at this stage.

The enzyme catalysing this de-phosphorylation is poorly investigated but is a pyrophosphatase similar to inorganic pyrophosphatase, ATPase, or thiamine pyrophosphatase. The reaction catalysed by all these enzymes can be written as in Figure 3.28.

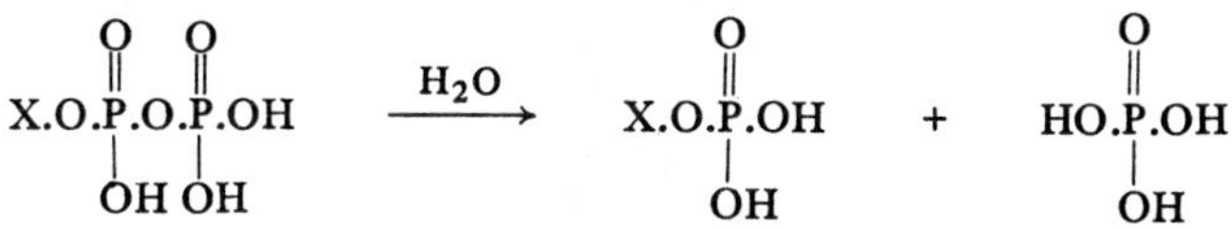

Organic pyrophosphate — Organic mono-phosphate — Inorganic ortho-phosphate

where X = H for inorganic pyrophosphatase
= AMP for myosin ATPase (as well as adenosine for granular ATPase)
= thiamine for thiamine pyrophosphatase

Figure 3.28. Mechanism of pyrophosphatases.

All these enzymes are activated by divalent cations, particularly Mg^{++} or Ca^{++}.

An important clue to the mode of action of bacitracin is the effect of slightly alkaline solution on antibiotic properties. Under these conditions, bacitracin *A* is slowly converted to bacitracin *F*, which is inactive. The change from A to F is an oxidative deamination affecting only one part of the molecule (Figure 3.29).

Loss of antibiotic activity by changing amine for carboxyl strongly suggests that the amine is involved in the mode of action. Bacitracin A, but not F, is an active chelation agent of divalent metal ions. The antibiotic is often included in oral preparations as the zinc chelate, which is tasteless, whereas free bacitracin is bitter. The chelates are also much more stable than the free antibiotic. How bacitracin chelates metal ions is

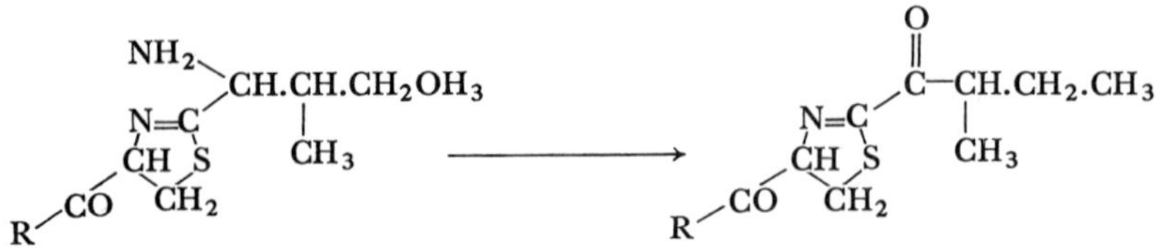

R = rest of molecule.

Figure 3.29. Conversion of bacitracin A to bacitracin F.

uncertain but by analogy with co-ordination complexes such as cuprammonium, EDTA, etc., one possibility is shown in Figure 3.30.

Displacement of a metal ion from chelates by another ion is well known for bacitracin F, zinc is displaced by nickel, nickel by copper and zinc by manganese. Relevance to the mode of action is not clear, but pyrophosphatases are activated by Mg^{++}

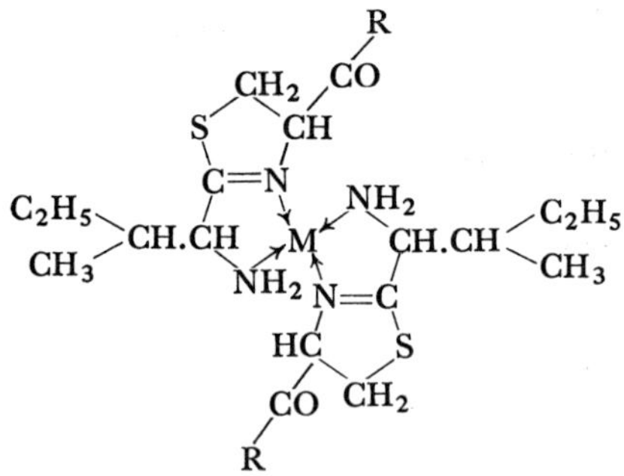

where M = a divalent metal ion
R = rest of bacitracin molecule

Figure 3.30. Bacitracin chelation.

or Ca^{++}, but inhibited by Cu^{++}, Mn^{++} or Zn^{++}. If bacitracin is able to transport inhibitory ions to pyrophosphatase, they may block active sites. Support comes from the discovery that antibiotic potency of bacitracin is reduced by EDTA (which would remove inhibitory ions released by the antibiotic), but enhanced by additions of Mn^{++}, Zn^{++} or Cd^{++}.

(v) Gentian Violet

The final compound inhibiting bacterial cell-wall biosynthesis is *hexamethylpararosaniline hydrochloride*; also known as

"Gentian Violet" or "Crystal Violet". It is a synthetic compound (*not* an antibiotic) used as an antiseptic, due to its selective action against gram-positive bacteria. Most commercial preparations contain varying amounts of penta- and tetramethyl derivatives.

The dye blocks cell-wall biosynthesis specifically at the condensation of L-alanine with UDP-muramic acid. The enzyme catalysing this reaction is unknown and the mode of action of the dye at molecular level is uncertain.

5. Miscellaneous Anti-metabolic Actions

Some antibiotics cannot be conveniently grouped into the previous modes of action. All have actions on metabolism of nucleic acids, proteins and cell walls, but these appear secondary to other effects. The important *primary* sites of action are respiration and oxidative phosphorylation; two closely related processes.

Sideromycins are a complex group of iron-containing antibiotics, with a tightly-bound Fe^{+++} ion held by three hydroxamic acid residues, to peptides. The general structure around the iron is shown in Figure 3.31.

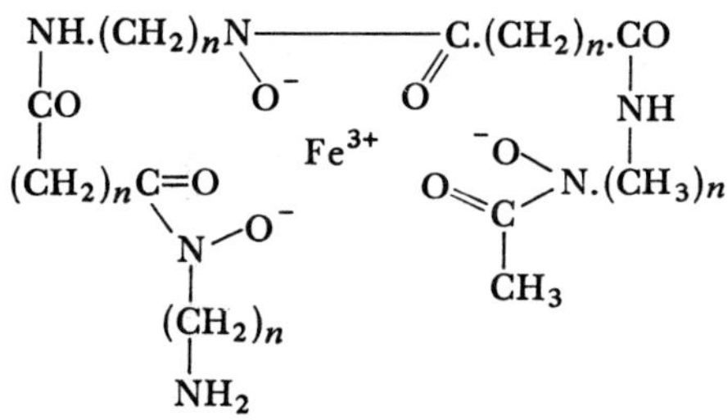

where n = 2, 3, 4 or 5

Figure 3.31. Sideromycins.

Iron is removed from sideromycins by treatment with 8-hydroxyquinoline, without damage to the rest of the molecule. The iron-free compounds re-combine only with ferric ions.

Two sideromycins used clinically are *albomycin* and *grisein* both of which are mixtures of several related compounds. Albomycin is a cyclic peptide consisting of 3 moles of L-serine and 3 moles of N^5-acetyl-N^5-hydroxyl-L-ornithine (see below), which supplies hydroxamate residues for binding iron. There is

$$\text{Hydroxamate residue } CH_3.CO.\underset{HO}{\underset{|}{N}} \vdots CH_2.CH_2.CH_2.\underset{NH_2}{\underset{|}{CH}}.COOH$$

also one 3-methyl-cytosine residue, linked to peptide by sulphate and perhaps esterified to hydroxyl of a serine.

Sideromycins at bacteriostatic concentration only slowly block biosynthesis of DNA and RNA, though protein formation stops earlier. For facultative bacteria, sideromycins inhibit growth strongly in the presence of oxygen, but have less effect in anaerobic conditions. Oxygen consumption by bacteria is much reduced, but rises if the antibiotic is removed by washing (it binds only weakly). Biosynthesis of porphyrins is unaffected, and haemin and catalase are formed in the presence of the antibiotics. The precise site of action is still uncertain, but it is likely that sideromycins inhibit respiratory enzymes, perhaps by interfering with cytochrome-dependent systems. These antibiotics cannot accept or donate electrons as their iron is permanently in the ferric state.

A chemically unrelated antibiotic is the blue pigment, *pyocyanine*, which is structurally similar to riboflavin. Pyocyanine readily undergoes oxidation and reduction and its redox potential is similar to those of flavin enzymes. It is a broad spectrum compound, inhibiting growth of bacteria, fungi, and protozoa, as well as mammalian cells. The first biochemical response to pyocyanine is increased oxygen uptake, gradually followed by a decrease as cells are killed. Pyocyanine offers an alternative pathway of electron transport from NAD- and NADP-linked dehydrogenases to oxygen, without passing through flavins or cytochromes. A simplified version is shown in Figure 3.32 for comparison with the situation in pyocyanine-treated bacteria.

As the normal electron-transport pathway is bypassed by pyocyanine, oxidation becomes uncoupled from phosphorylation, so production of ATP is much reduced.

Several other antibiotics act specifically on the electron-transport pathway. Table 3.19 summarizes these effects.

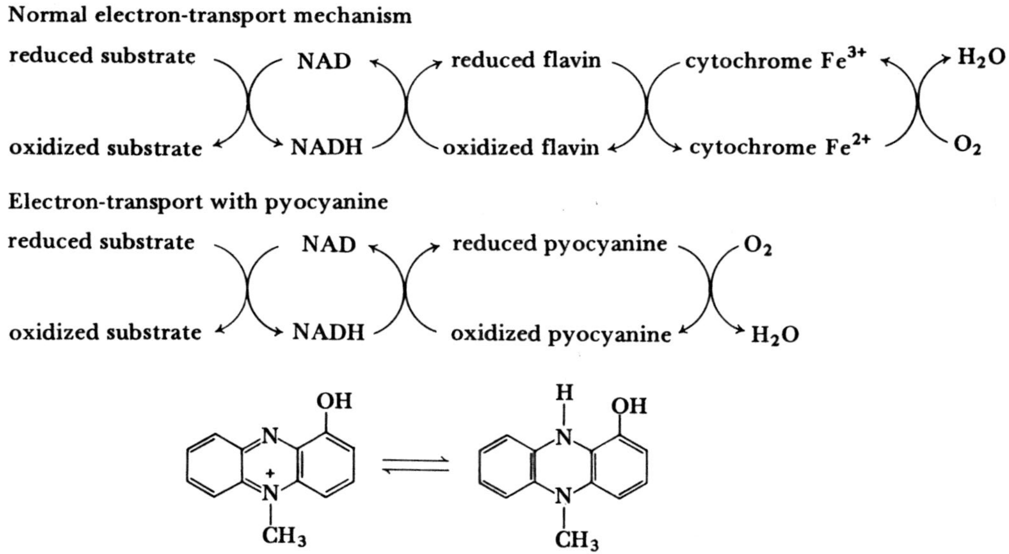

Oxidized pyocyanine   Reduced pyocyanine

Figure 3.32. Effect of pyocyanine on electron transport.

Table 3.19. Antibiotics acting on electron transport mechanisms

Antibiotic	Action
Actinomycin A	Inhibits electron transport from cytochrome *b* to cytochrome c_1.
Oligomycin	Inhibits oxidation of nicotinamide and flavin coenzymes, but *only* when coupled to phosphorylation.
Aurovertin	As for oligomycin, but probably acts at a slightly different site.

Inhibition of electron transport necessarily leads to inhibition of oxidative phosphorylation, but the reverse is not true. A number of antibiotics, such as gramicidins and valinomycin, uncouple phosphorylation without inhibiting electron transport.

6. Specificity and Resistance

We began this section by listing most of the commonly used antibiotics in groups, depending upon their spectrum of activity. Now that modes of action have been covered, there are two important biochemical questions to be asked:

(i) What determines the antibiotic spectrum of a compound?
(ii) Why do sensitive micro-organisms sometimes become resistant to particular antibiotics?

These questions cannot be answered in a completely satisfactory manner, but a considerable amount of information is available in some cases.

The problem will be approached from the basic principle of biochemical pharmacology, namely that in the vast majority of cases a compound acts by combining with specific receptors. We have seen that receptors for antibiotics can be DNA, RNA, various enzymic and non-enzymic proteins, etc. Clearly, an antibiotic cannot be expected to act on micro-organisms lacking its receptors, so that the presence or absence of its receptors in different classes of micro-organisms will determine its spectrum of activity. Similarly, genetic alteration of micro-organisms so that particular receptors are lost, offers an explanation for the development of resistance.

Table 3.20. Antibiotic activities and resistance development

Receptors	Antibiotics	Activity spectrum	Resistance mechanisms
DNA	Griseofulvin	Only fungi with cellulose walls; it probably cannot pass through fungal chitin or bacterial cell walls	Not understood, but not genetic as rapidly lost in absence of griseofulvin.
	Mitomycins	Many bacteria, viruses and human neoplasms	Genetically determined permeability changes and release of inactivating factors.
	Cycloheximide	Very broad spectrum	Genetically determined permeability change (?), without inactivation.
30 S ribosome	Streptomycin	Broad spectrum	Genetically determined change in ribosome structure.
	Tetracyclines	Broad spectrum	Genetically determined decreased permeability.
50 S	Chloramphenicol	Broad spectrum	Genetically determined decreased permeability and increased degradation.
	Erythromycin	Mainly grampositive bacteria; most other organisms are impermeable	Genetically determined change in ribosome structure.
Membrane sterols	Amphotericin	Most fungi and some protozoa, but not bacteria; requires sterols in cell membranes which are absent in bacteria.	Genetically determined (low-level) resistance by decreased sterols in membranes (?) or increased degradation.
	Nystatin	As amphotericin	Genetically determined resistance, but not cross-reacting with amphotericin.
Enzymes of cell wall biosynthesis	Bacitracin	Gram-positive bacteria; other organisms may lack the specific enzyme-receptor	Not understood, but probably not genetically determined as rapidly lost in absence of the antibiotic.
	Ristocetin	Most gram-positive bacteria and mycobacteria; other organisms may lack the specific receptor	Rare and not understood but probably a genetically determined change in permeability.

Table 3.20—continued

Receptors	Antibiotics	Activity spectrum	Resistance mechanisms
	Cycloserine	Broad spectrum, but more effective against gram-positive bacteria	Genetically determined large increase in activity of receptor enzymes (up to 800% of susceptible strains).
	Penicillins	Most gram-positive and some gram-negative bacteria: receptor enzyme more abundant in former, while β-lactamase is high in latter	Genetically determined increase in β-lactamase (penicillinase) activity.
Respiratory	Albomycin	Gram-positive bacteria and some gram-negative (mainly coli)	Not understood, but an unstable resistant population rapidly reverts in the absence of the antibiotic; excretes very low levels of glutamic acid.
	Oligomycin	Various fungi, excluding yeasts; receptors require mitochondria which are absent from bacteria; fungal selection is probably due to permeability	Rare and not understood.
Enzymes of oxidative phosphorylation	Valinomycin	Many fungi and mycobacteria, with some gram-positive and negative bacteria; presence of mitochondria favours activity	Rare: probably due to increased degradation.
	Gramicidin	Most gram-positive and a few gram-negative bacteria	Rare and not understood.

The second major factor is that a sensitive micro-organism must not only possess the specific receptor for a particular antibiotic, but the antibiotic must be able to reach it. If the antibiotic is unable to penetrate to its receptors, due to poor permeability, or is destroyed by enzymes before it has time to penetrate, it will have no effect.

We will consider some of the major antibiotics and attempt to explain the biochemical basis of their activity spectra, together with resistance mechanisms, by considering their major receptors. This is done in Table 3.20.

It will be seen that most bacteria possess a considerable ability to mutate to antibiotic-resistant forms. These mutations affect abilities of sub-cellular particles, such as ribosomes, to bind antibiotics, or decrease cell wall permeability to antibiotics, or increase synthesis of enzymes catalysing breakdown of antibiotics. Alternately, some micro-organisms respond by greatly increasing the amount of antibiotic receptor so that despite the formation of antibiotic–receptor complexes, the concentration of free receptors is biologically significant.

While antibiotics are powerful medicinal substances, they should always be seen against their biochemical background, as anti-metabolites secreted by micro-organisms effective against other micro-organisms, but whose therapeutic effects are determined by the ability of pathogens to change:

(i) the chemistry of their receptors by mutation.
(ii) cell wall permeability to the antibiotic.
(iii) the rate of secretion of enzymes degrading the antibiotic.

It seems very likely that broad spectrum antibiotics must also act on human cells, while narrow spectrum compounds will be continually susceptible to the development of resistant strains.

IV. ANTI-METABOLITES FROM PLANTS

1. General

In a previous Chapter (p. 53) we saw that many pharmaceuticals are derived from tissues of higher plants. Indeed, the preparation of herbal medicines is as old as mankind. Plants also provide the bulk of human foods, and aside from obviously important crops such as rice, maize, wheat and potatoes, there are hundreds of edible species of root vegetables, fruits, mushrooms, and leaves. In this section we consider anti-metabolites of clinical importance found in plants used either as human food, or for preparation of medicines.

A discussion has been given of microbial anti-metabolites that may contaminate plant foods, such as citreoviridin in rice (p. 225), aflatoxins in groundnuts (p. 228), botulinum exotoxin (p. 201), etc. Such anti-metabolites are not produced by the plant and so are excluded from the following account.

Toxins of plant foods represent a very mixed group of substances. Some are of considerable medical importance in particular countries, e.g. lathyrogens of vetch in India, while others contribute to an unknown extent, e.g. goitrogens of the cabbage family. The distinction between food and medicinal plants becomes blurred when toxins of rhubarb, castor beans, nutmeg, etc. are considered, for these are sometimes used for medicinal purposes.

As in previous sections of this chapter, emphasis will be given to the chemistry of plant toxins, their receptor molecules in human tissues, and the biochemical actions they induce. We begin by discussing various toxins indigenous to plants used for human food, then of food plants used as medicines, and finally, of strictly medicinal plants.

2. Antimetabolites of Food Plants

(i) Phytoagglutinins

The discovery that plant tissues contain substances capable of agglutinating human blood was surprising, but even more interesting was the finding that some plant substances are specific for group A or group B, and do not cross-react. Immunochemistry of blood agglutination will be outlined here. Considering only ABO blood groups, all humans possess, on their erythrocytes, polysaccharides of type A (group A), or type B (group B), or both (group AB), or neither (group O). These polysaccharides are antigenic and their protein antibodies are known as "isoagglutinins". Antibody to polysaccharide A is isoagglutinin α, while that for polysaccharide B is isoagglutinin β. If isoagglutinin α comes into contact with blood containing erythrocytes carrying polysaccharide A, it cross-links cells into clumps. Isoagglutinin β has similar action on erythrocytes carrying polysaccharide B. Persons of blood group A have isoagglutinin β in their plasma, group B have isoagglutinin α, group AB has neither, while group O has both.

Phytoagglutinins are substances resembling isoagglutinins. Some resemble α, other β, but mostly are non-specific and agglutinate cells carrying either A or B polysaccharides. Specific phytoagglutinins are used in haematology laboratories to determine blood groups of donors and patients.

Phytoagglutinins are sometimes called "lectins", or just simply "plant haemagglutinins". The phytoagglutinin of a particular plant is often given a specific name and Table 3.21 lists phytoagglutinins from some food plants.

Table 3.21. Phytoagglutinins from potential food plants

Plant	Family	Phytoagglutinins	M.W.
Castor bean (*Ricinus communis*)	Euphorbiaceae	Ricin	*c.* 40 000 (monomer) *c.* 80 000 (diamer)
		Ricin D	62 000
Kidney bean (*Phaseolus vulgaris*)	Leguminoseae	PHA	128 000
Black bean (*Ph. vulgaris*)	Leguminoseae	Phaseolotoxin A	126 000
Yellow bean (*Ph. vulgaris*)	Leguminoseae	YWBH	121 000
Jack bean (*Canavalia ensiformis*)	Leguminoseae	Concanavalin A	71 000
Soybean (*Glycine max*)	Leguminoseae	Glycinotoxin	105 000
Hyacinth bean (*Dolichos lablab*)	Leguminoseae	Dolichotoxin	?
Rosary pea* (*Abrus precatorius*)	Leguminoseae	Abrin	?

* Rarely used as food.

A publication appearing in 1964 listed over 500 species of plants known to contain haemagglutinins. Most were members of Leguminoseae or Euphorbiaceae, and toxins were confined, in the main, to seeds, roots and bark. Only a few plants in this list were major human foods, though a few types of beans are widely grown and eaten.

First description of a phytoagglutinin was of *ricin*, from castor beans, in 1889 and "agglutinin" was coined to describe this and related toxins in 1897. Only later was this term extended to its generalized use in modern immunology. While agglutination of human erythrocytes holds major medical interest in these substances, it was found at an early stage that some phytoagglutinins are even more active against red blood cells of non-human species. The lowest dilution of castor bean extract that agglutinates human erythrocytes is about 1 in 800, and the same is true for carp blood. However, rabbit erythrocytes agglutinate with a dilution of 1 in 8000, horse blood with 1 in 16 000, and pigeon blood with a dilution as low as 1 in 32 000. There are few reliable data on acute toxicity of phytoagglutinins, but LD_{50} by the intra-muscular route for ricin is about 0.1 g/kg for rabbits and about 30 g/kg for goats. Toxicity by the oral route is very much less.

Specificity of phytoagglutinins is curious. Substances acting on erythrocytes of several animal species agglutinate human blood of any group (e.g. ricin), but several specific phytoagglutinins are known which act only on *one* human blood-group and have little or no action on blood of other species (e.g. phytoagglutinin of lima beans acts only on human group A). As agglutination reactions require a high degree of immunochemical specificity, it is exceedingly difficult to explain evolution of plants containing such very specific agglutinins.

Human poisoning by phytoagglutinins is rare, but well-documented epidemics have occurred. The most toxic phytoagglutinin is ricin and products containing castor beans are usually associated with individual and epidemic poisoning. Partially cooked castor bean flakes added to other foods have had lethal consequences. Children are particularly susceptible and death has occurred after ingestion of as few as *five* castor beans. Workers using fertilizers containing pressed castor beans may absorb or inhale sufficient ricin to become very ill. Acute illnesses following meals containing uncooked legumes, such as runner beans, or kidney beans, have also been described and are due to phytoagglutinins. Flour from various beans containing phytoagglutinins is sometimes added to wheat flour for bread and cake manufacture, but has not been shown to be dangerous, though the practice is questionable.

Very little is known of the chemistry of phytoagglutinins, beyond the fact that they are proteins with molecular weights ranging from 60 000 to about 120 000. Few have been obtained in a pure state and there is considerable confusion between different laboratories on properties of isolated products. Ricin occurs as a monomer (M.W. 40 000), or diamer (M.W. 80 000), depending on methods of extraction. Claims have been made that toxicity of ricin is only partly explained by agglutination and that some preparations are devoid of agglutinin, yet retain toxicity. PHA from kidney beans is known to stimulate mitosis, as well as to act as haemagglutinin. Whether the two properties are shared by the same substance is not known.

Hydrolysates of purified phytoagglutinins yield mixtures of amino acids together with glucosamine and other sugar derivatives. Most phytoagglutinins are glycoproteins (or mucoproteins), but whether carbohydrate is involved in the agglutination reactions is not known though there is none in concanavalin A. Small amounts of lipid occur in the purer preparations. Most phytoagglutinins contain very little, if any, cystine, but other amino acids are typical of plant proteins.

Attempts to prepare active fragments of phytoagglutinins have met with limited success. A glycopeptide fragment of YWBH (from yellow beans) and another from agglutinin of Robina bark, retain similar haemagglutination activity to parent proteins.

Presumably, receptors of phytoagglutinins are blood-group polysaccharides of erythrocyte membranes, with which they combine; similar to human isoagglutinins. The result is macroscopic and microscopic damage to numerous tissues, due to defective blood supply, capillary blockage, ischaemia, etc. These in turn produce numerous secondary changes in tissue biochemistry. Lesions seen in experimental animals following parenteral administration of phytoagglutinin solutions depend more on the amount of toxin given than its source.

Interaction of phytoagglutinins with erythrocytes is a reversible process, for at about 56°C, adsorbed toxin is released into solution from cells. A similar effect is seen at acid pH. In fact at high temperatures, or low pH, phytoagglutinins fail to agglutinate red cells *in vitro*. Estimates of binding capacity for phytoagglutinin molecules by a single erythrocyte suggest about 400 000 sites per cell, each capable of binding one phytoagglutinin molecule. It is likely, however, that only a few molecules are required per cell to initiate agglutination.

Binding sites for phytoagglutinins occur on cells of many tissues, other than erythrocytes, but the role of extravascular binding of toxins in human poisoning is uncertain.

Attempts to identify the chemical bonds involved in phytoagglutinin binding to receptors have revealed little. Treatment of erythrocytes *in vitro* with sialidase (an enzyme catalysing hydrolysis of sialic acid-containing mucopeptides) protects against agglutination, both by phytoagglutinins or influenza virus. However, mucoprotein preparations which protect erythrocytes against virus agglutination fail to alter effects of phytoagglutinins.

Several phytoagglutinins containing less than 10% of lipid were nevertheless totally inactivated by incubation with lipases.

The practical significance of phytoagglutinins is small, despite their widespread distribution in human food. Being proteins, they are denatured by heat and lose their toxic properties. It is only when uncooked, or partly cooked foods containing phytoagglutinins are eaten, that poisoning results. Fortunately, the dose required to produce an effect is very much greater by the oral than the parenteral route. This is partly due to inactivation of phytoagglutinins by digestive enzymes, and partly to their high molecular weight, which impedes absorption. The degree of digestion of phytoagglutinins varies considerably. Ricin resists the action of pepsin and trypsin, and in one famous experiment, 10% of a purified ricin preparation was recovered intact after being incubated with pancreatic juice for 5 months. PHA of kidney beans is similarly resistant, but glycinotoxin is rapidly inactivated by either pepsin or trypsin.

Interestingly, while phytoagglutinins usually act as *antibodies* to antigenic polysaccharides or erythrocytes, in some circumstances they act as *antigens*. Animals can be immunized against phytoagglutinins by treatment with sub-lethal doses of toxins, or heat-inactivated preparations. Antibodies to phytoagglutinins then appear in serum and injection of serum from an immunized into a non-immunized animal confers protection against quite large doses of toxin. The onset of phytoagglutinin poisoning is slow and animals can be protected by injections of immune sera for up to 6 hours *after* injection of a lethal dose of ricin. This could be used as a treatment of human poisoning by castor bean products, though specific anti-serum is rarely available.

Most phytoagglutinins can be "toxoided" by treatment with formaldehyde, the resulting product still inducing antibody formation, but having lost its toxic properties. As antibodies to toxoids combine with, and inactivate untreated phytoagglutinins, this demonstrates that antigenic sites of phytoagglutinin molecules are different to haemagglutination sites, but that combination of phytoagglutinin with its antibody masks *both* sites.

(ii) Goitrogens

Nutritional goitre is usually caused by inadequate iodine intake, but occasionally arises from eating plant foods containing thyroid anti-metabolites, known as goitrogens. These range from simple compounds, such as inorganic thiocyanates or alkyl isothiocyanates, to complex thioglycosides. A list is given in Table 3.22.

With the exception of the first simple compounds, all other substances in Table 3.22 are β-thioglucosides with the general structure shown in Figure 3.33. Aside from compounds listed, at least 50 others have been detected in extracts of plants of the Brassica family. Generally, each species contains a mixture of several thioglucosides, with one compound predominating. Thioglucosides are not themselves goitrogens but must first be converted to a goitrogenic compound by an enzyme found in plant, but not in animal tissues. In contrast, inorganic thiocyanate and allyl isothiocyanate are directly goitrogenic and do not require enzymic activation.

The enzyme of plant tissues which activates thioglucosides is known as "thioglucosidase", but its actions are complex. The reaction catalysed is a simple hydrolysis of the C–S bond between the glucose residue and sulphur atom, with liberation of glucose, but the aglycone undergoes a spontaneous rearrangement with loss of the sulphate group (Figure 3.34). Of three possible final products, isothiocyanates are usually formed in greatest amount. The matter is further complicated, however, by the nature of the *R*-grouping, which sometimes participates in the rearrangement. This is seen in Figure 3.35, which compares products formed by action of thioglucosidase on sinigrin and progoitrin. The ratio of various possible products is determined by temperature and other physical variables, but under most conditions likely to occur during the preparation of human foods, the products are as shown above. As thioglucosides are

Table 3.22. Goitrogens of food plants

Goitrogen	Structure	Occurrence
Sodium thiocyanate	$Na^+ (S{-}C{\equiv}N)^-$	Many *Brassica*
Allyl isothiocyanate	$CH_2{=}CH.CH_2.N{=}C{=}S$	Some *Brassica*
Sinigrin	$CH_2{=}CH.CH_2.C.S{-}G$ $\Vert$ N | $O.SO_3^-$	Many *Brassica*
Progoitrin	OH | $CH_2{=}CH.CH.CH_2.C.S{-}G$ $\Vert$ N | $O.OS_3^-$	Turnips, rape, rutabaga, rubsen, naverte
Glucotropaeolin	$C_6H_5.CH_2.C{:}S{-}G$ $\Vert$ N | $O.SO_3^-$	Garden cress
Gluconasturtiin	$C_6H_5.CH_2.CH_2.C.S{-}G$ $\Vert$ N | $O.SO_3^-$	Turnips, horseradish, rape, kale
Glucoalyssin	$CH_3.SO.(CH_2)_4.C.S{-}G$ $\Vert$ N | $O.SO_3^-$	Rape
Gluconapin	$CH_2{=}CH.CH_2.CH_2.C.S{-}G$ $\Vert$ N | $O.SO_3^-$	Kale, cabbage, sprouts, broccoli, cauliflower
Glucobrassicin	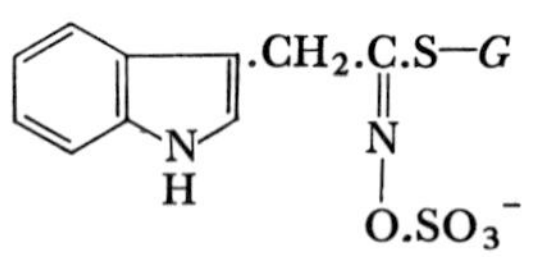	Many *Brassica*
4-Methylthio-3-butenyl glucosinolate	$CH_3.S.CH{=}CH.(CH_2)_2.C.S{-}G$ $\Vert$ N | $O.SO_3^-$	Radish

Where $.S{-}G$ = 1β-thioglucose.

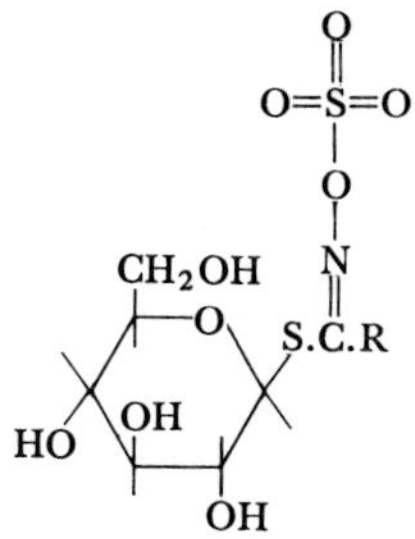

Figure 3.33. General structure of β-thioglucosides.

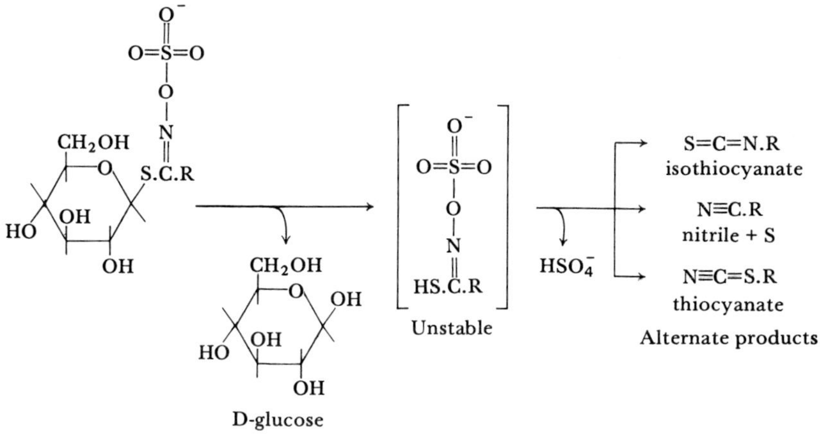

Figure 3.34. Hydrolysis of β-thioglucosides.

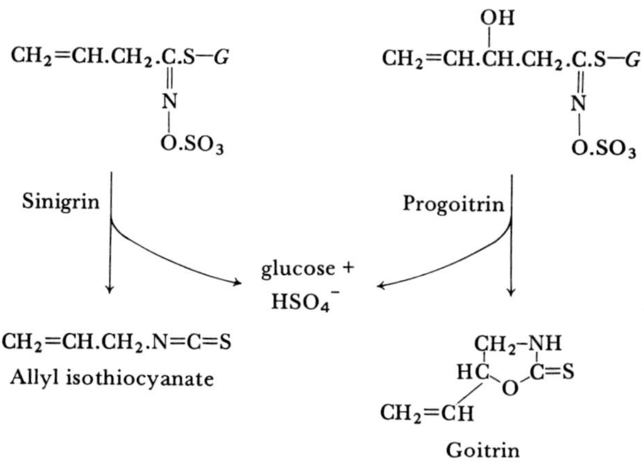

Figure 3.35. Comparison of hydrolysis of sinigrin and progoitrin.

not goitrogenic, it has often been thought that preparation of food containing progoitrins may affect the amount of goitrogenic metabolite formed. For example, if whole plants are boiled intact, the enzyme is inactivated and thioglucosides are not broken down, as they would be if plants were chopped or crushed before cooking (or eaten raw). While this is true, it is of no practical significance, for intestinal bacteria possess thioglucosidases to activate thioglucosides. Pure progoitrin given by mouth is recovered from urine largely as goitrin, but not if antibiotics are given first to eliminate intestinal flora.

Experiments with goitrin, allyl isothiocyanate, and other products formed from thioglucosides of Brassicas, show them to be potent goitrogens. The most active is goitrin, which occurs as two stereoisomers, *R*- and *S*-, depending on configuration of the hydroxyl- bearing C atom of progoitrin from which it is formed. Both isomers are equally active and have about 133% of antithyroid property of an equal weight of the potent synthetic drug, propylthiouracil. As the latter suppresses thyroid in adults at doses of about 50 mg daily, it is obvious that small concentrations of progoitrin in food could have serious effects. Fortunately, concentrations of progoitrin in the edible portions of Brassicas tend to be low, though seeds contain up to 30 mg/g.

Inorganic thiocyanates are goitrogenic, but are far less active than goitrin, daily doses of up to 200 mg KSCN being required for thyroid suppression. However, cabbage contains up to 10 mg thiocyanate per 100 g portion fresh vegetable, while thiocyanate, as well as goitrin, is found in milk from cows fed on cabbage, rape, kale, etc.

The mode of action of inorganic thiocyanate on the thyroid is different to that of goitrin or other goitrogenic derivatives of thioglucosides. Thiocyanate ions compete with inorganic iodide for uptake by specific receptors of the thyroid gland, so that additional iodide counteracts the action of thiocyanate. In contrast, goitrin (in common with propylthiouracil) blocks coupling of iodotyrosines, as well as iodination of MIT, so that its actions are not reversed by additional iodide. Both types of goitrogen nevertheless have a similar result on the thyroid and produce compensatory hyperplasia and hypertrophy, due to uncontrolled stimulation by thyrotrophin. The clinical result is similar to iodine-deficiency goitre.

Table 3.23. Iodine effect on plant goitrogen toxicity

Plant food	Goitrogen present	Reversed by iodide
Raw soybean (partly inactivated by heat	Unknown	Yes
Groundnuts	Arachidosides (glycoside pigments)	Yes
Cashew nuts	Anthocyanins	Yes
Almonds	Anthocyanins	Yes
Areca nut	Anthocyanins	Yes
Onions	*n*-Propyl disulphide	Yes
Cassava	Unknown	?

Goitrogenic activity has occasionally been reported for plant foods lacking thioglucosides. Little is known of the compounds responsible but Table 3.23 summarizes available information.

The action of anthocyanin pigments, and related compounds, on the thyroid appears to be by competition for "active iodide". Some of these pigments have phenolic rings that interact with iodide to yield iodinated derivatives, which are then excreted. These pigments are only mild goitrogens and large amounts are needed to produce an effect. The presence of unidentified goitrogens in such widely used human foods as soybeans and cassava, however, is alarming and requires more investigation.

It is particularly difficult to assess the role of plant goitrogens in the aetiology of human goitre. Iodine deficiency occurs in most geographical areas where simple goitre is common. However, it is on populations where iodine intake is marginally inadequate that goitrogens would be expected to have the most effect; so that *both* iodine-deficiency and goitrogen intake may be of equal importance. There is no doubt that certain goitrous areas, such as Finland, grow large amounts of Brassicas for human food, and also for feeding to dairy cows. In areas of Finland where goitre is endemic, milk contains 50 to 100 μg/l goitrin and up to 8 mg/l thiocyanate. While these levels of goitrogen are insufficient to induce thyroid hypertrophy, their ingestion over a period of years could lead to cummulative biochemical damage to the gland. Interestingly, thiocyanate ions compete with iodide for secretion into milk, so that the higher the thiocyanate content of milk, the lower the iodide. This could be a further factor determining human iodide intakes in goitrous regions.

The incidence of goitre in endemic regions of under-developed countries is often over 60% of the population, and is usually associated with a high incidence of cretinism. The importance of establishing the role of plant goitrogens in these areas can hardly be over-emphasized.

While the epidemiological significance of goitrogens on a large scale is still doubtful, occasional isolated cases of human goitre have been traced beyond reasonable doubt to ingestion of goitrogens. Some cases are due to food fads, leading to ingestion of truly collosal amounts of brassicas, while others are children who have eaten brassica seeds. Soybean products used as milk substitutes for infants are occasionally goitrogenic, even when prepared from heated soybeans. Only occasional batches of soybean are goitrogenic and thyroid enlargement arises only in infants with low iodine intake. It is reversed by changing to cow's milk, or by giving additional iodides.

(iii) Cyanogens

Inorganic cyanide ion (CN^-) is a potent metabolic poison, inhibiting electron transport by blocking cytochrome oxidase, and combining with haemoglobin to prevent oxygen uptake. The lethal oral dose for adults is about 200 to 250 mg NaCN or KCN, which is equivalent to about 50 to 130 mg CN^-. Exposure to HCN vapour at 200 ppm is usually fatal in about 3 minutes.

Chronic poisoning, by regular exposure to sub-lethal doses of inorganic cyanide, would be expected to produce anoxic damage to many tissues, particularly central and peripheral nerves with high oxygen demand. The clinical outcome would be impaired vision, hearing, etc., and defective intellectual performance. There would also be muscular weakness and, perhaps, some paralysis.

The reason for doubt over the clinical presentation of chronic cyanide poisoning is that the condition is exceedingly difficult to diagnose, for humans are constantly exposed to low levels of cyanide from industrial sources, tobacco smoke, and, as we shall see, certain plant foods. Cyanide is metabolized by humans to

$$CN^{--} + S_2O_3^{--} \xrightarrow{\text{rhodanase}} CNS + SO_3^{--}$$

Figure 3.36. Action of rhodanase.

thiocyanate, a reaction catalysed by rhodanase: a widely distributed human enzyme (Figure 3.36). Cyanide ions also bind avidly to methaemoglobin to form cyanhaemoglobin by exchange of OH by CN. A minor pathway of cyanide detoxification is through reaction with disulphide bridges of cystines to give cystine and 2-iminothiazolidine-4-carboxylic acid (Figure 3.37).

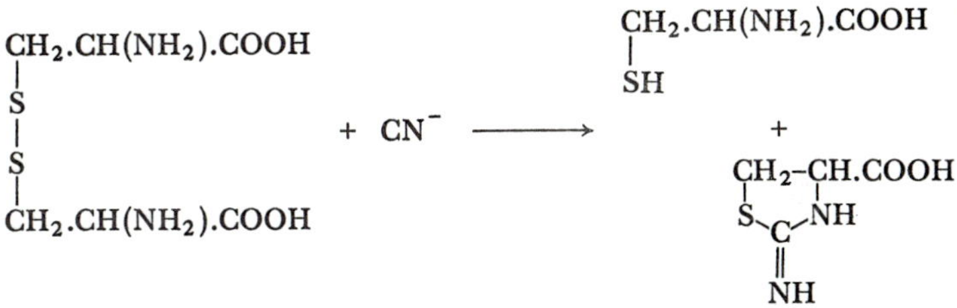

Figure 3.37. Minor pathway for cyanide detoxication.

As cyanide in low doses is rapidly metabolized to thiocyanate, persons suffering from chronic cyanide poisoning would be expected to have high tissue levels of thiocyanate, but, as we have just seen, thiocyanate is present in many foods, so that its presence in tissues is a poor criterion of exposure to cyanide. The normal range of thiocyanate in human saliva is reported as 26 to 270 mg/l, while milk from cows (*not* fed Brassicas) is 1 to 2 mg/l. Thiocyanate is even a normal constituent of CSF (30 to 290 μg/100 ml).

Organic cyanides are considerably less toxic than inorganic cyanides, but are subject to enzymic hydrolysis and liberation of HCN. As for thioglucuronides, enzymes catalysing activation of organic cyanides of plant tissues are confined to plants and absent from humans. They may, however, be present in intestinal bacteria.

Plant organic cyanides are glucosides of various cyanide-containing aglycones. They are widely distributed in the plant kingdom, but only three compounds are found in plants used as human food. These are listed in Figure 3.38. Dhurrin is hydroxy-amygdalin, and both compounds contain two glucose residues, while linamarin has only one. All three compounds are β-glucosides and are hydrolysed by β-glucosides found in plant

Cyanogenetic glucoside	Structure	Occurrence
Amygdalin	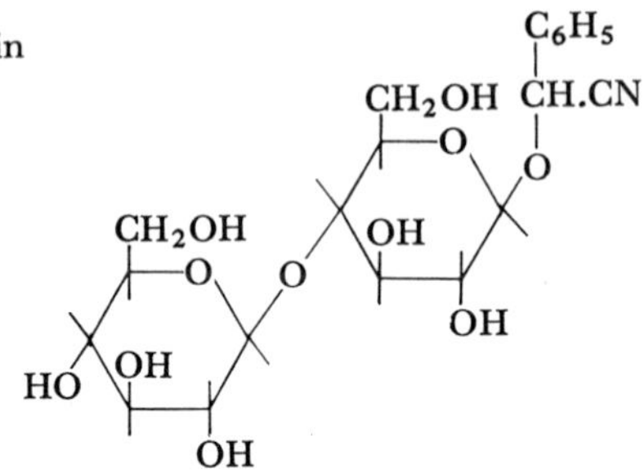	Almond Lemon Lime Apple Pear Cherry Apricot Plum
Linamarin	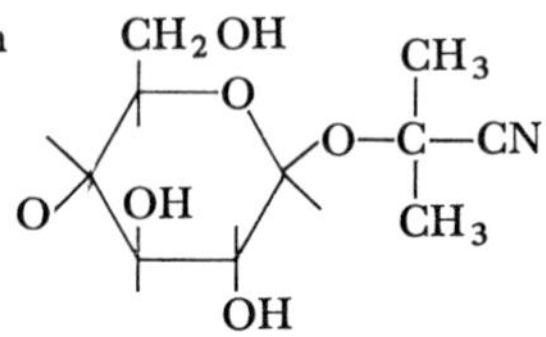	Cassava Linseed (flax) Sweet potato Yam Peas Lima beans
Dhurrin	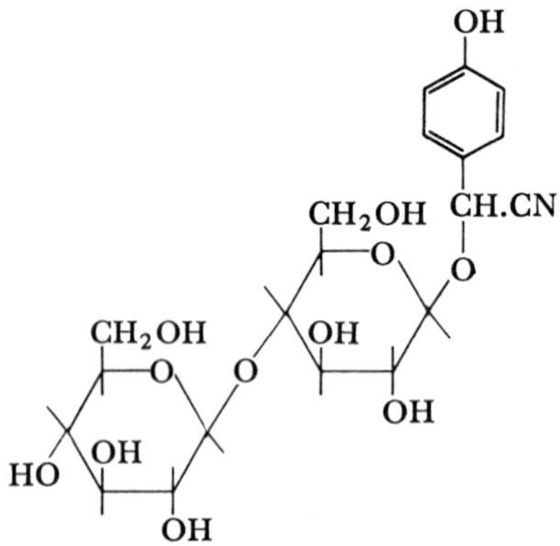	Sorghum Maize Millet Bamboo Sugar cane

Figure 3.38. Occurrence and structures of cyanogenetic glucosides.

tissues. Animal enzymes catalysing hydrolysis of β-glucoside linkages, such as β-amylase, β-glucuronidase, lactase, etc. are ineffective. For amygdalin and dhurrin, the two glucose residues are removed one at a time by the enzyme. The net result in each case is release of cyanide-containing aglycone, together with glucose. The linamarin aglycone is highly unstable and immedi-

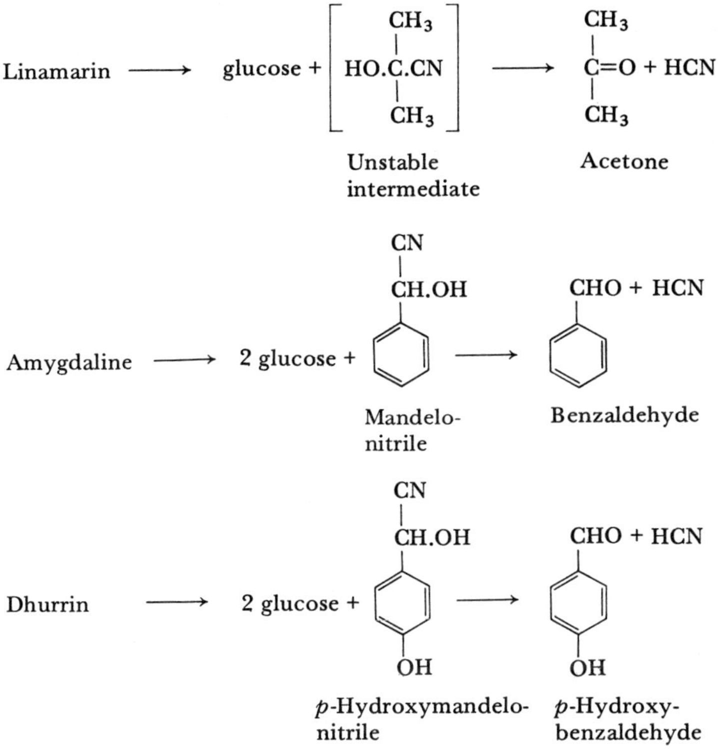

Figure 3.39. Hydrolysis of cyanogenetic glucosides.

ately decomposes, while the cyanohydrin aglycones from the other two compounds slowly hydrolyse (Figure 3.39).

β-Glucosidases of plant tissues are extra-cellular; glycosides are intracellular. While cellular integrity of plant tissues is maintained, HCN is not liberated, but bruising, crushing, cutting, etc., all bring enzyme and substrate together. Soaking in water also aids final hydrolysis and release of HCN boiling rapid inactivates β-glucosidase. The glycosides may also by hydrolysed non-enzymically by any weak acid, though they are poorly water-soluble and very little hydrolysis occurs below 60°C.

While normally considered as a water-solution, HCN is a volatile liquid with a boiling point of 26°C, so that poisonous vapours form over plants containing cyanogens particularly if the plants are damaged and stored under hot, humid conditions.

As for goitrogens, there is little definite evidence either way on the effects of chronic poisoning by cyanogens, though there are well documented cases of acute poisoning. Bruised cassava has been known to be poisonous since at least 1605 when the first recorded death was published. Deaths following ingestion of *Phaseolus lunatus* (lima or butter beans) have been reported many times from various countries. General Gordon's horses were poisoned in the Sudan by eating Egyptian vetch, and this incident led to the extraction and identification of linamarin.

Table 3.24. Cyanides in food plants

Plant foodstuffs	HCN yield (maximum–mg/100 g)
Bamboo shoots (tips of shoots)	800
(immature stems)	300
Lima beans (black Puerto Rican)	300
(white Burmese)	210
(white American)	10
Almonds	250
Sorghum (immature)	250
Cassava (root cortex)	245
(whole root)	55
(stem)	110
Haricot beans	5

Table 3.24 lists yields of HCN from some human foods known to contain cyanogens. These are maximum values, so that yields from most samples of these plants will usually be lower. Highest concentrations of cyanogens are found in growing tips of plants, though the compounds are distributed throughout all plant tissues.

It will be seen that portions of most of these plants contain potentially fatal doses of HCN. That fatal poisonings are not common is probably due to methods of food selection and preparation adopted over long periods of time by peoples who include these plants in their normal diets. Bruised or damaged plants are rejected, while even good specimens are stored for only short periods. Plants are well washed and peeled where possible. Cooking is thorough and the water is discarded. Some countries have introduced legislation forbidding the sale of foods containing more than a particular concentration of cyanogen (USA = 20 mg/100 g).

There is considerable medical interest in knowing whether populations who regularly eat cyanogen-containing foods develop chronic poisoning, despite precautions. Tobacco amblyopia occurs in heavy smokers and is suspected of being an induced damage to optic nerves by HCN of tobacco smoke. Amblyopia is relatively common in some African countries where cassava is a staple food, and occurs even in non-smokers. It is often associated with ataxia, nerve deafness and muscular weakness. Patients with this syndrome usually also have signs of nutritional defects, so it may have a multiple aetiology with cyanogens as one factor. There are many features in common between these patients and those with Leber's disease, a rare inherited optic atrophy, thought to be due to defective rhodanase and failure to detoxify cyanide.

(iv) Lathyrogens

Lathyrism is one of the oldest recorded human diseases, arising only in populations eating seeds of *Lathyrus* species (vetches). It was described by Hippocrates, who was aware of the cause. Despite this, lathyrism is a major health problem on the Indian sub-continent and in parts of North Africa. Occasional cases from Europe and other parts of Asia are reported. A survey published in 1958 on a rural Indian population showed 25 000 cases of lathyrism out of a total of 634 000 people.

The biochemistry of human lathyrism has been somewhat complicated by the discovery that toxic vetches fed to experimental animals produce symptoms quite different to those seen in human patients. In man, lathyrism is a degeneration of the spinal cord, leading to spastic paralysis of the legs and other neurological lesions. In animals, the plants and their anti-metabolites induce skeletal and mesenchymal lesions, without much effect on nerves. It seems likely that laboratory animals are sensitive to a lathyrus toxin acting principally on bone, which has little effect in humans. Conversely, humans are seriously affected by lathyrus nerve toxins which have little action in animals.

The *osteo*lathyrogen of vetches appears to be β-(γ-L-glutamyl)-aminopropionitrile (Figure 3.40) whereas the *neuro*-lathyrogens (which cause the human disease) appear to be β-*N*-oxalyl-L-α,β-diaminopropionic acid together with L-α,γ-diaminobutyric acid and γ-*N*-oxalyl derivative (Figure 3.41).

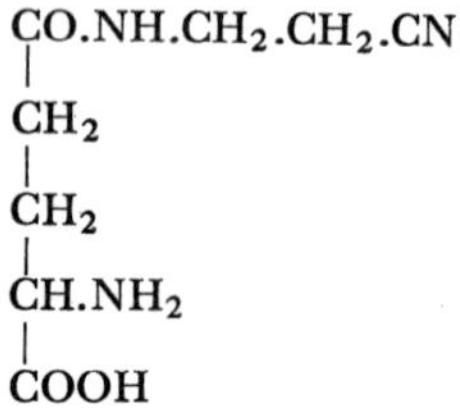

β-(γ-L-glutamyl)-aminopropionitrile

Figure 3.40. An osteolathyrogen.

The anti-metabolites present in particular vetches depend on species and variety, together with stage of growth and method of cultivation. The total lathyrogen content of seeds ranges from about 0.2 to 2.0%. In parts of India, it is customary for vetch seeds to make up between a third and a half of the diet. At this level of intake, symptoms of lathyrism develop after about 3 to 6 months and are most common in young men and, to a lesser extent, in young women. Assuming a daily food

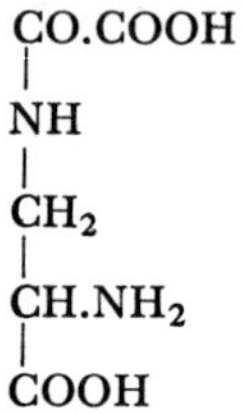

β-*N*-oxalyl-L-α,β-diaminopropionic acid

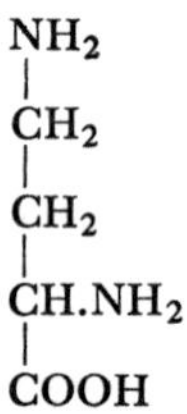

L-α,γ-diaminobutyric acid

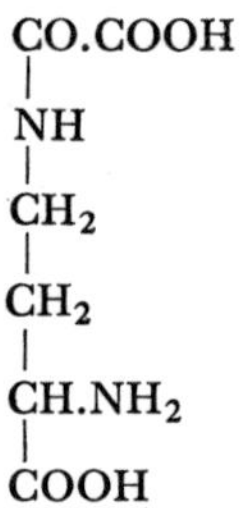

γ-*N*-oxalyl-L-α,γ-diaminobutyric acid

Figure 3.41. Some neurolathyrogens.

intake of about 2000 kcal, with an average calorific value of 6 kcal/g and moisture content of 20%, the daily intake of lathyrogens for these people is about 4 g daily (i.e. about 65 mg/kg).

Oral administration of neurolathyrogens to rats produces CNS lesions only at very high levels (680 mg/kg). Neural lesions develop in young chicks following parenteral administration of similar doses, but adult animals seem resistant. Similar resistance is seen in adult rats, dogs and guinea-pigs, but young animals are susceptible. It is possible that transport across the blood-brain barrier is poor in older animals, though not in humans, so that spinal degeneration occurs only in young animals and in man.

The mode of action of neurolathyrogens is unknown and the absence of satisfactory animal models makes investigation difficult. A large number of synthetic compounds are osteolathyrogens in animals, but this is probably irrelevant to the human disease. It will be noted that two of the three compounds thought to be human lathyrogens are *N*-oxalyl derivatives of L-amino acids not found in human proteins. In day-old chicks, the synthetic compounds in Figure 3.42 are neurotoxins.

N-oxalyl-glycine	$HOOC.CO.NH.CH_2.COOH$
N-oxalyl-β-alanine	$HOOC.CO.NH.CH_2.CH_2.COOH$

Figure 3.42. Synthetic neurotoxins.

Considering the importance of glycine as an inhibitory transmitter in the spinal cord, it is tempting to suggest that neurolathyrogens act as analogues interfering with inhibitory transmission. One of the earliest symptoms of human lathyrism is spastic contraction of calf muscles, accompanied by exaggerated knee and ankle reflexes, that could well be explained by suppression of spinal inhibition. It is more difficult, however, to explain spinal degeneration leading to paralysis and loss of control over the lower limbs.

(v) Protease Inhibitors

Numerous plants contain substances inhibitory to proteolytic enzymes of the human digestive tract (pepsin, rennin, trypsin, chymotrypsin, carboxypeptidase, etc.).

Food containing inhibitors reduces digestion and absorption of dietary protein, with increased excretion of faecal nitrogen. Chronic administration in animals results in pancreatic hypertrophy.

Table 3.25 lists some common food plants known to contain protease inhibitors. In general, these have been tested only against trypsin and their action on other digestive enzymes is often in doubt. Chymotrypsin is inhibited by anti-metabolites from soybean and potato, but not by that from wheat. Pepsin is not inhibited by any of these three substances.

Table 3.25. Protease-inhibitors of food plants

Plants containing protease-inhibitors	Present in	Inactivated by heat
Groundnuts	Seeds	No
Soybeans	Seeds	Yes
Garden peas	Seeds	?
Lima beans	Germ	Yes
Oats	Seeds	No
Barley	Seeds	No
Wheat	Germ	Yes
Maize	Seeds	No
Rice	Seeds	Yes
Beetroots	Roots	Yes
Turnips	Roots	Yes
Potatoes	Roots	Yes
Sweet potatoes (yams)	Roots and leaves	Yes
Coffee	Seeds	Yes

Purified preparations of protease inhibitors contain polypeptides of molecular weight 14 000 to 23 000. These chemically combine with protease molecules, to form inactive complexes. We have seen elsewhere that trypsin catalyses hydrolysis of peptide bonds formed from the carbonyl of either arginine or lysine with the amine of any other residue. Catalysis requires interaction of the hydroxyl at the enzyme active centre (-Asp-Ser-Gly-) with the peptide (see p. 151). Inhibition of trypsin by protease inhibitors has been studied in detail and the outlines of chemical interactions involved are known. Soybeans contain several trypsin-inhibitors, one of the best known being the "Kunitz inhibitor" a polypeptide of 198 amino acid residues with an N-terminal aspartyl and a C-terminal leucyl. The molecular weight is about 24 000, which is similar to that of trypsin. The inhibition is competitive and is greatly reduced by low molecular weight artificial enzyme substrates, such as *N*-benzoyl-L-arginine ethyl ether. The enzyme–inhibitor complex has a very low dissociation constant. Kunitz inhibitor molecules possess two disulphide bridges, both of which are essential for activity; though curiously they can be opened and

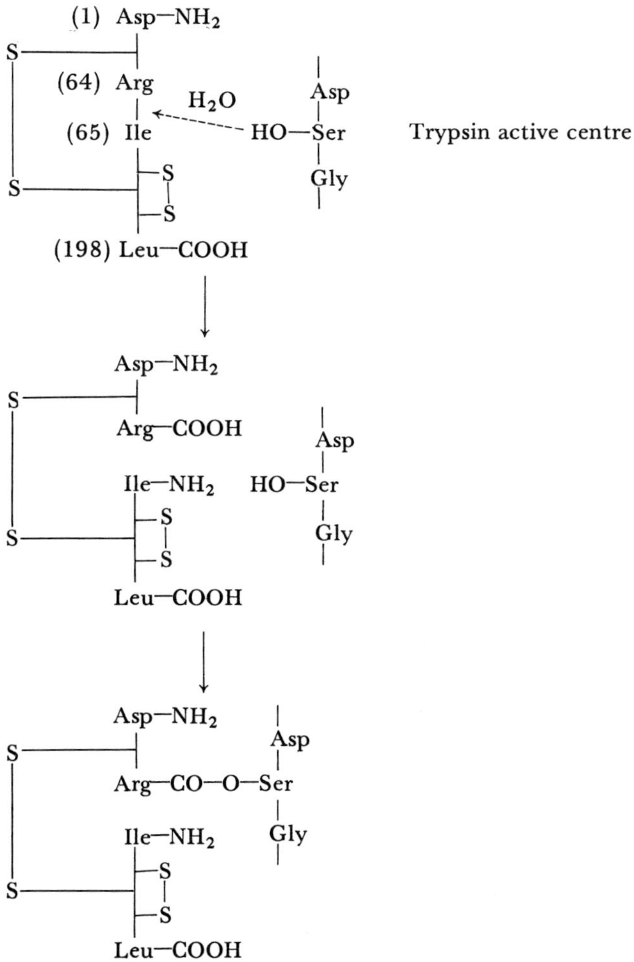

Figure 3.43. Kunitz inhibitor.

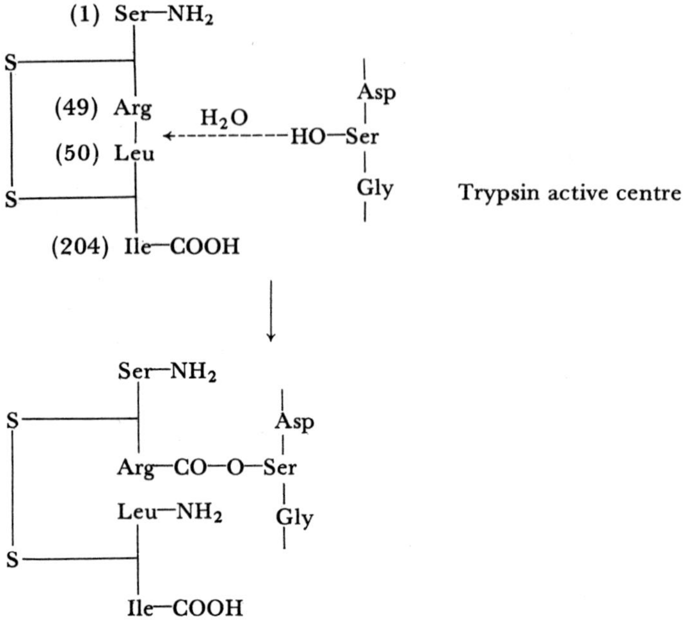

Figure 3.44A. Maize protease-inhibitor.

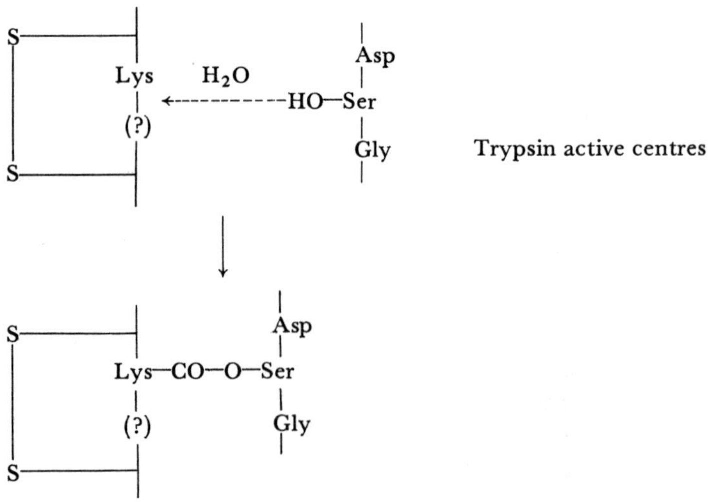

Figure 3.44B. Lima bean protease-inhibitor (Six related polypeptides with 76, 77, 84, 89, and 93 amino acid residues).

reclosed *mis-matched* and still yield an active inhibitor. Interaction of inhibitor with enzyme occurs in two stages, shown in Figure 3.43.

The active centre of trypsin catalyses the hydrolysis of an arginyl-isoleucyl bond of the inhibitor with formation of a new C-terminal arginyl residue. Carboxyl of this new residue then spontaneously esterifies the seryl-hydroxyl and effectively blocks the enzyme active centre. A similar reaction occurs with protease inhibitors of maize (Figure 3.44A), and of lima beans (Figure 3.44B), though in the latter case it is a lysyl bond which is split and a lysyl carboxyl which esterifies the seryl of trypsin.

If trypsin is treated with DIPPF, which specifically combines with seryl of the active centre, the enzyme no longer combines with Kunitz, maize or lima bean inhibitors.

The active centre of trypsin also requires a nearby histidyl residue, which will specifically combine with TLCK. Trypsin treated with this inhibitor also cannot combine with protease inhibitors, even though the seryl group remains free. This demonstrates that only enzymically active trypsin can combine and that the combination involves an enzymic change in the inhibitors. Modification of ϵ-amino groups of lysyl residues of Kunitz and maize inhibitors has no effect on their combination with trypsin, but completely prevents interaction of lima bean inhibitors, for a lysyl bond is involved only for the latter.

Confirmation of this mechanism comes from chemical reduction of inhibitor–trypsin complexes, so that disulphide bridges are broken. As inhibitor molecules are linear peptides, which have been split by the enzyme, two shorter fragments are released, while the ester bond to the enzyme hydrolyses spontaneously. The Kunitz inhibitor gives one peptide with 64 amino acid residues (aspartyl-N terminal, arginyl-C terminal) and another with 134 residues (isoleucyl-N terminal, leucyl-C terminal), while the maize inhibitor yields one of 49 residues (seryl-N terminal, arginyl-C terminal) and a second of 155 residues (leucyl-N terminal, isoleucyl-C terminal).

The action of other protease inhibitors is not so well understood. The inhibitor of potatoes is particularly active against chymotrypsin, but will not combine with enzyme pre-treated by DIPPF to block the seryl active centre. However, it does combine with chymotrypsin pre-treated by TLCK, which blocks the histidyl residue also required for activity.

Another very active chymotrypsin inhibitor is Bowman-Birk inhibitor from soybeans: a polypeptide of M.W. 20 435 that differs from Kunitz inhibitor by being insoluble in acetone. B-B inhibitor combines with trypsin to form a stable complex that will also combine with chymotrypsin to give a three-component complex. Similarly, chymotrypsin–B-B inhibitor complex will also combine with trypsin, but an inhibitor molecule will not combine with two molecules of either enzyme alone. This strongly suggests two separate, unrelated binding sites on the inhibitor; one specific for trypsin and the other for chymotrypsin, despite similarities between the active centres of the two enzymes. However, the specificity of chymotrypsin is low (unlike trypsin) and its greatest catalytic action is on peptide bonds formed from carboxyls of aromatic amino acid residues (which are unaffected by trypsin). It is possible that the trypsin site of B-B inhibitor is similar to that of Kunitz inhibitor, but that the chymotrypsin site is elsewhere on the molecule and involves an aromatic amino acid residue. Treatment of B-B inhibitor, following interaction with trypsin, by carboxypeptidase B (which specifically removes C-terminal arginyl residues), destroys ability of the inhibitor to combine with trypsin, but leaves combination with chymotrypsin unaffected. In contrast, treatment of inhibitor, following interaction with chymotrypsin, by carboxypeptidase B has no action on combination with either enzyme. Probable mechanisms involved in these reactions are set out in Figure 3.45.

Several other protease inhibitors are also "double-headed" like B-B inhibitor. The lima bean protease-inhibitor with 89 amino acid residues (M.W. 9667) behaves in a very similar manner, though the other 5 inhibitors from this plant do not. Mung bean protease-inhibitor (M.W. *c.* 8000) is unusual in possessing two trypsin-binding sites, though it does not combine with chymotrypsin. The inhibitor from black-eyed peas (M.W. 16 923) combines with one mole of trypsin and *two* of chymotrypsin.

The biological effect of protease inhibitors is to reduce the amount of proteolytic activity in the lumen of the gut, so that utilization of dietary protein is impaired. Inhibitors in ingested foods combine with intestinal proteases, particularly those of pancreatic origin, to form stable complexes, which pass down the gut with undigested food and are decomposed by bacteria

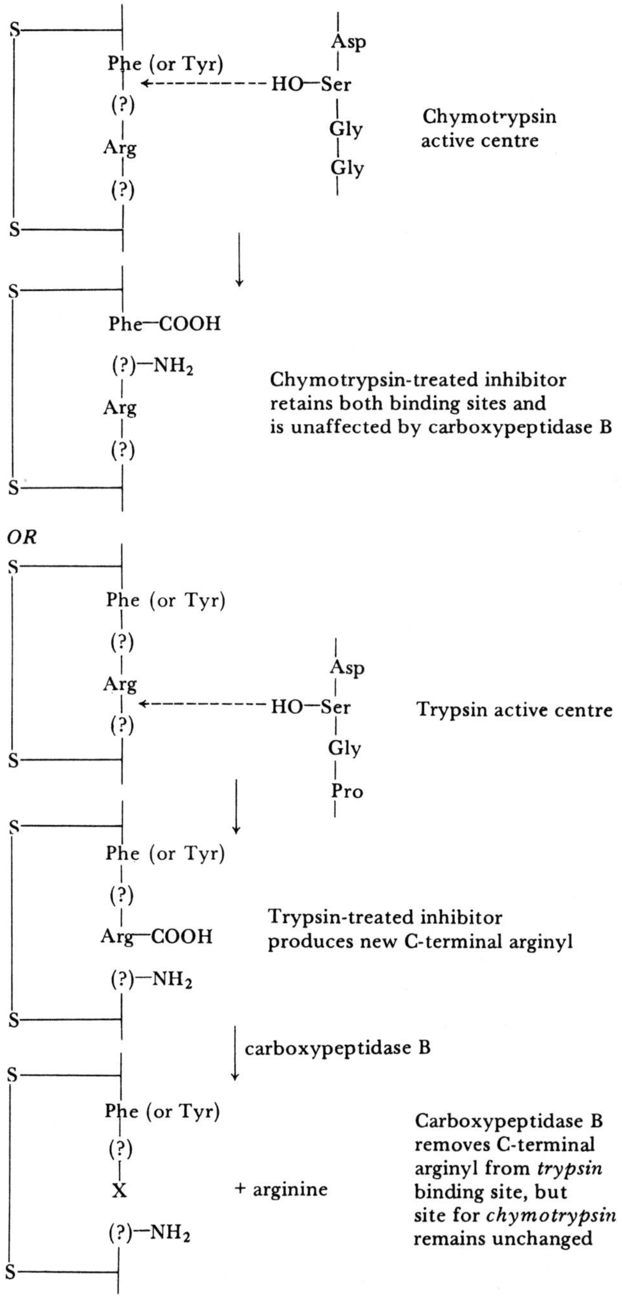

Figure 3.45. Bowman-Birk inhibitor.

of the colon. Only inhibitors not inactivated by cooking, or present in foods eaten raw, are of significance in human nutrition. These are protease inhibitors of groundnuts (which are present mainly in skins), oats, barley, and maize. On the basis of animal experiments, the major effect of protease inhibitors is to reduce availability of essential amino acids present in these foods, so that young animals fed diets marginally adequate in essential amino acids undergo restriction in growth if protease inhibitors are added. A growth restriction occurs even when essential amino acids of the food are free compounds, rather than in proteins. This suggests some toxic action unrelated to inhibition of digestive enzymes.

It has been mentioned that pancreatic hypertrophy occurs in laboratory animals (rat and chick, but *not* calf or pig), following chronic ingestion of protease inhibitors. The mechanism of pancreatic stimulation is not clear, but may involve increased release of intestinal hormones which act on the pancreas (e.g. secretin or pancreozymin-cholecystokin). Increased pancreatic secretion leads to protein loss from the body in the form of pancreatic enzymes, for these are normally digested and their amino acids reabsorbed from the small intestine. The presence of protease inhibitors prevents this return. As adults secrete about 700 ml of pancreatic juice daily, containing about 3.5 g protein (mainly proteolytic enzymes) the loss by combination with inhibitors together with simultaneous reduction in efficiency of digestion of dietary protein can precipitate essential amino acid deficiencies.

The significance of protease inhibitors in human disease is still uncertain, and it is not known if pancreatic hypertrophy occurs in humans eating foods containing these anti-metabolites.

(vi) Favism

This haemolytic anaemia is closely related to primaquine-sensitivity, and arises in susceptible individuals following a meal containing fava beans (*Vicia faba*); a widely-grown crop in many parts of the world. Haemolytic episodes have also been reported occasionally following garden peas, or after exposure to various pollens.

Like primaquine-sensitivity, favism is an inherited metabolic disease involving glucose-6-phosphate dehydrogenase (G6PD) isozymes of erythrocytes. However, the two conditions are not identical, for fava beans do not induce a haemolytic anaemia in Negroes sensitive to primaquine, though primaquine produces anaemia in non-Negroes sensitive to fava beans. Susceptible individuals show great variation in severity of responses to fava beans, some developing a severe crisis with haemoglobinuria and jaundice, while others show only a mild anaemia. Onset of the disease is rapid and symptoms occur within 24 hours of eating

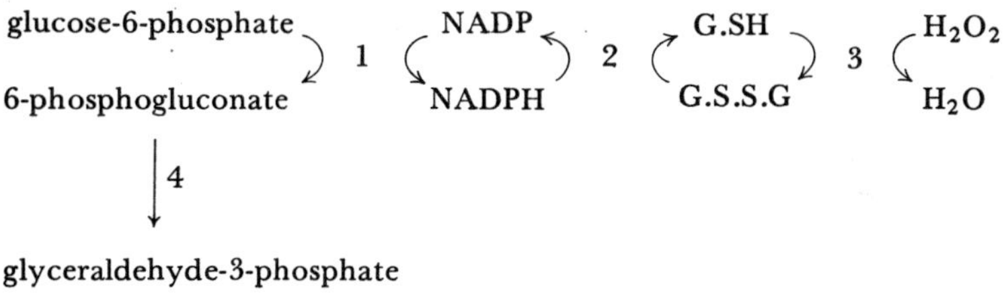

where, 1 = G6PD
2 = glutathione reductase
3 = glutathione peroxidase
4 = remainder of pentose phosphate pathway.

G.SH = reduced glutathione
G.S.S.G = oxidized glutathione

Figure 3.46. G6PD in erythrocyte biochemistry.

fava beans. In children the disease can be fatal, though blood transfusion is of great value. An attack is self-limiting and rarely lasts for more than 48 hours, for only old erythrocytes (aged 63 to 76 days) are affected by anti-metabolites from the beans. Once these cells are destroyed, no further haemolysis occurs, and the subject is relatively resistant to fava beans until a further aging erythrocyte population has developed.

The principal pathway of glucose metabolism by erythrocytes is via the pentose phosphate pathway, for which the first enzyme is G6PD. Its coenzyme is NADP, which is constantly regenerated as the oxidized form by glutathione reductase. The net effect is for G6PD to maintain cell concentrations of reduced glutathione, the major substance protecting sulphhydryl groups of essential enzymes and membrane proteins from oxidation and inactivation. Interrelationships are shown in Figure 3.46.

The gene for erythrocyte G6PD is on X-chromosomes, so

that males are always homozygous, but females may be heterozygous for different isozymes. Isozymes of G6PD are readily separated by electrophoresis at pH 8.6, and several different fractions have been identified. About 20 isozymes are now known, though many are very rare. The commonest are isozymes A and B, with A being the fastest-moving electrophoretic component. The genetic locus for G6PD is usually written "*Gd*", so that normal male phenotypes are Gd^{A} and Gd^{B}, while women may be Gd^{A} Gd^{A}, Gd^{B} Gd^{B}, or Gd^{A} Gd^{B}. In all populations the commonest isozyme is A, but Negroes show a high proportion of B. In West Africa about 56% of men are Gd^{B}, while U.S. Negroes males have an incidence of about

Table 3.26. G6PD isozymes

Isozyme	Electrophoretic mobility compared to isozyme B	K_m G6P (pM)	K_m NADP (pM)	Heat stability
A	Fast	50–78	2.9–4.4	Normal
B	—	50–78	2.9–4.4	Normal
Canton	Fast	20–36	2.0–2.4	Slightly reduced
Mediterranean	Normal	18–26	1.2–1.6	Reduced
A^-	Fast	50–78	2.9–4.4	Normal

66%. Isozymes A and B are equally active and of the same stability. They differ in substitution of an aspartyl residue in A by asparaginyl in B, which explains why isozyme A moves faster to the anode on electrophoresis.

The commonest defective isozymes are A^- (Gd^{A-}), Canton (Gd^{Can}), and Mediterranean (Gd^{Med}). The first of these is seen only in Negroes, the second in S.E. Asia, while the third occurs in Europe and North Africa around the Mediterranean. Men possessing a defective isozyme have low erythrocyte G6PD activity: less than 20% of normal for Gd^{A-}, less than 24% for Gd^{Can}, and less than 7% for Gd^{Med}. Properties of the various isozymes are summarized in Table 3.26.

The Canton and Mediterranean isozymes have reduced catalytic abilities, probably due to amino acid substitution near to the active centre. The nature of the defect in isozyme A^- is not known, but it is much less stable *in vivo* than either A or B: having a half-life of only 13 days, compared to the normal 62 days. As erythrocytes lose their nuclei at an early stage, they have no mechanisms to replace defective enzymes.

The situation in women is more complicated than might be expected. Activity of G6PD in erythrocytes from $Gd^{A}Gd^{A}$ women is the same as Gd^{A} men, despite the presence of two genes (on each X chromosome) in female cells. A likely explanation is that some mechanism "switches off" one gene, but leaves the other fully active. In homozygous women, this is of no consequence and half the erythrocytes have G6PD molecules derived from a gene of the mother's X chromosome, while the other half have enzyme derived from father's gene. Women who are heterozygous (e.g. $Gd^{A}Gd^{A-}$) have a mixed population of erythrocytes, some with defective G6PD and some with normal enzyme. As the average life-span of erythrocytes with defective G6PD is reduced from the normal of about 100 days to about 76 days, activity of enzyme in erythrocytes from a woman heterozygous for a defective isozyme is slightly more than half the mean of activity between normal women and those homozygous for the defect. For example, the relative G6PD activities in erythrocytes from women who are $Gd^{B}Gd^{B}$, $Gd^{B}Gd^{Med}$ or $Gd^{Med}Gd^{Med}$, are 100, 62, 5. While favism occurs in women homozygous for defective *Gd* alleles, it is rare in heterozygotes and takes a mild form.

Various defective isozymes differ considerably in medical significance. Isozyme A^{-} begins as a fully active enzyme, gradually becoming inactive but isozymes Canton and Mediterranean are *formed* as poorly-active enzymes, present in erythrocytes from the start. It would be expected that *all* erythrocytes with defective G6PD should suffer damage to sulphhydryl groups of essential proteins, due to inability to maintain protective levels of glutathione. It would also be expected that isozymes defective throughout the life of erythrocytes produce more trouble than those inactivated slowly. Unfortunately, correlations of this type have proved impossible. Mediterranean isozyme is associated less with favism than would be expected, though populations in particular regions show a higher incidence than others, despite similar patterns of isozymes and consumption of fava beans. Nevertheless, the apparent resistance of Gd^{A-} Negroes to fava beans is probably due to relatively higher G6PD activity than for some other isozymes, combined with significantly lower haemolytic activity of fava toxins than drugs like primaquine.

The toxins of fava beans are probably derived from glycosides shown in Figure 3.47, for the aglycones are haemolytic *in vitro*, particularly to sensitive erythrocytes. The glycosides are hydrolysed on crushing or cutting the plant, by β-glucosidase, The enzyme is inactivated by heating, but the glycosides are stable and are hydrolysed by intestinal bacteria.

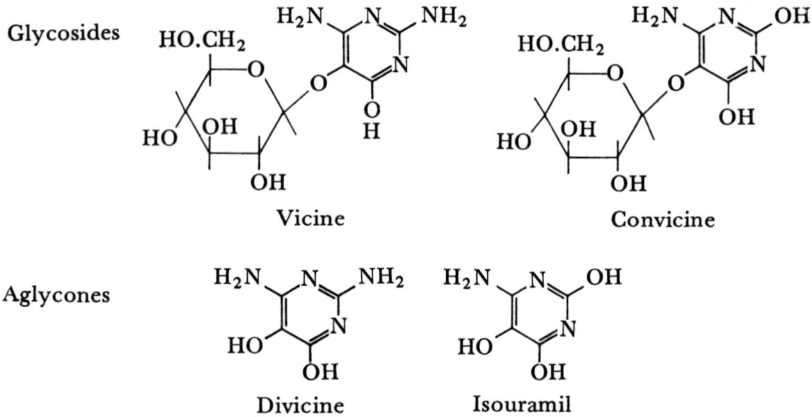

Figure 3.47. Glycosides and aglycones of fava beans.

If these aglycones are favism toxins, their mode of action is probably due to non-enzymic oxidation of reduced glutathione. The concentration of glutathione in normal erythrocytes is remarkably high (about 70 mg/100 ml), but is much reduced in erythrocytes from subjects with any defective G6PD isozyme. Addition of fava aglycones to erythrocyte suspensions produces a rapid oxidation of glutathione that is carried to completion. In the absence of oxygen there is no effect. As the glycosides are without activity, it is likely that the C-5 hydroxyl group is involved: probably in a simple redox system of the type shown in Figure 3.48.

The probable chain of events in favism is as follows:

(1) An inherited defective G6PD isozyme leads to either
  (i) low G6PD activity in erythrocytes *throughout* their life

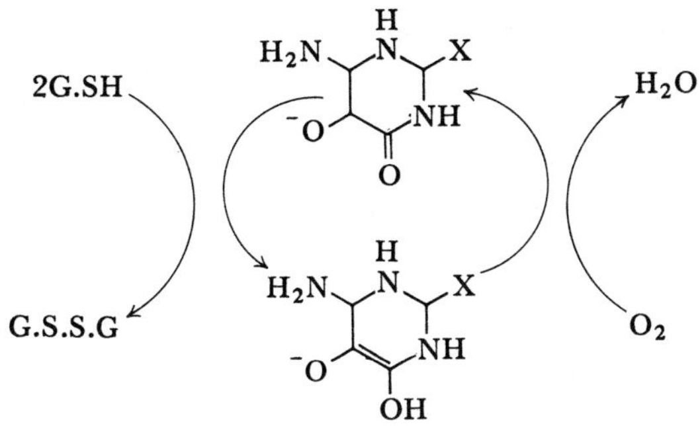

X = NH_2 divicine
= OH isouramil

Figure 3.48. Glutathione oxidation by vicia aglycones.

or

(ii) *development* of low G6PD activity in erythrocytes due to instability of the enzyme protein.

(2) Low G6PD activity reduces metabolism of glucose via its major pathway in erythrocytes: the pentose phosphate pathway.

(3) Limitation of glucose metabolism decreases ATP concentration.

(4) The ratio of reduced to oxidized NADP decreases.

(5) The ratio of reduced to oxidized glutathione decreases.

(6) The concentration of toxic endogenous oxidizing agents, such as H_2O_2, increases.

(7) Conversion of haemoglobin to methaemaglobin increases.

(8) Activity of erythrocyte sulphhydryl-dependent enzymes (e.g. hexokinase, NADPH-diaphorase, phosphomoesterase, pyrophosphatase, etc.) decreases.

(9) Cummulative damage occurs in erythrocyte cell walls.

(10) The average circulation survival-time of damaged cells is significantly decreased and they are removed at an earlier age by the reticuloendothelial system and spleen.

(11) Ingestion of fava beans brings divicine and isouramil into the circulation and contact with erythrocytes.

(12) The aglycones catalyse oxidation of reduced glutathione, the concentration of which is already low, particularly in older erythrocytes.

(13) The metabolic effect of aglycones on the already damaged cells leads to an abrupt failure to maintain integrity of erythrocyte membranes so that haemolysis occurs.

Toxins other than divicine and isouramil may occur in fava beans. There is evidence for a non-specific phytoagglutinin, though its activity is weak. 3,4-Dihydroxyphenylalanine (dopa) is present in fava bean pods at about 250 mg/100 g mainly as free compound, but with small amounts of β-glucoside. Dopa is another agent promoting oxidation of reduced glutathione; presumably after conversion to dopaquinone, which then reacts non-enzymically with glutathione (Figure 3.49). However, the

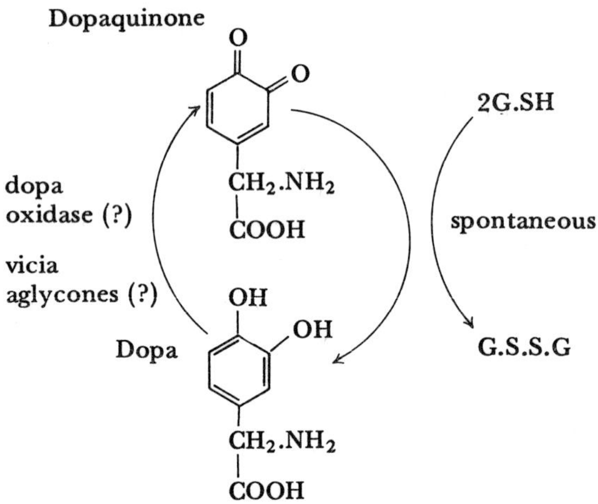

Figure 3.49. Oxidisation of glutathione by dopa.

plasma concentration required to produce this effect requires the amount of dopa equal to 0.5 kg fava pods, which seems somewhat excessive. More significantly, oxidation of reduced glutathione by fava aglycones is synergized by dopa to a highly significant degree.

From the medical viewpoint, favism is unlikely to be controlled while fava beans are grown and eaten by populations with a relatively high incidence of G6PD defective isozymes. Research to select plant strains forming only low amounts of haemolytic anti-metabolites seems worth while, but only a widespread education programme is likely to reduce the incidence of this serious and sometimes life-threatening disease.

(vii) Miscellaneous Plant Food Toxins

(*a*) *Allergens*

A detailed discussion of drug-allergy is given in another section, while general immunochemical principles of antigen–antibody reactions are discussed elsewhere. Here only one or two biochemical aspects of allergic reactions to plant foods are mentioned.

Allergens are an unusual toxin (or antimetabolite) for they produce undesirable effects only in sensitized people and have no action in others. The allergic response varies from rhinitis and urticaria, to asthma, eczema and anaphylactoid shock. Onset of symptoms can be immediate and severe (even fatal), or delayed and moderate.

While drugs are allergic as haptens, food allergens are high molecular weight proteins with true antigenic properties. There is also evidence that plant mucosaccharides act as allergens. Almost nothing is known of the chemistry of these substances, while there is controversy over their heat-stability. In Table 3.27 a list is given of common plant foods known to produce allergic responses in sensitized patients.

Table 3.27. Allergens in plant foods

Plant	Allergens	Heat-stability	Common symptoms
Wheat	Albumins	Destroyed above 120°C	"Baker's asthma" and "Baker's eczema"
Maize	Insoluble protein	Very stable	Anaphylaxis
Rice	Mainly in husk	Unstable	Eczema
Buckwheat	? (Very potent)	?	Asthma, anaphylactoid shock
Soybean	Insoluble proteins	Very stable	Anaphylaxis
Tobacco	?	Unstable	Asthma, eczema
Groundnut	Protein of skin	Very stable	Eczema, asthma
Castor bean	Insoluble proteins	Moderately stable	Asthma, anaphylactoid shock
Strawberry	Soluble protein on outer fruit surface	Unstable	Urticaria
Coffee	Soluble protein	Unstable	Headaches, g.i. distress, eczema

Almost any plant food is a potential allergen for its proteins are foreign to humans, and patients sensitive to one food are usually allergic to others. Skin testing with heated and unheated extracts of food often reveals many unsuspected allergies.

(*b*) *Saponins*

Several β-glycosides of plant sterols, or the structurally related triterpenes, are potent haemolytic agents found in over 400 plant species; including common foods, such as spinach, beetroot, asparagus, tea, soybean, yam, etc. Potency of these compounds is usually expressed as a "haemolytic index" (HI), so that 100 g of a saponin inducing complete haemolysis of 1.0 ml erythrocytes (under standard conditions) is arbitarily said to have an HI of 20 000. HI values of most saponins are rather lower than this value, but soybean saponin has an HI of about 9000 and represents about 0.5% of the beans (i.e. 500 mg/100 g). Hence saponin from 100 g soybeans has the potential to haemolyse about 50% of erythrocytes in an adult man.

Fortunately, despite haemolytic potency, widespread distribution and high concentrations in foods, saponins are very rarely involved in human diseases. The reason is two-fold: saponins are poorly absorbed from the intestine, while haemolytic activity is lost by binding to plasma proteins. Some toxic saponins have LD_{50} about 100 mg/kg by the oral route, so that an average man would need to ingest about 7 g saponin to have a 50% risk of death; but this requires eating at least 1.4 kg of food containing 0.5% saponin, or 7.0 kg containing 0.1%; both of which are very unlikely.

Oral saponins have not been implicated in human deaths from haemolytic crisis, although animals are killed by high doses. However, death is related to severe intestinal inflammation, rather than a systemic action. Intravenous administration of saponins produce severe haemolysis, but only when a dose is sufficient to saturate saponin-binding sites of plasma proteins, as well as overcome protective actions of plasma lipids.

Action of saponins on erythrocyte membranes is related to their potent detergent effect, which disrupts essential lipids. Saponins are also potent foaming agents, and form large quantities of strong, stable foam when their solutions are aerated. Ruminants grazing alphalpha grasses have a high intake of saponin which can fill the rumen with foam and lead to death from pressure symptoms ("bloat"). Saponins are also lethal to fish by altering gill permeability.

(*c*) *Hormone-like Toxins*

Many substances in food plants have hormonal effects on humans and, in certain circumstances, produce serious diseases, or even death. Long-term effects of low level intake of these substances are unknown.

Blighia sapida grows in many tropical countries and ingestion of its fruits leads to what, in Jamaica, is called "vomiting sickness". In Jamaica, the plant is called *ackee*, and in Nigeria, *isin*. The condition is very serious, particularly in children, and

Hypoglycin A

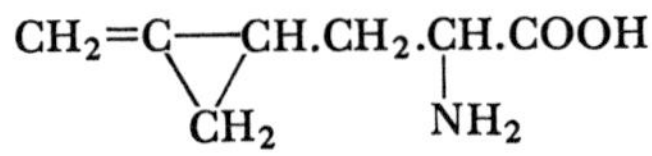

β-(Methylenecyclopropyl)-alanine

Litchitoxin

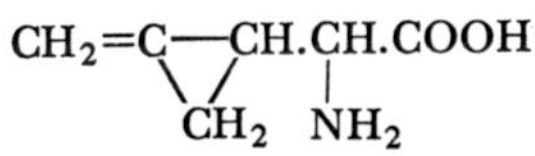

α-(Methylenecyclopropyl)-glycine

Figure 3.50. Hypoglycaemic plant toxins.

the initial vomiting, which may not occur for several hours following ingestion of fruits, is followed by convulsions, coma, and death.

The plant contains two hypoglycaemic toxins, known as hypoglycins A and B, A being the more active. The structure of A is shown in Figure 3.50, together with that of a related compound, also with hypoglycaemic properties, occurring in litchi ("lychee") seeds. Oral or parenteral administration of these substances produces a marked decrease in blood glucose and depletion of liver glycogen. These toxic fruits when eaten by hungry or malnourished people, who have low fasting blood glucose, produce a more marked effect than in healthy subjects.

Chronic administration of hypoglycins at sub-lethal doses to rats induces fatty liver, while they are highly teratogenic during the first 5 days of gestation in pregnant rats.

Hypoglycaemic action of these compounds is obscure, but litchitoxin is a leucine antagonist. Both may be metabolized *in vivo* to 1-methylenecyclopropane-2-carboxylic acid; an inhibitor of fatty acid β-oxidation. A number of natural amino acids, including leucine, stimulate pancreatic release of insulin, augment insulin hypoglycaemia, and decrease hepatic gluconeogenesis, so that plant hypoglycins may have similar, though more potent, effects.

Many foods contain small amounts of the vasoactive hormones, serotonin and noradrenaline; both of which are

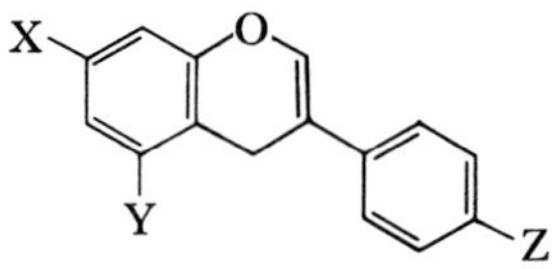

Compound	X	Y	Z
Genistein	OH	OH	OH
Daidzein	OH	H	OH
Biochanin A	OH	OH	OCH_3
Prunetin	OCH_3	OH	OH

Figure 3.51. Estrogenic isoflavones.

potent vasoconstrictors and increase blood pressure. Serotonin is found in bananas, plantains, pineapples and watermelons, while noradrenaline occurs in bananas and plantains. The concentration of serotonin in bananas (*c.* 3 mg/100 g) is such, that in regions of Africa where they are eaten in abundance, daily intake of serotonin may be as high as 30 mg. There is a high incidence of carcinoid–like heart disease among Africans living in these regions, for which serotonin from bananas could be an aetiological factor.

It has been mentioned elsewhere that some plants contain weakly estrogenic compounds of non-steroidal structures (p. 66). Most are isoflavones and structures of several are given in Figure 3.51. All these compounds have the usual physiological effects of estrogens when administered to laboratory animals.

Genistein and daidzein occur in soybeans, biochanin A in clover and Bengal gram, and prunetin in plums, cherries and related fruits. Most isoflavones are heat-stable and there is little loss on cooking. It is not known whether any of these substances is active in humans, but numerous cases of infertility in sheep have been traced to ingestion of clover containing biochanin A.

(*d*) *Hepatoxins*

Probably the most important hepatoxic anti-metabolite derived from food plants is ethanol, but this will not be discussed here. Nor will any further mention be made of hepatoxic fungal metabolites found in food (see p. 226).

Pyrrolizidine alkaloids occur in several plant species and produce considerable liver damage if given to laboratory animals. Most plants of the genus *Senecio* contain these alkaloids and have been conclusively related to human liver disease in several countries, including South Africa and Jamaica. While plants of this genus are not intentionally grown for food they often contaminate fields of wheat, or other grains, and so are harvested with these crops if insufficient care is taken. Bread made from the resulting flour may contain toxic concentrations of alkaloids. An alternative source of the poison is from infusions of dried leaves ("bush tea") of many plants picked at random, which are commonly prepared by certain rural populations, particularly of the Caribbean region.

The various alkaloids induce hepatic cirrhosis and necrosis, with occlusion of hepatic veins. They are also suspected carcinogens. The structure of retrorsine, which occurs in many species of *Senecio*, is shown in Figure 3.52, together with the numbering system from the pyrrolidizine ring. Considerable investigation of the relationship of chemical structure to hepatoxic activity has been undertaken and the following is a summary of essential features:

(1) C-1,2 unsaturation.
(2) C-1 hydroxymethyl, esterified.
(3) C-7 hydroxyl, unesterified.
(4) Branched mono- or di-carboxylic acid esterified to the C-1 hydroxymethyl: the acid may contain 4, 5 or 6 carbon atoms.

Hepatoxic actions by compounds not meeting these criteria such as retrorsine are due to conversion to toxic metabolites *in vivo* (i.e. lethal synthesis).

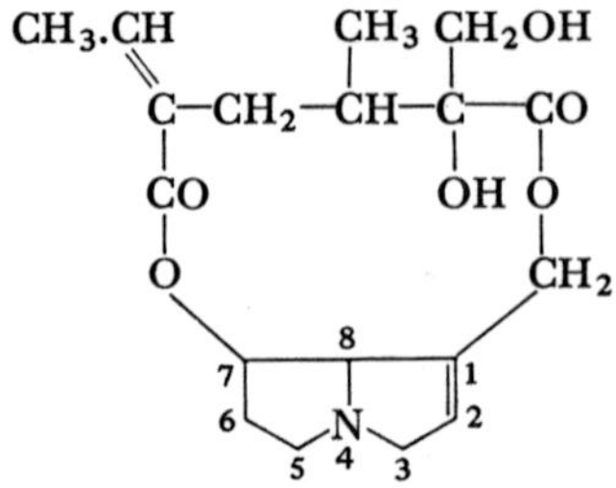

Retrorsine

Figure 3.52. Pyrrolidizine alkaloid of Senecio.

Neurotoxin

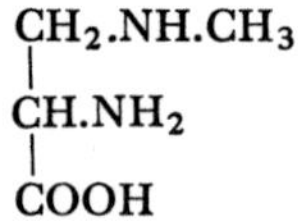

α-Amino-β-methylaminopropionic acid

Hepatoxin

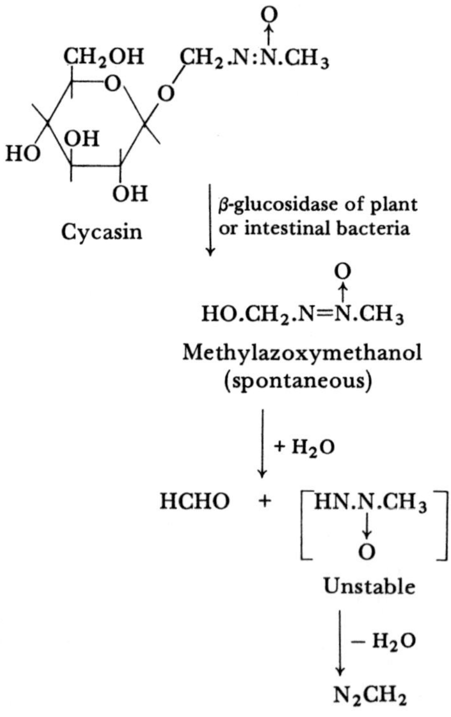

Figure 3.53. Toxins of cycads.

Other hepatoxins occur in cycads, palm-like plants grown widely in many tropical and sub-tropical regions. They are commonly used as a source of starch, for thickening soups or used in pastry. Leaves of the young plant are eaten as a vegetable. Most people using cycads for food, appreciate the toxic nature and attempt to extract and sun-dry the plants before use. There is little doubt that cycads contain several chemically unrelated anti-metabolites, including a neurotoxin (α-amino-β-methylaminopropionic acid), allergens, as well as hepatoxins.

The hepatoxin of cycads is believed to be the β-glucoside of methylazoxymethanol ("cycasin"); this compound is a potent liver carcinogen in rats, though effects of very low doses in humans are unknown. The structure is shown in Figure 3.53, together with its metabolism to the aglycone. While methylazoxymethynol is itself carcinogenic, it is unstable in aqueous

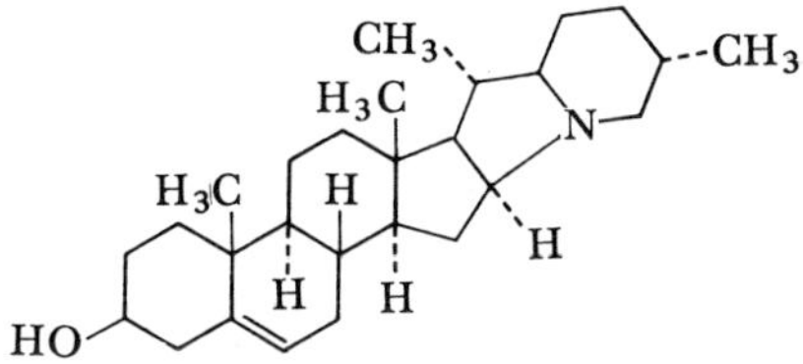

Figure 3.54. Solanine.

solution and decomposes to diazomethane, a well-known alkylating agent of both nucleic acids and proteins. The aglycone is also mutagenic.

Mention must also be made of solanine (Figure 3.54), a water-soluble alkaloid found in potatoes and destroyed by boiling, but not by baking. Concentrations are highest in young potatoes and skin, and the compound is a potent inhibitor of cholinesterase. Fatal poisoning has followed meals of green potatoes containing 25 to 45 mg/100 g solanine, while mild to serious intoxications are relatively common. The average solanine content of peeled potatoes in about 5 to 10 mg/100 g.

3. Antimetabolites of Food Plants Used Medicinally

We have already mentioned one or two food plants used as medicines, such as castor beans, which are used whole, or as castor oil, as a purgative. This action is lubricant related mainly to glycerides of ricinoleic acid which are poorly absorbed and so have a good action. Toxins of castor (ricin and allergens) are absent from the oil, which is a common vehicle for intramuscular injections, particularly for steroid hormones.

Several other medicinal products acting as purgatives or laxatives are derived from food plants, such as olive oil, mucilagenous extracts like agar and carrageenans, together with extracts of figs, rhubarb, or prunes. Some of these are members of the small class of drugs without molecular receptors (see p. 103) in that they assist passage of materials along the gut by physical, rather than chemical, processes. They cannot be considered anti-metabolites.

Agar and carrageenans are complex sulphated galactose polysaccharides which are not digested to any appreciable extent in human intestines. They are widely used in foods for thickening effects, while their value in constipation lies in increasing faeces bulk by absorbing water, so that peristaltic reflexes of the gut wall are activated by pressure. Intact and degraded carageenans have been used to treat peptic ulcers, as they decrease by an unknown mechanism the volume of gastric acid produced by a standard dose of histamine. Sub-dermal implants of carrageenans induce granulation and are sometimes used in the assay of anti-inflammatory actions of corticosteroids.

The action of fruits mentioned above is biochemical, rather than physical. Figs, rhubarb and prunes are all members of the large group of plants which contain anthraquinone purgatives (non-food plants of this group include senna, cascara, aloes, and podophyllum), and they are frequently used as mixed extracts. Medicinal effects are complicated by the large number of substances present. A summary is given in Table 3.28.

The active purgatives in each case are β-glycosides of substituted anthranol (10-hydroxyanthracene), but little is known of the chemistry or mode of action at molecular level. All are irritant to mucous membranes and increase intestinal peristalsis. Unlike many other plant glycosides, their aglycones are much less active than the parent compounds. Many plant purgatives also contain rutin.

The presence of a proteolytic enzyme (ficin) in figs probably contributes little to medicinal action of the plant, but other enzymes from food plants are used therapeutically. The commonest are papain from pawpaws (*Carica papaya*) and bromelain from pineapples (*Ananas comosus*). Both are probably mixtures of several proteolytic enzymes with similar activity. They are endopeptidases with a sulphydryl group at the active centre, which requires activation by reducing agents. They catalyse hydrolysis of peptide bonds between α-amino of lysyl or arginyl residues and carboxyl of most other amino acids. Both enzymes have been used in the treatment of chronic gastritis and dyspepsia, as enemas for impacted faeces, as well as in soft tissue injury and oedema.

Table 3.28. Active constituents of plant purgatives

Plant	Constituents	Biological effects
Rhubarb (*Rheum palmatum*)	Anthranol glycosides rhein glycosides	Purgatives
	Oxalates	Decrease absorption of Ca^+ and other ions
	Tannins	Astringent
Figs (*Ficus carica*)	Anthranol glycosides	Purgatives
	Ficin	Proteolytic enzyme
Prunes (*Prunus domestica*)	Anthranol glycosides	Purgatives
	Hemicelluloses	Demulcent

The use of carrageenans in peptic ulcer was mentioned above. Another food plant used for this indication is liquorice (roots of *Glycyrrhiza glabra*). One active ingredient is glycyrrhizinic acid, a glycoside that yields glycyrrhetic acid (also called "enoxolone") on hydrolysis. The aglycone has anti-inflammatory properties and other cortisol-like actions in humans. It has been used therapeutically in conditions such as rheumatoid arthritis, various skin diseases, and as the disodium succinate ester ("carbenoxolone") for gastric ulcer. Structures are shown in Figure 3.55.

The cortisol-like action is presumably due to binding to corticosteroid-receptors, as the aglycone bears a structural resemblance to cortisol though requires reduction of 11-oxo group. As glucocorticoids are usually ulcerogenic, the beneficial effect of these compounds in gastric ulcer is curious and their

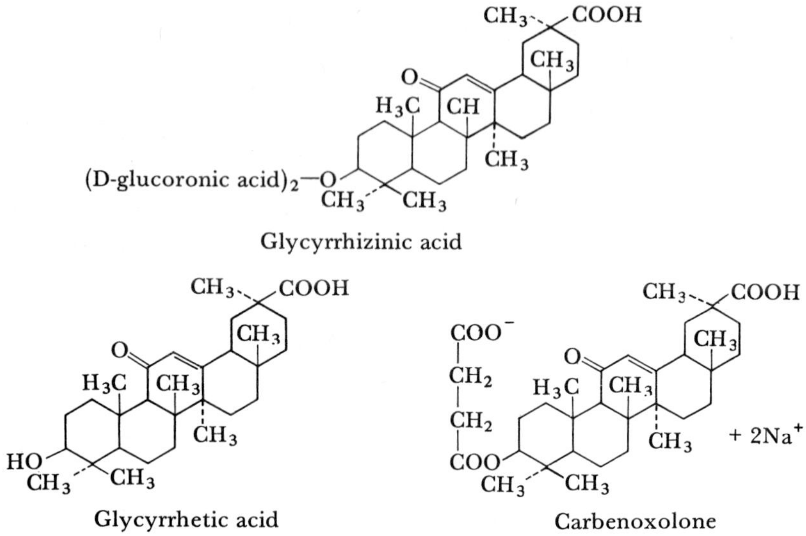

Figure 3.55. Medicinal compounds from liquorice.

mechanism is uncertain; it may involve stimulation of mucous secretion.

Many plant oils have non-specific irritant actions on skin or mucous membranes, and this forms the basis for their use as rubefacients, counter-irritants in neuralgia or muscle pain, or even appetite-stimulants. Table 3.29 lists some food plants and their essential oils used for such purposes. The mode of action is unknown, but presumably involves stimulation of release of kinin mediators of inflammation.

Most anti-metabolites of food plants used medicinally discussed above, have local, rather than systemic actions, as well as being rather unspecific. An important group of centrally acting drugs from food plants are methylated purines, such as caffeine, which act as CNS stimulants and diuretics. Table 3.30 sets out structures of these compounds, their occurrence and principal actions in humans.

Due to low solubility, these compounds are little used as more soluble synthetic derivatives, such as aminophylline (theophylline ethylenediamine), are available.

These anti-metabolites produce two quite different types of effect. The first is on brain, where stimulation of higher centres leads to wakefulness and increased mental activity. There is also

Table 3.29. Medicinal oils of plant foods

Food	Ingredients of oils	Comments
Garlic *(Allium sativum)*	Allyl propyl disulphide diallylsulphide	Sometimes used as diuretic and in bronchitis, but can be lethal to children.
Cayenne pepper *(Capsicum minimum)*	Capsaicin	Very powerful irritant, painful in the eye.
Horseradish *(Armoracia rusticana)*	Allyl isothiocyanate (see p. 277)	–
Aniseed *(Pimpinella anisum)*	Anethole (1-*p*-methoxyphenyl-1-propene) *p*-methoxybenzyldehyde *p*-methoxybenzoic acid	Mild carminative.
Ginger *(Zingiber officinale)*	Oleoresin	–
Lemon *(Citrus limon)*	Citral $((CH_3)_2C{=}CH.(CH_2)_2.\underset{CH_3}{\underset{\vert}{C}}{=}CH.CHO)$	Suspected vitamin A antagonist.
Nutmeg *(Myristica fragrans)*	Myristicin 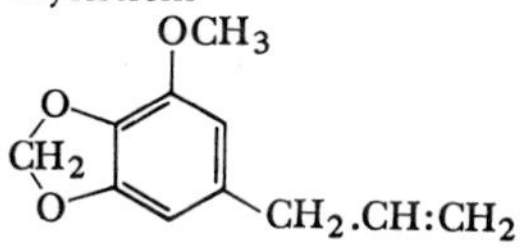	Potent toxin inducing stupor, epileptiform convulsions and other serious side-effects.

stimulation of respiratory and vasomotor centres. The second is peripheral, causing arteriolar vasodilation and increased diuresis by inhibition of renal tubular resorption. Caffeine and theophylline show both effects while theobromine shows only the second.

In humans, caffeine is metabolized *in vivo* partly to theophylline and partly to 1,7-dimethylxanthine. Further *N*-demethylation yields 7-methylaxanthine and 1-methylxanthine. Theobromine is metabolized to 7-methylxanthine and 3-methylxanthine. Whether these demethylations are involved in the mode of action is uncertain. From differential actions of the three compounds, it appears that a 1-methyl group is required for action on the CNS. Effects on blood vessels are reduced in excised organs, so that circulatory actions are partly centrally controlled, but low concentrations of caffeine are

Table 3.30. Plant methyl-purines

Compound	Chemical structure	Occurrence	Principal actions
Caffeine (1,3,7-trimethyl-xanthine)	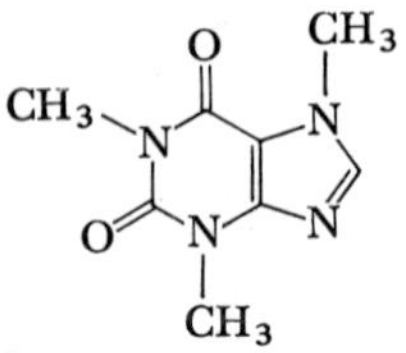	Tea (1 to 4%) Coffee (1 to 2%) Cocoa (trace)	CNS stimulant, mild diuretic, coronary dilator.
Theobromine (3,7-dimethyl-xanthine)	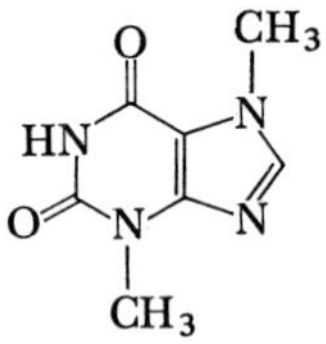	Cocoa (1 to 3%)	Mild diuretic, coronary dilator (no action on CNS).
Theophylline (1,3-dimethyl-xanthine)	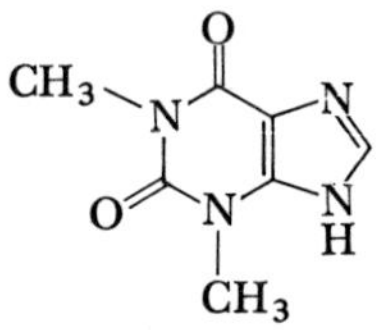	Tea (*c.* 1%)	CNS stimulant, moderate diuretic and coronary dilator.

inhibitory to constrictor effects of adrenaline. Similarly, effects on the heart involve a combination of vagus nerve stimulation and direct dilator action on coronary vessels. A 1-methyl group reduces direct action on peripheral and coronary arteries. As adrenaline is metabolized (and inactivated) primarily by *O*-methylation, it is tempting to suggest that plant methyl-purines act as methyl donors for this reaction.

From chemical structures these compounds might act as purine anti-metabolites, and disrupt biosynthesis of nucleic acids, but there is no evidence that this is their primary mode of action. Large doses of caffeine are mutagenic to animals and bacteria, and teratogenic to rats. These effects may involve DNA changes.

Many other plant food derivatives are used medicinally, but their biochemical effects are either uncertain or uninteresting. Starches (e.g. arrowroot), or pectins (from fruits) are used as absorbants of water in diarrhoea; intravenous infusion of sucrose reduces cerebral oedema, while mustard is a popular emetic. Treatment of hypercholesterolaemia by safflower oil has been attempted.

4. Anti-metabolites of Medicinal Plants

(i) Affecting the Nervous System

Many compounds from plants act in different ways on the human nervous system. Indeed most of the oldest medicines fall into this category (e.g. opium, atropine, ergot, etc.). Before discussing actions of these substances, a brief review of nervous system biochemistry is given in Table 3.31.

Table 3.31. Human nervous system

Major division	Part	Transmitters
Brain	—	Acetylcholine (+) Noradrenaline (+) Dopamine (–) γ-Aminobutyrate (–) Others
Spinal cord	—	Acetylcholine (+) Noradrenaline (+) Glycine (–)
Peripheral nerves	Voluntary and sensory (cranial and spinal)	Acetylcholine (+)
	Autonomic pre-ganglionic neurones	Acetylcholine (+)
	Autonomic post-ganglionic neurones (parasympathetic)	Acetylcholine
	Autonomic post ganglionic neurones (sympathetic)	Acetylcholine *or* noradrenaline

(+) synaptic stimulant (–) synaptic inhibitor.

An obvious way for anti-metabolites to alter nervous function is to disrupt metabolism of synaptic transmitter substances. These vary considerably in different parts of the body, but acetylcholine and noradrenaline are the commonest. A detailed discussion is given elsewhere on anti-cholinesterases, which inhibit enzymes catalysing breakdown of acetylcholine and so greatly enhance its effects (p. 151). Similarly, blocking action at the molecular level by acetylcholine antagonists interferes with binding of acetylcholine by its receptors (p. 155). The nervous system contains two classes of transmitter substances; one favouring transmission of impulse, the second inhibiting transmission. Tetanus toxin (p. 206) appears to prevent release of glycine, an inhibitory transmitter of the spinal cord, while methionine sulphoximine inhibits enzymes involved in biosynthesis of γ-aminobutyric acid, an inhibitory transmitter of brain.

Any of the following anti-metabolic actions will disrupt functions of the nervous system:

(a) Inhibition of transmitter formation
(b) Inhibition of transmitter break-down
(c) Blocking transmitter receptors
(d) Stimulating receptors in the absence of transmitter.

Disruption of neurone metabolism also occurs through mechanisms not involving transmitters. Lathyrus neurotoxins (p. 287) produce demyelination of nerve fibres, while neurotoxins of snake venoms act in a variety of ways (p. 335).

To assist appreciation of different biological effects of plant compounds acting on the nervous system, a summary is given in Table 3.32 of principal connections within the human autonomic nervous system. The role of acetylcholine and noradrenaline at various junctions is given.

Most organs with double autonomic innervation (sympathetic and parasympathetic) respond in opposite ways to impulses received via two nerves; one of which releases acetylcholine (parasympathetic) and the other noradrenaline (sympathetic).

We have mentioned that atropine is an acetylcholine antagonist, which at low doses blocks receptors for this transmitter. At high doses it penetrates autonomic ganglia to block transmission. The major actions of atropine are listed below:

(a) Dilation of pupils (blocks parasympathetic nerves, but allows sympathetic nerves to function)
(b) Stops saliva secretion (as (a))
(c) Dilates and dries bronchi (as (a))
(d) Relaxes intestinal spasms (as (a))
(e) Suppresses sweating (cholinergic sympathetic nerves to sweat glands are blocked)
(f) Increases heart rate (in high doses) (blocks vagus inhibition)
(g) Cutaneous dilation (as (e))

It will be seen that these follow logically from blocking of

Table 3.32. Human autonomic nervous system

Part	Nerves	Termination	Action
Cranial nerves (parasympathetic: acetylcholine only)	III	Ciliary muscles	Accommodates
		Iris	Constricts
	VII	Lachrymal gland	Stimulates
		Nasal mucosa	Stimulates
		Salivary glands	Stimulates
	IX	Salivary glands	Stimulates
	X	Heart	Rate slows
		Bronchi	Constricts
Spinal nerves (sympathetic: post-ganglionic neurones are *either* cholinergic *or* adrenergic)	T1	Iris (noradrenaline)	Dilates
	T1 to T5	Blood vessels and sweat glands of head and neck (acetylcholine)	Constricts vessels and stimulates glands
	T2 to T9	Blood vessels and sweat glands of arms (acetylcholine)	Constricts vessels and stimulates glands
	T2 to T6	Heart (noradrenaline)	Rate increases
	T5 to T9	Bronchi (noradrenaline)	Dilates
	T5 to T9	Adrenal medulla (acetylcholine)	Stimulates
	T4 to L2	Blood vessels and sweat glands of legs (acetylcholine)	Constricts vessels and stimulates glands
	T6 to L2	Gut (noradrenaline)	Inhibits (relaxes)
	L1 to L2	Urinary bladder (noradrenaline)	Relaxes
		Genitalia (noradrenaline)	Relaxes
Spinal nerves (parasympathetic acetylcholine only)	S2 to S4	Urinary bladder	Contracts
		Genitalia	Stimulates
		Gut	Stimulates

acetylcholine receptors in end organs of both parasympathetic and cholinergic sympathetic nerves. High doses interfere with acetylcholine receptors of the CNS or voluntary muscles and, in atropine poisoning, there is excitement and delerium, accompanied by partial paralysis of limbs. Death ensues from asphyxia, resulting from inhibition of the respiratory centre.

The chemical structure of atropine is shown in Figure 3.56. As extracted from plants (e.g. *Belladonna*, *Stramonium*, Mandragora, etc.) the compound is a racemate, but it is formed within plant tissues as a *laevo*-isomer (known as hyoscyamine); which undergoes racemization, partly within the plant, and partly during extraction. As atropine is (±)-hyoscyamine, it is interesting to compare relative actions in the body. Hyoscyamine is about twice as active as atropine on the pupil, heart rate, and suppression of sweat glands, but effects on intestines and (at high doses) CNS are the same. Hyoscyamine is less toxic than atropine, for the (−)-isomer is metabolized faster than the (+)-isomer. The products of

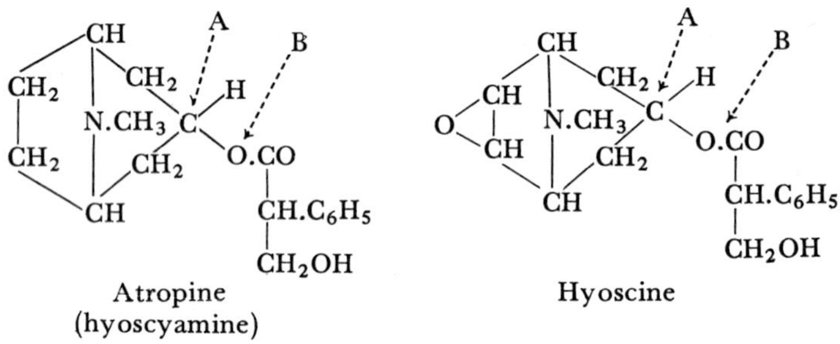

where A = asymmetric C atom
B = site of hydrolysis by esterases

Figure 3.56. Atropine and hyoscine.

metabolism are tropine and trophic acid (α-benzyl-β-hydroxy-propionic acid). Tropine is biologically inactive. The esterase responsible for the hydrolysis is absent from human blood, but is present in rabbit plasma. Both atropine and hyoscyamine are bound by plasma proteins, but can pass into the CNS. Ionized derivatives, such as atropine methyl sulphate, pass less easily and are preferred therapeutically.

Other plant compounds acting on acetylcholine receptors include (−)-hyoscine (also called "scopolamine"), which is metabolized *in vivo* to scopine and trophic acid. Its actions on pupils, intestines, sweating and salivation are similar to atropine, though its effect on heart rate is weak. Psychic actions are more pronounced at therapeutic doses, and it induces drowsiness and sleep, sometimes with hallucinations.

Atropine and hyoscine are efficient blocking agents for acetylcholine receptors of tissues with autonomic innervation, but their action on acetylcholine receptors in skeletal muscle is weak, except at high doses. In contrast, (+)-tubocurarine (an alkaloid of *Chondrodendron tomentosum* and major active ingredient of the South American arrow-poison known as "curare") acts primarily to compete with acetylcholine for receptors on "end-plates" in skeletal muscles, but has little action elsewhere. Anti-cholinesterases are inhibited by combining at the enzyme active site (p. 151) and so increasing survival time of acetylcholine liberated by nerve endings. The paralysing action of tubocurarine on muscle is counteracted by anti-cholinesterases (e.g. neostigmine), which increase the amount of acetylcholine in competition for receptors.

The dose required to produce significant effects on tissues with parasympathetic innervation is nearly twice that blocking skeletal muscle, while that for inhibition of cholinergic sympathetic innervation is five times as much. Very high doses completely block autonomic ganglia. Differences in sensitivity to acetylcholine antagonists are presumably related to structural differences in binding-sites of macromolecular receptors. The receptors for acetylcholine in autonomic ganglia are blocked by pentamethonium ions, but not by the longer molecules of decamethonium (p. 156). Conversely, decamethonium (or succinylcholine) blocks receptors of skeletal muscle end-plates on which pentamethonium has no action. Biological effects of atropine and tubocurarine strongly suggest the existence of a third type of acetylcholine receptor located in tissues with autonomic innervation, which is blocked by atropine, but not by tubocurarine, decamethonium, or pentamethonium; except at very high doses. Autonomic ganglia receptors are located on dendrites of post-ganglionic neurones, skeletal muscles receptors on end-plates, while the third type must be on cell surfaces of tissues of heart, blood vessels, iris, sweat and salivary glands.

Tubocurarine molecules (Figure 3.57) have two quaternary nitrogen atoms about 20 Å apart, which is approximately the

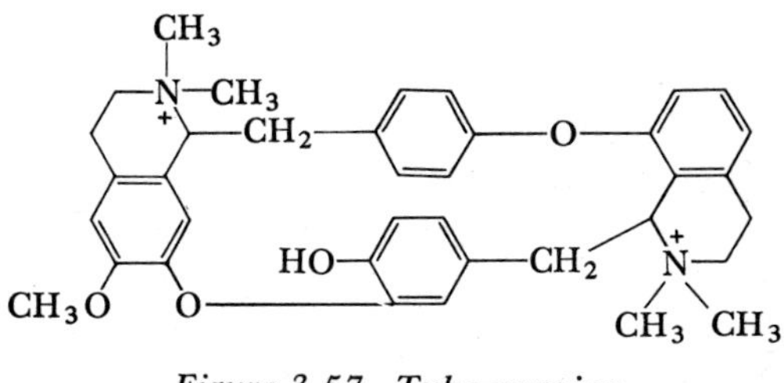

Figure 3.57. Tubocurarine.

separation in decamethonium and succinylcholine. The dimethyl ether of tubocurarine is between two and three times as potent as the parent compound. This difference is probably due to decreased metabolism of the dimethylether, for tubocurarine is rapidly cleared from blood and metabolized by both liver and kidneys. Only about a third can be recovered unchanged from urine. The nature of its metabolites is uncertain, but may be formed by breaks in the ring structure, so that spatial separation of the nitrogen atoms no longer matches the anionic sites of acetylcholine receptors.

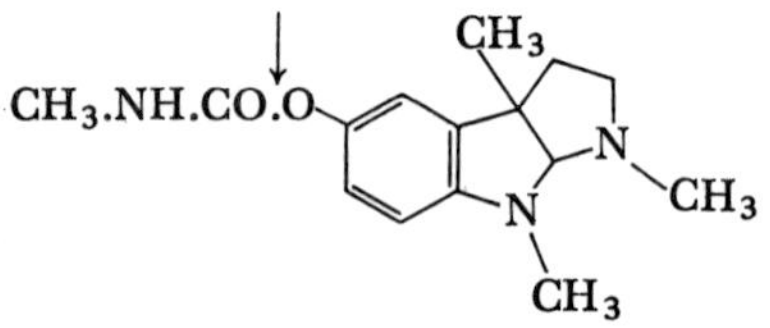

Physostigmine
(eserine)

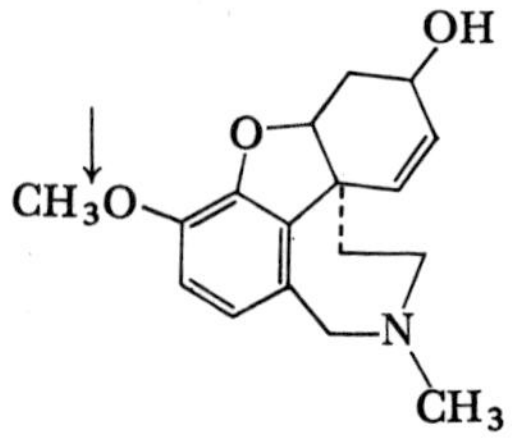

Galanthamine
(from *Galanthus woronowii*)

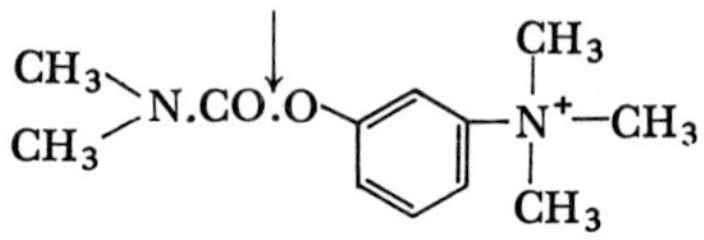

Neostigmine
(synthetic)

Figure 3.58. Anti-cholinesterase alkaloids.

Inhibitors of cholinesterases unblock the action of tubocurarine on muscle end-plates by inducing marked local increases in acetylcholine concentration. A discussion of the mode of action of synthetic anti-cholinesterases is given elsewhere (see p. 151), but several naturally occurring compounds have this action and have been used medicinally. One is physostigmine (also called "eserine"), an alkaloid of calabar bean (*Physostigma venenosum*). Its structure is shown in Figure 3.58, together with related compounds. Each is acted upon by cholinesterase enzymes at the points indicated to give a phenol and a substituted seryl-hydroxyl at the enzyme active centre. For physostigmine, the group transferred to the enzyme is

CH_3 . NH . CO–, for galanthamine it is CH_3–, while for neostigmine it is $(CH_3)_2$. N . CO–. The substituted seryl residues are very stable and the enzyme molecule, deprived of its active centre, is completely inhibited.

These compounds are exceptionally poisonous and have caused many deaths. The fatal oral dose for adults is about 10 mg of physostigmine. Calabar beans have long been used by African witch-doctors, but deaths have also occurred accidentally in laboratory and medical workers.

It could be expected that these alkaloids act at all three types of acetylcholine receptors (see above). By augmenting effects of cholinergic nerves they produce wide ranging effects. Pupils constrict, motility of gut increases, skin arterioles are dilated, while sweating and salivation are increased. Heart rate is slowed by small doses, but larger amounts block the vagus centre, so that heart rate increases due to withdrawal of vagus inhibition.

As atropine blocks acetylcholine receptors of tissues with autonomic innervation, it antagonizes the action of physostigmine at these sites, and so is a useful antidote in physostigmine poisoning. Atropine cannot antagonize effects of physostigmine at skeletal muscle end-plate receptors, though this can be achieved in laboratory animals by tubocurarine (very dangerous in man), with suppression of muscular twitchings and convulsions induced by physostigmine overdose.

A further class of plant alkaloids mimic the actions of acetylcholine. Three types of receptor for this transmitter occur in different tissues, which have different sensitivities for compounds blocking its action. Similarly, alkaloids are known which will combine with, and *stimulate* some acetylcholine receptors, but not others. Muscarine, the alkaloid of *Amanita muscaria* (fly agaric), stimulates receptors of tissues with autonomic innervation, but has no effect on receptors in autonomic ganglia or in skeletal muscle end-plates. Nicotine, a tobacco alkaloid, strongly stimulates autonomic ganglia, but has little action elsewhere. Pilocarpine, an alkaloid of *Pilocarpus microphyllus* (Jaborandi), acts on receptors of *both* autonomic ganglia *and* tissues with autonomic innervation, but even at toxic doses has only weak action on muscle end-plates. Compounds acting directly on end-plate receptors are rare, but veratridine and cevadine (*Schoenocauden officinale*) have some effect, as well as on acetylcholine receptors elsewhere. Structures of alkaloids with acetylcholine-like activity are shown in Figure 3.59.

A summary of medicinal plant alkaloids acting on acetylcholine receptors is given in Table 3.33. Reversal of veratridine effects is difficult due to alkaloid binding tightly to receptors; from which it is displaced with great difficulty. Large doses of tubocurarine produce a partial reversal of muscle spasm, but atropine is unable to reverse slowing of the heart rate. Both veratridine and cevadine also have complex actions on other receptors.

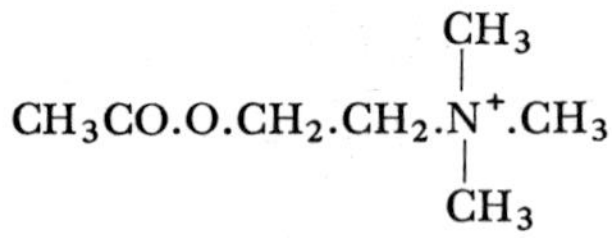

Acetylcholine

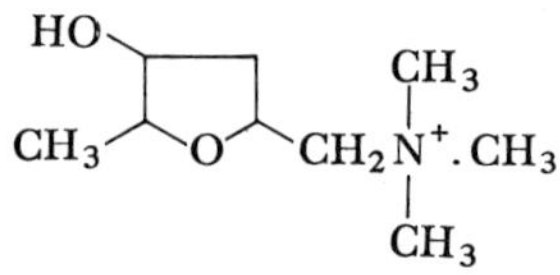

Muscarine

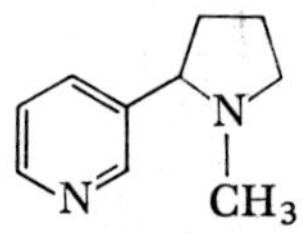

Nicotine

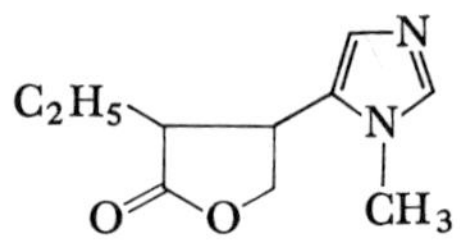

Pilocarpine

Figure 3.59. Acetylcholine-receptor stimulants.

Post-ganglionic fibres of some sympathetic nerves are adrenergic (release noradrenaline), even though they are activated in the ganglion by acetylcholine. It follows that action of compounds affecting ganglion receptors will be complex, for *both* cholinergic *and* adrenergic post-ganglionic fibres will be affected. For tissues with double autonomic innervation (see Table 3.32), effects of ganglion blocking compounds will be related to the changed balance between the two fibres.

For example, nicotine blocks autonomic ganglia and the pupils respond first by constriction (cholinergic effect), then by dilation (adrenergic effect), which may be explained as an initial partial inhibition of the adrenergic fibres, followed by a more intense inhibition of cholinergic fibres.

Alkaloids disrupting adrenergic functions are known, but the situation is complicated by interactions between adrenaline released by the adrenal medulla and noradrenaline released locally from sympathetic nerve endings. Differences in action of these two compounds are that the primary effect of adrenaline is as a wide-acting hormone influencing metabolism of many tissues, whereas noradrenaline is a locally acting substance transmitting impulses from sympathetic nerve endings to tissues of eyes, bronchi, heart, gut, bladder and genitalia. Noradrenaline is a consistent vasoconstrictor, whereas adrenaline dilates vessels of the heart and some other tissues.

Table 3.33. Alkaloids affecting transmitter-receptors

Alkaloid	Acetylcholine receptor affected	Antagonized by
Atropine, Hyoscine	Tissues with autonomic innervation (–)	Physostigmine
Tubocurarine	Skeletal muscle end-plates (–)	Physostigmine
Physostigmine, Galanthamine	Cholinesterases (–)	Atropine (autonomic) Tubocurarine (end-plates)
Muscarine	Tissues with autonomic innervation (+)	Atropine
Nicotine	Autonomic ganglia (+)	Pentamethonium
Pilocarpine	Tissues with autonomic innervation (+) and autonomic ganglia (+)	Atropine (autonomic) Pentamethonium (ganglia)
Veratridine, Cevadine	Skeletal muscle end-plates (+) and tissues with autonomic innervation (+)	Tubocurarine (end-plates)

Noradrenaline is stored in granules at nerve endings of adrenergic fibres and these have the capacity to reabsorb noradrenaline when a nerve impulse has passed. Noradrenaline is also inactivated by catechol-*O*-methyl transferase (C*O*MT) and monoamine oxidase (MAO). Both adrenalines in blood plasma are inactivated by C*O*MT, but MAO is responsible for inactivation of noradrenaline from nerve endings.

Adrenergic function may be altered by plant medicines which interfere with noradrenaline storage granules; promote synthesis of compounds in nerves unable to activate the noradrenaline

Compound	Occurrence	Chemical structure
Adrenaline	adrenal medulla	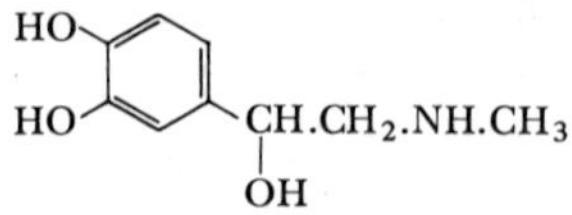
Noradrenaline	adrenergic nerve ends	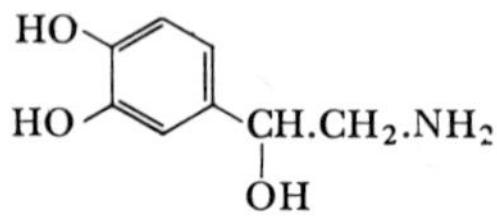
Reserpine	*Rauwolfia* sp.	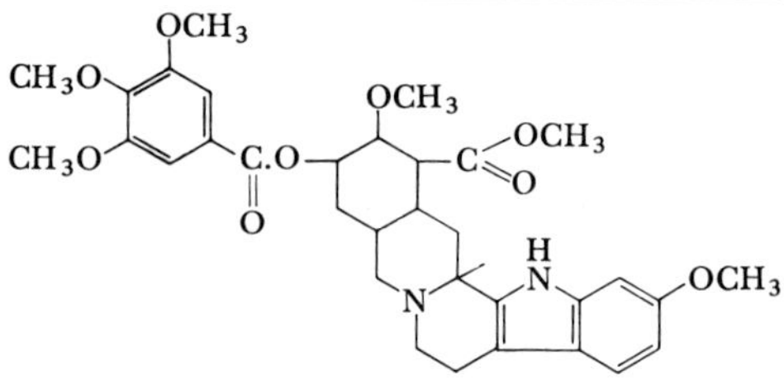
β-Phenylethylamine	Mistletoe	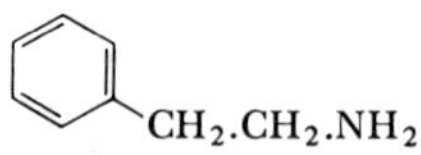
Tyramine	Many plants (also intestinal bacteria)	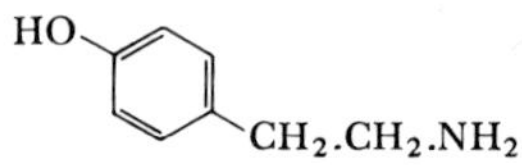
Dopamine	Banana	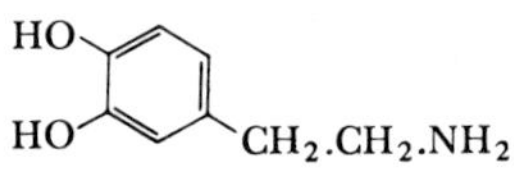
Hordenine	Barley	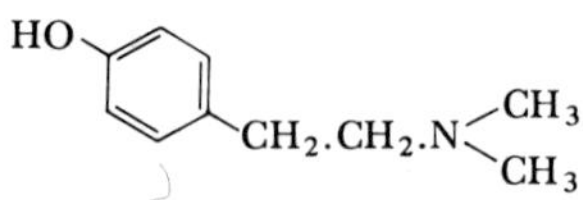
Mescaline	*Lophophora williamsii*	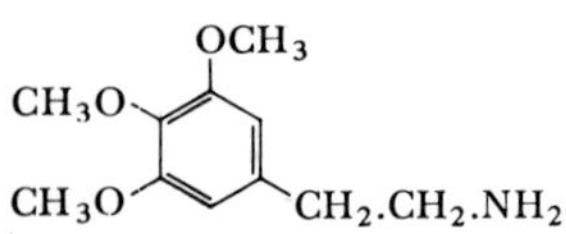
Ephedrine	*Ephedra* sp.	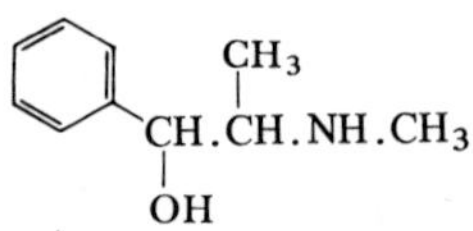

Figure 3.60. Alkaloids affecting catecholamines.

receptors; by acting as false transmitters or by blocking the receptors.

A widely used drug in the treatment of blood pressure is reserpine, an alkaloid originally derived from the Indian plant *Rauwolfia serpentia*. Reserpine also has actions on the CNS which make it valuable in some psychiatric states. Its action on blood pressure is via noradrenaline storage granules of sympathetic nerve endings, which gradually decrease in size and are prevented from re-forming. In this way little noradrenaline is released by the nerves and peripheral vasodilation predominates. This action of reserpine makes it a useful tool for investigation of effects of other compounds with adrenergic actions. For example the potent vasoconstrictor tyramine is devoid of effects in animals pre-treated with reserpine, which indicates that it acts through stimulation of noradrenaline release, rather than directly.

Plant alkaloids with chemical structures resembling adrenalines are shown in Figure 3.60.

Tyramine and β-phenylethylamine are converted to "false transmitters" within neurones, if present in sufficient amounts. They are converted to corresponding β-hydroxyl compounds, which are stored in nerve-end granules to be released by passage of an impulse, but are poor activators of noradrenaline receptors. Many MAO inhibitors serve to promote formation of false transmitters by allowing endogenously formed tyramine to accumulate in neurones. Cocaine is a further alkaloid affecting adrenergic nerve endings, though it has more pronounced actions elsewhere. It appears to sensitize noradrenaline receptors, as well as blocking uptake into storage granules.

Only brief mention has so far been made of effects of medicinal plant substances on brain, yet this is the primary action in many cases. Unfortunately, the biochemistry of interactions is often poorly understood. Interconnections between various parts of brain and spinal cord are very complicated and both excitory and inhibitory tracts are known. Table 3.34 lists some drugs of plant origin which act on the CNS.

Hallucinogens have attracted much attention, due to widespread mis-use of opiates, cannabis and LSD; as well as the possibility that knowledge of their actions may give an insight into the biochemistry of human mental disease. Serotonin is involved in several brain functions, but many hallucinogens have

Table 3.34. Medicinal plant compounds acting on CNS

Compound	Occurrence	Actions
Strychnine	*Strychnos nux-vomica*	Blocks spinal inhibitory transmitter (glycine) and induces convulsions.
Picrotoxin	*Anamirta paniculata*	Stimulates neurones of the medulla oblongata and induces convulsions.
Camphor	*Cinnamomum camphora*	Stimulates neurones of brain, mainly above optic thalami, but involving motor cortex and spinal cord.
Bulbocarpnine	*Corydalis cava*	Stimulates neurones of midbrain basal ganglia and induces convulsions.
Morphine	Opium poppy	Depresses higher functions of brain and induces catatonic rigidity through effect on midbrain.
Papavarine	Opium poppy	Relaxes involuntary muscles without affecting higher functions.
Cannabinols	*Cannabis sativa*	Hallucinogens affecting higher centres and producing motor excitation.
Lysergic acid (LSD)	Ergot	Hallucinogen stimulating higher centres.
Reserpine	*Rauwolfia* sp.	Suppresses nerves stimulating diencephalic centres to induce relaxation without narcosis—also acts on adrenergic peripheral nerves.
Cocaine	*Erythroxylon coca*	Stimulation and paralysis of many areas to induce excitement exhilaration and tremors, with loss of touch and pain senses.
Aconitine	*Aconitum napellus*	Stimulates brain centres to slow heart rate and respiration with peripheral anaesthesia and muscular convulsions.

an effect on its metabolism, though interactions are complex. LSD induces an increase in brain serotonin concentration, probably by reducing the rate of inactivation. Neurones containing serotonin in the mid-brain are blocked by LSD, but serotonin receptors in other areas are activated.

Mode of action of serotonin at the molecular level in brain is uncertain, but it may intercalate with base-pairs of DNA to promote interaction of template DNA with transcriptase enzymes. Compounds which prevent this intercalation, or which displace serotonin from the site, may be hallucinogens.

The mode of action of morphine-like analgesics is unknown, but interesting structure–activity relationships provide information about the receptors, whatever these may be. Chemical structures of various compounds in this series are shown in Figure 3.61.

Naturally occurring morphine is the D(−)-isomer and its enantiomorph, L(+)-morphine is devoid of analgesic properties. Acetylation of both hydroxyls yields heroin, which is marginally more active than morphine (though more addictive). Methylation of hydroxyl at C-3 yields codeine, with about one-tenth the analgesic potency of morphine; which it may possess through *in vivo* metabolism to the parent compound. Interestingly, both morphine and codeine inhibit microsomal oxidases (see p. 13) in liver of *male*, but not *female*, rats. This may be due to a greater inductive effect of testosterone than estrogens on these enzymes.

The synthetic analgesic, D(−)-levopharanol, lacks both C-6 hydroxyl and the oxygen bridge, but is fully active. Treatment of morphine with HCl produces an internal rearrangement to give apomorphine, which is devoid of analgesic activity, but is a potent centrally acting emetic.

The N atom of the piperidine ring of morphine is about 80% ionized at physiological pH and seems a likely candidate for binding to anionic sites of receptor molecules (compare with acetylcholine receptors, p. 155). Substitution of *N*-methyl by *N*-allyl produces D(−)-nalorphone, which is a potent *antagonist* of morphine analgesics. The alkaloid thebaine is found in opium at about 0.5% concentration and is exceedingly toxic, yet differs from codeine by only a methoxy group and an additional double bond in ring C. The most potent compound in the morphine series, with about 10 000 times the analgesic

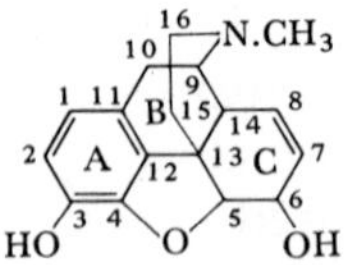

Morphine

(asymmetric at C-13, D(–) active, L(+) inert)

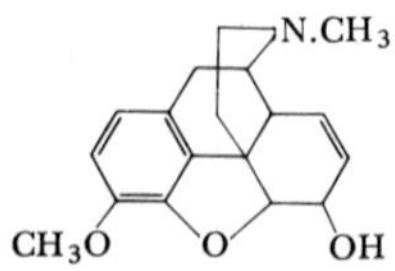

Codeine

(about 10% activity of D(–)-morphine)

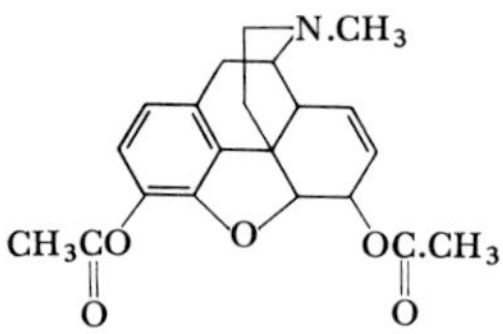

Heroin

(slightly more active than morphine, but more addictive)

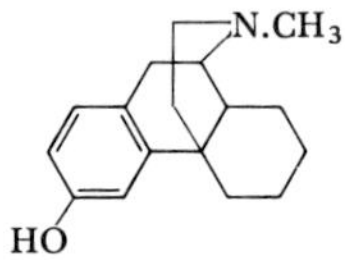

Levorphanol

(only D(–) active, with greater potency than morphine)

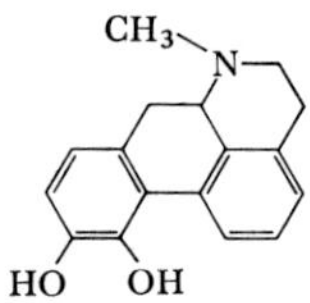

Apomorphine

(emetic, not analgesic)

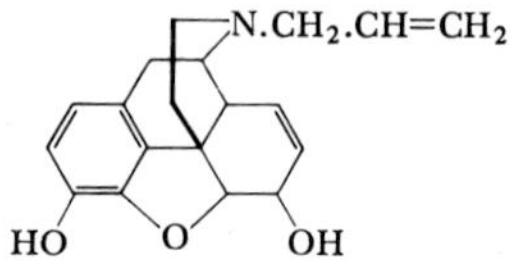

Nalorphine

(potent morphine antagonist, only D(–) active)

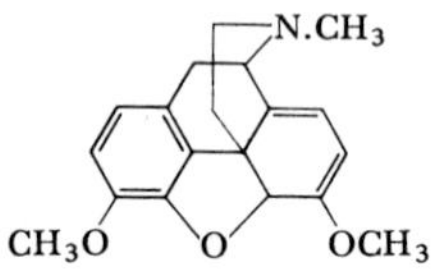

Thebaine

(highly toxic)

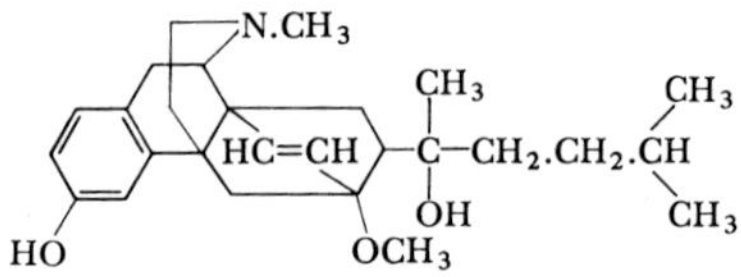

Synthetic derivative

(10 000 times as active as morphine)

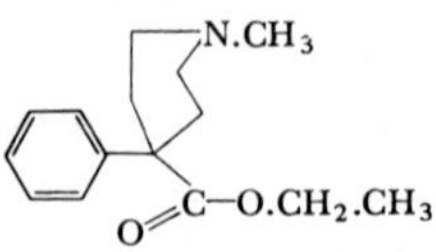

Pethidine

(synthetic, only D(–) active)

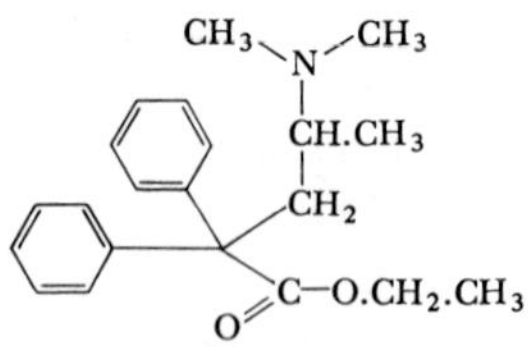

Methadone

(synthetic, only D(–) active)

Figure 3.61. Morphine analgesics.

activity of morphine, yet low toxicity, is the synthetic compound shown in Figure 3.61, and, like thebaine, has modifications to ring C.

Other synthetic compounds acting at the same site include pethidine and methadone. The latter probably occurs in a conformation with a pseudopiperidine ring, spatially similar to morphine.

It is likely that morphine receptors possess anionic sites to which N of the piperidyl ring links, while C-15 and C-16 lie in a nearby groove. An aromatic ring is needed to bind a second site of the receptor, but substituents on this ring are critical. Hydroxyl or acetyl at C-3 are allowed, but methoxy is not.

Precisely how combination of morphine with its specific macromolecular receptors of brain produces analgesic and other actions in humans is not known, but may involve biochemical changes to neurones supplying "pain centres", which are diffuse areas involving large tracts of thalamus and frontal cortex.

(ii) Affecting the Cardiovascular System

Medicinal plant compounds affecting the nervous system often have effects on heart and circulation by altering nervous control of these tissues. Other compounds have a direct action, but may also produce central effects. Cardiac glycosides of *Digitalis* (foxglove) and other plants are one such group.

These compounds act directly on heart muscle to increase tone, excitability and contractility although they also stimulate the vagus centre and slow impulses through the bundle of His. The result, at therapeutic doses, is a slower, more powerful beat.

Cardiac glycosides are steroidal aglycones linked to various sugars, including glucose, cyamarose and digitoxose. Sugar residues are not essential for activity. Glycosides extracted by chemical methods from plants are degradation products of native compounds. Table 3.35 lists the major cardiac glycosides, together with their aglycones and parent compounds. Chemical structures are given in Figure 3.62.

Essential structural features of aglycones acting on the heart are a 5-membered unsaturated lactone ring in the 17 position, a 14-hydroxyl, and *cis*-configuration of rings C and D. The more hydroxyls present, the more rapid the onset of action, though the shorter its duration (rapid conjugation).

Table 3.35. Cardiac glycosides

Plant	Native glycosides* in plant tissues	Glycosides† extracted	Aglycone‡ released by acid hydrolysis
Digitalis	Purpurea glycoside A	Digitoxin	Digitoxigenin
	Purpurea glycoside B	Gitoxin	Gitoxigenin
	Lantoside A	Digitoxin	Digitoxigenin
	Lantoside B	Gitoxin	Gitoxigenin
	Lantoside C	Digoxin	Digoxigenin
Strophanthus	K-strophanthoside	Strophanthin	Strophanthidin
	K-strophanthoside	Cymarin	Strophanthidin
	G-strophanthoside	Ouabain	Ouabagenin
	S-strophanthoside	Sarmentocymarin	Sarmentogenin

* Native glycosides

purpurea glycoside A	: digitoxyl-1,4β-D-glucose
purpurea glycoside B	: gitoxyl-1,4β-D-glucose
lantoside A	: acetyldigitoxyl-1,4β-D-glucose
lantoside B	: acetylgitoxyl-1,4β-D-glucose
lantoside C	: acetyldigoxyl-1,4β-D-glucose
K-strophanthoside	: strophanthyl-1,4β-D-glucose
G-strophanthoside	: ouabanyl-1,4β-D-glucose
S-strophanthoside	: sarmentocymaryl-1,4β-D-glucose

† Glycosides extracted

digitoxin	: digitoxigenyl-3β-(D-digitoxose)$_3$
gitoxin	: gitoxigenyl-3β-(D-digitoxose)$_3$
digoxin	: digoxigenyl-3β-(D-digitoxose)$_3$
strophanthin	: strophanthidyl-3β-D-cymarosyl-1,4β-D-glucose
cymarin	: strophanthidyl-3β-D-cymarose
ouabain	: ouabagenyl-3β-L-rhamnose
sarmentocymarin	: sarmentogenyl-3β-D-sarmentose

‡ See Figure 3.12.

Mode of action at molecular level is unknown, but cardiac glycosides should be compared to toad poisons.

A totally different group of plant medicines acting on the vascular system are various anti-coagulants. Most are coumarins (lactones of *cis-o*-hydroxy cinnamic acid) which are widely distributed in vascular plants. Structures of active compounds are given in Figure 3.63.

Both coumarol and dicoumarol are naturally occurring anti-coagulants, while warfarin and phenindone are related synthetic compounds. Coumarol occurs in clover and hay (it is responsible for their characteristic "sweet" odour), but is converted to dicoumarol by micro-organisms. Spoiled hay is known to cause haemorrhagic disease in cattle.

Aglycones

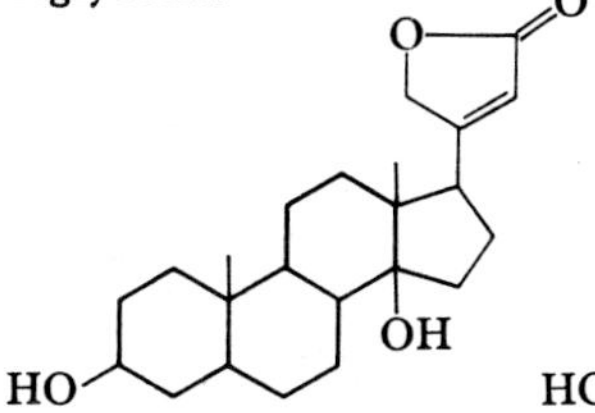

Digitoxigenin

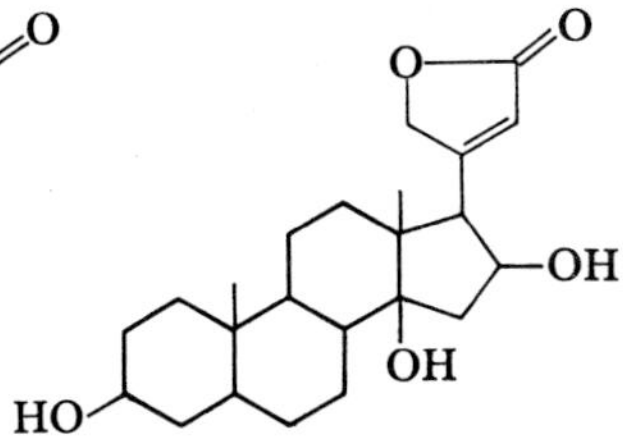

Gitoxigenin

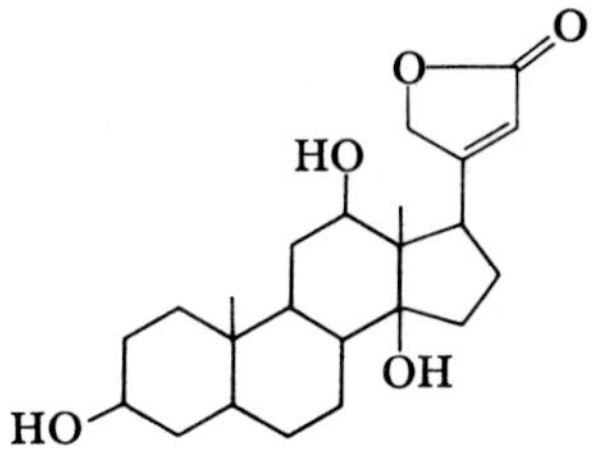

Digoxigenin

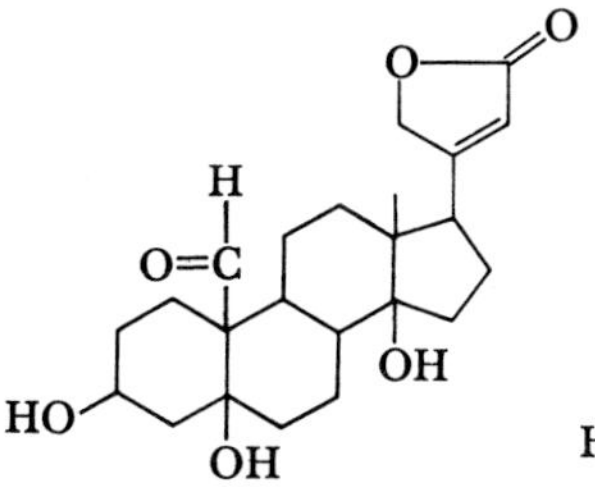

Strophanthidin

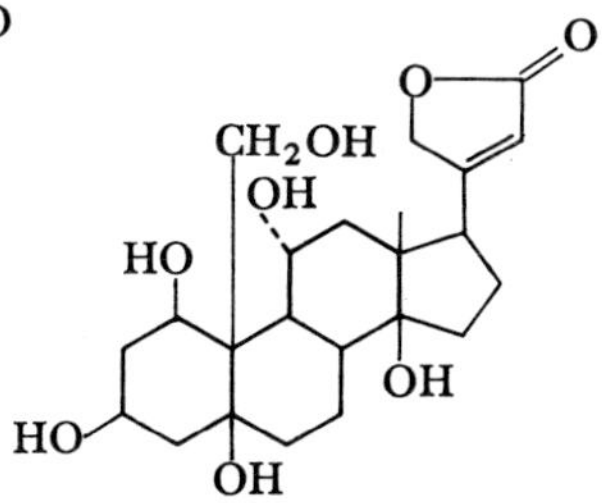

Ouabagenin

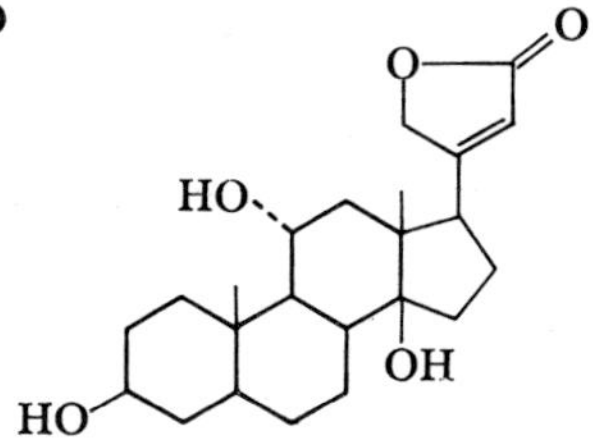

Sarmentogenin

Figure 3.62. Structures of cardiac glycosides.

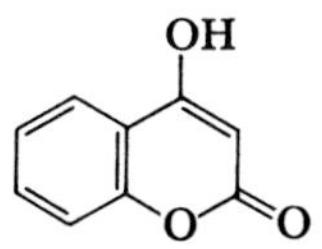

4-Hydroxycoumarin (coumarol)

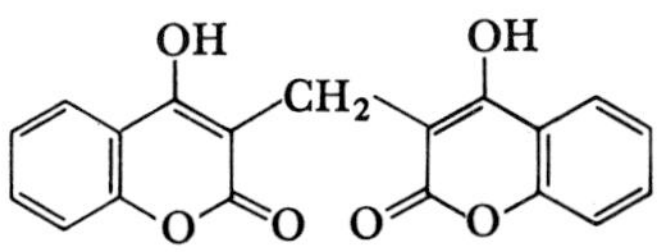

Bishydroxycoumarin (dicoumarol)

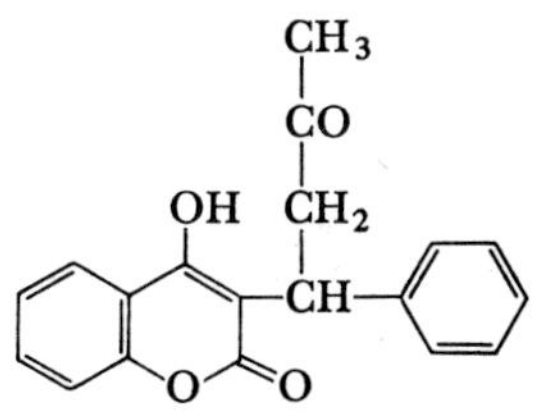

Warfarin (synthetic)

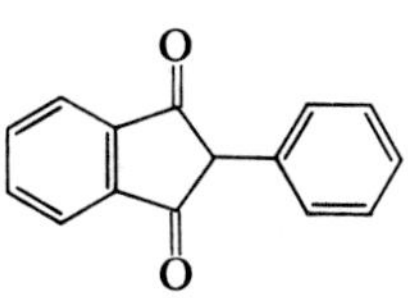

Phenindione (synthetic)

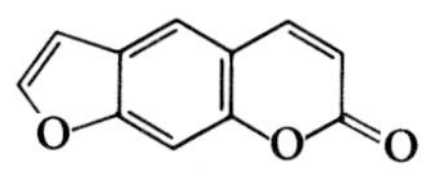

Psoralen

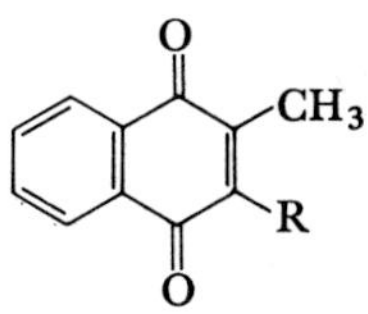

Vitamin K group

Figure 3.63. Coumarin anti-coagulants and related compounds.

Table 3.36. Miscellaneous plant medicines

Compound	Structure	Uses	Comments
Ergotamine	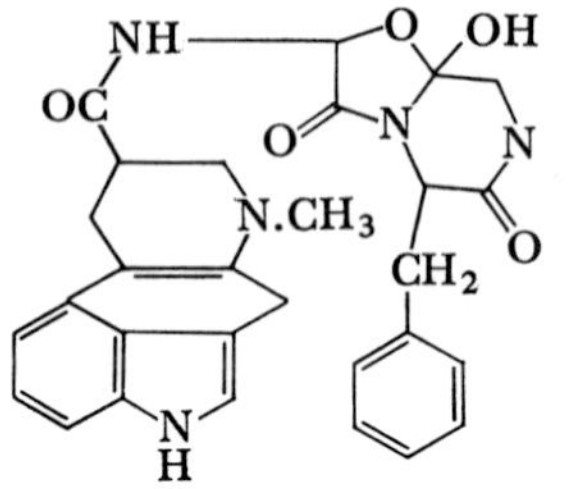	Migraine	Vasoconstrictor and α-adrenergic blocking agent.
Ergometrine	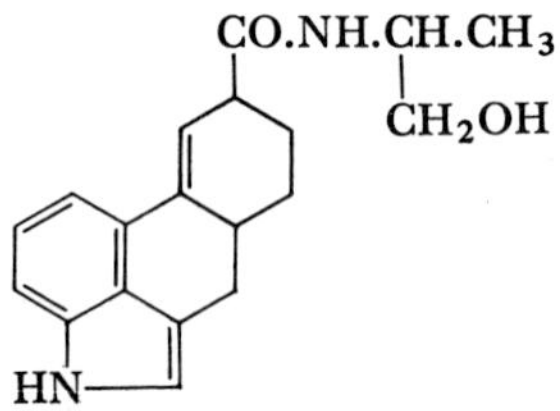	Post-partum haemorrhage	Powerful oxytocic action on uterus.
Colchicine	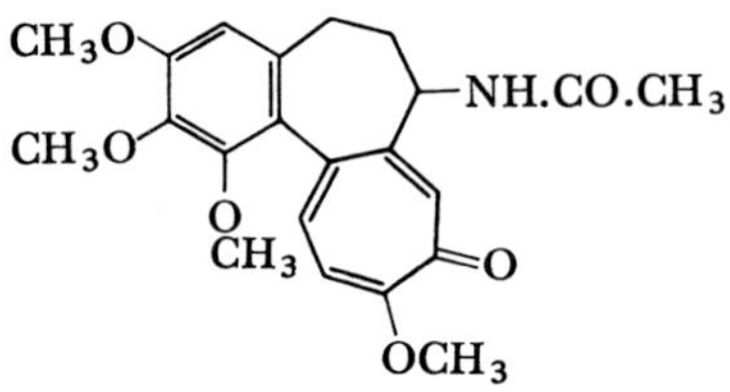	Gout	Uricosuric agent interfering with phagiocytosis producing inclusion of micro-crystals.
Tannic acids (e.g. *m*-digallic acid)	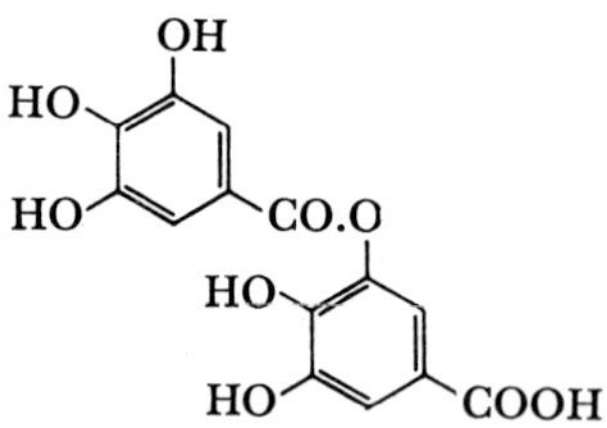	Astringent	Hepatoxins.
Rotenone	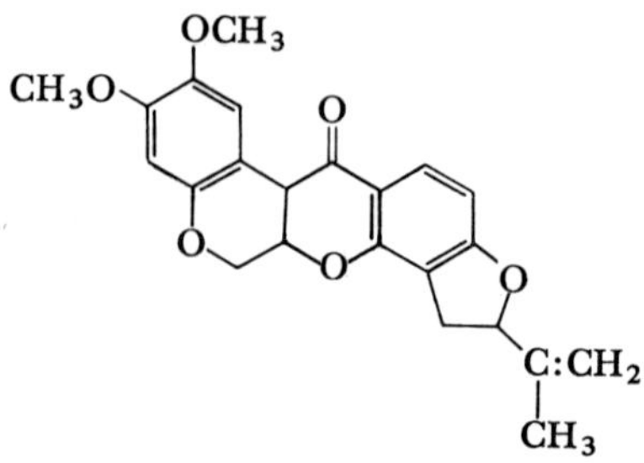	Insecticide of derris	Inhibits electron transfer between $NADH_2$ and flavin.

Psoralen is a structurally related compound, used in the treatment of vitiligo.

Coumarin anti-coagulants are vitamin K antagonists, which interfere with action of the vitamin in biosynthesis of blood coagulation factors. The action is reversed by additional vitamin K. As coumarins are inhibitors of coagulation factor biosynthesis, they are devoid of effect *in vitro*, but slowly decrease blood levels of various factors over a period of 4 to 7 days after ingestion.

(iii) Miscellaneous Other Anti-metabolites

Some plant medicinal compounds which have not so far been discussed are listed in Table 3.36, with comments of their chemistry and biochemistry.

V. SELECTED REFERENCES

1. Vogel R., Trautschold I. and Werle E. (1968) *Natural Proteinase Inhibitors*. New York: Academic Press.
2. Liener I. E. (editor) (1969) *Toxic Constituents of Plant Foodstuffs.* New York: Academic Press.
3. Ajl S. J., Kadis S. and Montie T. C. (1970) *Microbial Toxins.* New York: Academic Press, 6 volumes.
4. Uraguchi K. (1971) Pharmacology of mycotoxins, in *Pharmacology & Toxicology of Naturally Occurring Toxins* (H. Raskova, editor). Oxford: Pergamon Press, volume II, p. 143.
5. Raskova H. and Masek K. (1971) Microbial toxins, in *Pharmacology & Toxicology of Naturally-Occurring Toxins.* (H. Raskova, editor). Oxford: Pergamon Press, volume I, p. 3.
6. Raudonat H. W. (editor) (1965) *Recent Advances in the Pharmacology of Toxins.* Oxford: Pergamon Press.
7. Landy M. and Braun W (editors) (1964) *Bacterial Endotoxins.* New Jersey: Rutgers University.

Chapter 4

VENOMS

I. VENOMOUS ANIMALS AND PLANTS

Venoms are a distinct class of poisonous substances, in that they are complex mixtures of compounds occurring as a biological *secretion*. In this way they can be classified separately to poisonous substances that are merely constituents of biological tissues. This distinction can be clearly seen in the following examples: one from each of the major biological kingdoms.

Various molluscs are venomous, in that they possess specialized organs for the production of a poisonous secretion which can be forced into wounds made by the mollusc's teeth on animals it attacks. In contrast, various shellfish used as human food are occasionally contaminated by dinoflagellates, which contain highly toxic compounds capable of inducing paralysis and death. Yet these toxins cannot be regarded as venoms (of either the molluscs or the dinoflagellates).

Similarly, in the plant kingdom, many whole plants are toxic if eaten (e.g. aconite, bracken, etc), but are not venomous. In contrast, plants such as the stinging nettle and poison ivy produce poisonous secretions into specialized structures on their surfaces, which produce toxic effects in animals with whom they come into contact.

Organisms with specialized venom apparatus are sometimes called *phanerotoxic*, while poisonous animals lacking such apparatus are named *cryptotoxic*.

In the animal kingdom, venomous animals can possess a wide variety of specialized apparatus for the production and administration of their poisons. Among vertebrates the commonest type of venom apparatus is some modification to the teeth, jaws and salivary glands, so that venom is administered by biting. A similar situation occurs in arachnids, though most venomous insects rely on modifications to their urino-genital tract and accessory organs, so that venom is administered by a sting. Other types of venomous structures include modified spines, or specialized regions of an exoskeleton.

Venomous plants are rare, but mostly rely on delicate spines which rupture and penetrate on contact with skin, so that venom spreads on and in the epidermis.

Envenomation of humans occurs on a wide scale throughout the world. Perhaps the commonest type is insect stings, though these are rarely serious. In contrast, bites by venomous reptiles and spiders are common causes of death in certain parts of the world, especially tropical regions. Death and serious consequences are also frequently reported following envenomation by marine animals. It has been estimated that the annual death-rate from the latter cause alone, is probably about 200 persons per year, though reliable estimates are difficult to formulate.

We will deal with venomous snakes, insects, and spiders in this chapter, making only brief mention of other organisms.

II. SNAKE VENOMS

1. Classification and Distribution

Snakes are the most abundant venomous reptiles, though other reptiles, such as the guila monster of Mexico, are also dangerous. Snakes belong to the order Ophidiae, which is sub-divided into four families. Venomous snakes occur in all four families, but those of the Aglyphae and Opistoglyphae families are of little importance to humans. The latter have no specialized venomous teeth, but some produce a salivary venom, so that this is administered into the skin by diffusion into punctures and abrasions produced by the teeth. Opistoglyphae snakes possess grooved poisonous teeth, but these are situated posterior to the superior maxillary, so that venom is not injected by these teeth into large animals bitten by these snakes. In contrast, both of the other families, Proteroglyphae and Solenoglyphae, possess specialized venomous teeth sited anterior to the superior maxillary, so that venom readily enters a wound through these teeth. In the Proteroglyphae the poisonous teeth are motionless and grooved, wherease the Solenoglyphae possess mobile, needle-like poisonous teeth.

Proteroglyphae are divided into two groups: Hydropheidae, or sea-snakes, of which there are 16 genera, and Elapidae, or cobras, of which they are 41 genera. Similarly, Solenoglyphae are divided into: Viperidae, or vipers, of which there are 10 genera, and Crotalidae, or pit-vipers, of which there are 6 genera. There are about 400 species of venomous snakes in all, mainly belonging to these two families.

It has long been recognized that certain patterns can be seen in the toxic symptoms induced in humans by venomous snakes, depending on the type of snake. Generally speaking, patients bitten by Proteroglypherae develop neurological damage, while those bitten by Solenoglyphae have signs of vascular changes and altered blood coagulation. A major exception to this rule is the South American Rattlesnake (*Crotalus durissus terrificus*) which although a Solenoglyphae, induces primarily neurotoxic signs in its victims.

Table 4.1 summarizes the major signs and symptoms developing in patients bitten by various types of snake.

While it is apparent from clinical observations that different snake venoms induce different pharmacological and pathological changes, and so must contain different types of poisons, there is also the problem of the *quantity* of venom injected by the snake. This will depend on many factors, including the type and age of snake, the interval between attacks and the rate of venom formation. It is likely that a prolonged attack, in which the snake makes a series of bites, will be likely to inject significantly more venom than a single brief attack.

Some estimate of the maximum venom capacity of snakes can be gathered by removing all the venom from healthy snakes in captivity. Experiments of this type show that sea-snakes contain only about 12–15 mg dried venom per animal, whereas some vipers (e.g. *Bitis gabonica*) may yield up to 500 mg dried venom per snake. Elapidae and Crotalidae are generally intermediate in amounts of venom (38–84 mg and 21–66 mg respectively), though occasional species yield larger amounts than others (e.g. *Bothrops jararacussu*, a crotalid, yielded 104–330 mg).

Toxicity of dried snake venom obtained from healthy animals in captivity is usually measured as LD_{50} s.c. in mice. Results are summarized in Table 4.2. It will be seen that the most toxic venom requires about 0.12 mg/kg to give a 50%

Table 4.1. Signs and symptoms of snake-bite

Type of snake	Major signs and symptoms
Hydropheid	Little pain at site of bite. Muscular stiffness, flaccid paresis, absence of reflexes. Myoglobinuria. Faecal and urinary incontinence. Respiratory failure and death.
Elapid	Painful bite. Drowsiness and lassitude. Gradual loss of consciousness. Respiratory difficulties. Tachycardia and weak pulse. Excessive salivation and abundant mucus secretion in air ways. Convulsions (occasionally). Coma and respiratory failure.
Viperid and Crotalid (except South American)	Burning, painful bite. Oedema and lymphangitis around bite. Petechia, vesiculation and haematoma. Haematemisis, haematuria, melaena and epistaxis. Faecal and urinary incontinence (in severe cases). Hypotension, shock and circulatory collapse: death.
South American Crotalid	Little pain at site of bite. Impaired vision. Respiratory difficulties. Palpebral ptosis. Death from respiratory failure.

chance of death in mice. If human toxicity is of the same order, this would mean about 8.4 mg for a 70 kg man. This would seem a reasonable amount of venom to be transferred by a healthy snake in a single bite, and the fatality rate for snakes with highly toxic venom is often as high as 50%. However, a snake with a less toxic venom (e.g. 3 mg/kg) would have to inject about 25 times this amount (i.e. 210 mg), which exceeds the amount of venom present in most species. It seems likely that the relatively low death-rate from the bite of most venomous snakes is simply due to the inability of the snake to inject sufficient venom into a man.

Table 4.2. Toxicity of dried snake venoms

Family	Species	Toxicity (mg/kg) LD_{50} (mouse, s.c.)
Hydrophediae	*Laticauda semifasciata*	0.5
Elapidae	*Bungarus multicinctus*	0.16
	Naja naja	0.31
	Naja haje	0.9
	Hemachatus haemachates	1.1
Viperidae	*Bitis lachesis*	0.4
	Bitis gabonica	0.8
	Causus rhumbeatus	1.2
Crotalidae	*Crotalus durissus terrificus*	0.12
	Bothrups jararaca	0.8
	Crotalus scutulatus	1.8
	Trimeresurus flavoviridis	3.0

For South Africa, where snake-bite is relatively common, the incidence is given as 240 per million people per annum, yet the death-rate from snake-bite, despite the presence of many highly venomous species in this region, is only about 4 per million people per annum. This implies a death-rate of 4 per 240 persons bitten, or 1.7%.

Factors contributing to the chances of death from snake-bite, aside from the species of the snake, are as follows:

(1) The amount of venom injected; which in turn depends upon
  (a) the amount of venom in the snake
  (b) the number of bites
  (c) the thickness of clothing at the site of bites
  (b) the thickness and penetration of subcutaneous fat.
(2) The amount of poisonous substances within the venom; this appears to vary widely from snake to snake of the same species, perhaps depending upon
  (a) the size and age of the snake
  (b) time since it last bit
  (c) physiological factors, such as time of day, state of nutrition, parturition, mating, etc.
(3) The physiological state of the victim is also important, particularly:
  (a) age (children under 5 are a poor risk)

(b) health and nutritional status
(c) previous exposure to snake venoms (there is some evidence of reduced susceptibility with increasing exposure)

(4) Finally the speed with which treatment is given, and the type of treatment, greatly affect mortality. Anti-sera must be specific and given in sufficiently high doses at an early stage.

2. Chemical Composition

(i) Low Molecular Weight Compounds

It is immediately apparent from what has been said, that snake venoms must vary considerably in the nature of their toxic components from species to species. There are two overall toxic actions that can be recognized: neurotoxic and vascular-toxic. However, like all animal secretions, snake venoms are very complex mixtures of many different substances, any one of which may contribute to the overall toxic action. It is unlikely, therefore, that there is a single neurotoxin, or a single vasculartoxin, common to all venoms inducing one particular type of action. It is more likely that each venom contains a number of components, each contributing to the overall effect.

While it is possible that inorganic compounds could contribute to toxic actions, this is not likely. Nevertheless, some strange results have been reported for various snake venoms. For example the approximate concentrations of some metallic elements reported for snake venoms and some other body fluids have been compared and it has been claimed that amounts of aluminium, copper and zinc in venom are about 10 to 100 times higher than would be expected. However, the analytical method used was emission spectroscopy, which is not very reliable for biological specimens, while contamination is very easy in trace element work. On the present evidence, these findings should be regarded sceptically.

Of low molecular weight organic compounds, a most surprising finding has been the discovery by several workers of high concentrations of adenosine and adenosine phosphates in elapid venoms. The highest concentrations so far reported are about 26% of the total solids. Unfortunately, while adenosine is known to produce a variety of pharmacological effects (e.g. hypotension, heart-block, etc.), it seems unlikely that it plays any important role in snake-bite due to elapids, for these species rarely produce any fall in blood pressure. The presence of serotonin (5-hydroxytryptamine) and adrenaline has been occasionally reported for some snake venoms, but does not seem to be a regular constituent for most.

These largely negative findings on low molecular weight compounds indicate that most of the toxicity lies in high M.W. substances.

(ii) Proteins

About 90% of the solid constituents of most snake venoms is protein, some of which is enzymically active. The enzymes will be considered separately.

It is difficult to distinguish chemically between polypeptides and small proteins, except by arbitary definitions. A common procedure is dialysis through a cellophane membrane: high M.W. components (i.e. greater than about 45 000 M.W.) are retained, but lower M.W. substances pass through. It has been known for many years that the majority of the toxic activity of snake venoms is associated with proteins, but there is a marked difference between the various families of snakes on the dialysis behaviour of these toxins. For venom from Proteroglyphae, toxins readily diffuse through dialysis membranes, but almost none of the Solenoglyphae toxins can pass. This indicates that the former are polypeptides or small proteins, while the latter are high M.W. proteins.

Protein components of tissues and secretions are usually classified according to their similarity to plasma proteins. Using this system the results shown in Table 4.3 were obtained for protein constituents of venom from an elapidae (Proteroglyphae) and a viperidae (Solenoglyphae). It will be seen that

Table 4.3. Protein constituents of some snake venoms

Family	Group	Species	Percentage of total Albumins	 Globulins	 Polypeptides
Solenoglyphae	Viperidae	*Vipera russellii*	22	23	50
Proteroglyphae	Elapidae	*Naja naja*	40	20	28

the polypeptide content of the former (which contains the toxins) is significantly lower than the latter.

There have been many attempts to characterize venom proteins by electrophoretic separation. In every case, 5 to 10 components have been reported, depending upon the species examined and the conditions of the electrophoresis. Similar results have been obtained by immunological fractionations, and most snake venoms contain 5 to 10 antigenic protein fractions.

Many attempts have been made to obtain "pure" protein fractions from various venoms. In earlier days, methods used included salt fractionation, solvent fractionation, and ultra-filtration; often combined with dialysis, heat and pH treatment. A number of relatively pure toxic protein fractions were obtained, one of the most notable being "crotoxin", a crystalline product from the venom of *Crotallus durissus terrificus* (American rattlesnake). More recently, fractionation has been achieved by combinations of ion-exchange chromatography on resins and Sephadex-gel filtration.

Many of these protein fractions proved to be enzymes, but a few were not. Enzymes as toxic components of venom will be

Table 4.4. Non-enzymic protein toxins of snake venoms

Group	Species	Toxins	Approx. M.W.	Comments
Hydropheid	*Laticauda semifasciata*	Erabutoxin a	6750	Neurotoxin polypeptide N-terminal Arg 61 AA residues.
		Erabutoxin b	6770	Neurotoxin polypeptide N-terminal Arg 61 AA residues.
Elapidae	*Naja* sp.	(Neurotoxin)	?	Only one neurotoxin has been separated in most species, but M.W. has been reported from 2500 to 11 000.
Crotalidae	*Crotalus durissus terrificus*	Crotoxin	29 900	Can be separated into 2 components of similar size. Neurotoxin.
		Crotamine	*c.* 12 000	Basic peptide N-terminal Tyr C-terminal Gly. Neurotoxin.

dealt with below; here we will mention some non-enzymic toxic proteins and their possible mode of action. Table 4.4 summarizes some information on these substances.

Amino acid sequences, or partial sequences, are available for some of the purified non-enzymic toxins of snake venom, but little is yet known of functional groups involved in their biological activities. The hydropheid neurotoxins, erabutoxin *a* and *b*, differ by a single amino acid residue. Erabutoxin *a* contains a aspartyl residue where *b* contains a histidyl residue. Crotamine has no free sulphhydryl groups, but these are formed by mild reduction, during which process toxic properties are decreased.

Table 4.5. Pharmacological effects on non-enzymic protein toxins

Toxins	Principal actions
Hydropheid neurotoxins	1. Block neuromuscular transmission, without affecting nerve conduction or ability of muscles to contract (curare-like). 2. High concentrations may induce direct muscle damage.
Elapid neurotoxin	1. Block neuromuscular transmission, without affecting nerve conduction or ability of muscles to contract (curare-like). 2. Respiratory inhibition by blocking impulses to respiratory muscles: respiratory centre is not directly affected. 3. Block sensory nerve endings. 4. Produce heart blocks.
Crotalid neurotoxins	1. Block neuromuscular transmission, but may induce direct contraction of smooth muscle fibres. 2. Have little, if any, cardiac actions.

Non-enzymic toxins of venom are known to have a variety of pharmacological actions, though none seem to be directly involved in haemolytic effects. Table 4.5 summarizes the principal actions.

(iii) Enzymes

For many years it was believed that all snake venoms owed their toxic properties to their constituent enzymes. While this is now known to be untrue, it is nevertheless likely that enzymes contribute the major toxins in many species of venomous snakes.

Approximately 15 different enzymic activities have been detected in snake venoms and are thought to be related in different ways to poisonous actions. In some cases it is uncertain whether one enzyme catalyses several related reactions, or whether they are due to separate enzyme proteins. Table 4.6 lists the major enzymes of snake venoms and their occurrence.

Each of these enzymes will be briefly discussed, and its contribution to the toxicity of different types of venom will be mentioned.

(1) *L-amino acid oxidase:* This enzyme has been detected in venom from most types of poisonous snake, though hydropheid venom does not appear to have been examined. The longer chain amino acids (e.g. arginine, histidine) are oxidized more readily than the shorter chain compounds (e.g. serine, alanine), though all naturally occurring L-amino acids, together with various synthetic analogues (e.g. 2-aminohendecan-1,11-decarboxylic acid) are acted on to some extent. Snake venom L-amino acid oxidase is a flavoprotein similar to the mammalian hepatic enzyme, but the enzyme in venom is very much more active. It is not known what contribution, if any, this enzyme makes to venom toxicity. Its widespread distribution in venoms suggests that it must play some role, but the products of the enzymic reaction, 2-oxoacids, are not toxic, nor is sufficient enzyme administered by a snake bite to have any significant effect on plasma amino acid concentrations in the victim. Likewise, the amount of ammonia released would be rapidly dealt with by the victim's tissues.

(2) *Phospholipase-A:* The major substrates are lecithins, though cephalins (phosphatidylethanolamines) are also acted upon. The product formed by this snake venom enzyme is the corresponding α-lysophosphatide (unlike the α-toxin of Clostridia; see Figure 3.1). Studies with both natural and synthetic lecithins have established that the catalytic action is confined to hydrolysis of the ester linkage on C-2 of the glycerol residue, C-1 esters being unaffected. The chain-length and degree of unsaturation of the fatty acid residue at C-2 seem irrelevant to the action of this enzyme. Snake-venom phospholipase-A has been prepared in high purity from both *Naja naja*

Table 4.6. Major enzymes of snake venoms

Enzyme and classification	Reaction catalysed	Occurrence
1. L-amino acid oxidase (L-amino acid-O_2 oxidoreductase; EC 1.4.3.2)	L-amino acid + O_2 = 2-oxoacid + NH_3 + H_2O_2	Elapids Viperids Crotalids
2. Phospholipase-A (phosphatide acylhydrolase; EC 3.1.1.4)	Phosphatide + H_2O = α-lysophosphatide + fatty acid	Hydropheids Elapids Viperids Crotalids
3. Phospholipase-B (lysolecithin acylhydrolase; EC 3.1.1.5)	Lysolecithin + H_2O= glyceryl-1-phosphoryl-choline + fatty acid (also catalyses same reaction as phospho-lipase-A)	Elapids Viperids Crotalids
4. Acetylcholine esterase (acetylcholine acylesterase; EC 3.1.1.8)	Acetylcholine + H_2O = acetate + choline	Elapids
5. 5′-Nucleotidase (5′-ribo-nucleotide phosphohydrolase; EC 3.1.3.5)	Ribonucleotide + H_2O = ribonucleoside + phosphate	Elapids Viperids Crotalids
6. Phosphodiesterase (ortho-phosphoric diester phospho-hydrolase; EC 3.1.4.1)	Polyribonucleotide or oligodeoxyribonucleotide + H_2O = nucleoside-5′-phosphates	Elapids Viperids Crotalids
7. Hyaluronidase (hyaluronate glycanohydrolase; EC 3.2.1.36)	Hydrolysis of β-1,3-linkage between D-glucuronate and 2-acetamido-2-deoxy-D-glucuronate and 2-acetamido-2-deoxy-D-glucose residues in hyaluronic acids	Elapids Viperids Crotalids
8. Alkaline phosphatase (ortho-phosphoric monoester hydrolase; EC 3.1.3.1)	Many organic phosphates + H_2O = base + phosphate	Crotalids
9. Peptidases (peptide hydrolases; EC 3.4.4.?)	Peptides + H_2O = shorter peptides + amino acids	Elapids Viperids Crotalids
10. NADase (dinucleotide pyro-phosphatase; EC 3.6.1.9)	NAD + H_2O = adenylic acid + niacinamide-5′-ribo-nucleotide	Elapids
11. ATPase (ATP phosphohydrolase; EC 3.6.1.8)	ATP + H_2O = adenylic acid + pyrophosphate	Elapids Viperids Crotalids
12. ADPase (ADP phosphohydrolase; EC 3.6.1.?)	ADP + H_2O = adenylic acid + orthophosphate	Viperids Crotalids
13. Blood-coagulation factors:		
(a) Thrombin-like	Fibrinogen = fibrin + peptides	Elapids Crotalids
(b) Factor-X activators	Factor-X = active-factor-X	Viperids

and *Crotalus adamanteus*; the M.W. from both species was about 32 000, but at least two closely related isomeric forms were present. Ca^{2+} ions are required for maximum activity, while divalent heavy metal ions, e.g. Pb^{2+}, are inhibitors. Activity is also inhibited by diisopropylfluoro phosphate, so that the active site probably contains a seryl residue. Free SH groups are probably required also, for iodoacetamide is also an inhibitor, though *p*-chloromercuribenzoate is not.

Phospholipase-A occurs in the venom of all poisonous snakes so far studied and is probably intimately concerned in the induction of toxic effects. Lysolecithin, and related lysophosphatides, are potent cytolytic agents which solubilize lipid suspensions by micelle formation. For this reason they are haemolytic poisons, due to their solubilizing action on lipids of erythrocyte membranes, and also neurotoxins in that they induce demyelination.

Very low phospholipase-A activity occurs in the human central nervous system, but the rate of lysolecithin production is so low as to be of little significance in normal people. However, lecithins and other phosphatides are normal constituents of human blood, occurring in both plasma and cells. Large amounts of lysolecithins will be formed in the blood of a snake-bite victim. The effects of this enzyme, however, will depend not only on the amount injected by the snake-bite, but also on the amount of phospholipase-B that is present. The latter enzyme catalyses the further hydrolysis of lysolecithins to inactive products. It is possible that the degree of haemolysis following a snake-bite is at least partially determined by the relative amounts of these two enzymes.

(3) *Phospholipase-B:* This enzyme acts on lecithin in the same way as phospholipase-A to catalyse the formation of an a-lysolecithin, however, the latter compounds will also serve as substrates for a further reaction to hydrolyse the second fatty acid residue and yield glyceryl-1-phosphorylcholine. The enzyme has been obtained in a relatively pure state from elapid venoms. It has a pH optimum about 8–9, is very heat-stable, and is activated by Ca^{2+} ions.

Relative amounts of phospholipase-B in venoms vary considerably. Elapids usually show the highest activity, but some species have only low to moderate amounts. Some crotalids have moderate activity, but others have very little. Viperid venom also shows low activity.

(4) *Acetylcholinesterase:* A major difference between elapids and other poisonous snakes is the presence of high concentrations of this enzyme in their venom. Despite the fact that acetylcholine is the most important synaptic transmitter, purified acetylcholinesterase from elapid venom is *not* neurotoxic at doses similar to those in which it must enter the body of a snake-bite victim. The venom enzyme is not a true acetylcholinesterase, for it catalyses the hydrolysis of a range of choline esters, and also of some simple glycerides. It is inhibited by eserine (physostigmine) (see p. 320).

(5) *5′-Nucleotidase:* This enzyme catalyses the hydrolysis of 5′-phosphate groups from 5′-monophosphates of purine- or pyrimidine-containing nucleosides. Both ribonucleotides and 2-deoxyribonucleotides will serve as substrates, but the former are much better than the latter.

This enzyme is also widely distributed in snake venoms and has been purified about 1000-fold. It is activated by Mg^{2+}, but not by Ca^{2+}, and is inhibited by heavy metal ions such as Zn^{2+}. Chelating agents, such as EDTA, which remove the Mg^{2+} cofactor are also inhibitors.

The role of 5′-nucleotidase in venom toxicity is uncertain. There are small amounts of mononucleotides in human tissues, including blood, of which the most abundant are adenosine-containing compounds. Free adenosine in relatively large amounts is known to produce transient falls in blood pressure by peripheral vasodilation; it also increases coronary flow, but interferes with cardiac condition to produce bradycardia and partial A-V block. It is doubtful, however, if the amount of adenosine produced by 5′-nucleotidase following snake-bite is sufficient to induce any of these effects.

(6) *Phosphodiesterase:* Purified preparations of this enzyme from snake-venoms have found wide application in nucleic acid research, for they catalyse the hydrolysis of both RNA and DNA starting at the terminal nucleotide sequence with a free 5′-hydroxyl. The result is a mixture of nucleoside-5′-monophosphates.

Because the enzyme has found this application, it has been extensively purified from venoms. Most preparations appear to contain 2 or 3 isomeric forms. Full activity requires Mg^{2+} ions but Ca^{2+} is ineffective. It is not inhibited by most heavy metal ions, with the exception of Cu^{2+}. EDTA or other chelating agents are also inhibitors.

Most phosphodiesterase preparations from venom will not act on double-stranded DNA, though they readily attack denatured DNA. While polynucleotides containing the usual purine and pyrimidine bases are readily hydrolysed, unusual bases, such as hydroxymethylcytidine, prevent the enzymic reaction.

Phosphodiesterases catalysing the formation of 5′-nucleotides from nucleic acids are known as "exonucleases". Some snake venoms have also been shown to contain "endonucleases", which catalyse formation of 3′-nucleotides. Crotalid endonuclease acts on both RNA and DNA, but elapid endonuclease is specific for RNA. Both are activated by Mg^{2+}.

The role of phosphodiesterases in venom toxicity may be to supply 5′-nucleotides to serve as substrates for 5′-nucleotidase, so that amounts of cardiovascular toxins, such as adenosine, reach significant levels.

(7) *Hyaluronidases:* We have seen in the previous chapter (p. 201) that some bacterial exotoxins (e.g. of *Streptococcus pyogenes*) are hyaluronidases, and act as "spreading factors" by promoting tissue breakdown and rapid dissemination of the bacteria and their other toxins through the blood stream. Hyaluronidases of snake venoms probably play an analogous role, and occur in the venoms of most poisonous snakes.

Little is known of the chemical and physical characteristics of these enzymes from venom, though they are assumed to be similar to hyaluronidases isolated and studied from other sources, e.g. mammalian testes.

(8) *Alkaline Phosphatase:* Crotalid venom, though apparently not venom of other poisonous snakes, contains a non-specific phosphatase with a pH optimum at about 9·5. A large number of organic phosphates can serve as substrate, including adenosine-5′-phosphate, adenosine-3′-phosphate, ATP, FMN, niacinamide-mononucleotide, and *p*-nitrophenylphosphate.

The enzyme has been obtained in a relatively pure state. It is activated by either Mg^{2+} or Ca^{2+}, but inhibited by chelating agents or heavy metal ions.

Again the function of this enzyme in venom toxicity is uncertain, but it may contribute to the formation of adenosine from tissue nucleic acid break-down products.

(9) *Peptidases:* All snake-venoms contain several enzymes catalysing the hydrolysis of peptide linkages. It has been known for at least a century that snake venoms dissolve meat and can

disintegrate gelatin cubes or fibrin clots. More recent investigations have concentrated on the specificity of the peptidases. Most mammalian enzymes of this type show clear specificity in the substrates on which they can act. For example, trypsin catalyses the hydrolysis of peptide linkages in which the carbonyl is an arginyl or lysyl residue, while chymotrypsin and pepsin catalyse the hydrolysis of peptides containing aromatic amino acids, such as phenylalanine. Venoms of most viperids and crotalids readily hydrolyse synthetic substrates containing arginine (e.g. benzoyl-arginine ethyl ester or *p*-toluene-sulphonyl-arginine methyl ester), but have little or no effect on synthetic substrates containing aromatic amino acids (e.g. *N*-benzoyl-tyrosine ethyl ester or *N*-acetyl-tyrosine ethyl ester). It is therefore concluded that snake venoms of these groups contain trypsin-like enzymes, though as no inhibition occurs following treatment with soybean trypsin-inhibitor (see p. 290) or ovomucoid, they are not identical to trypsin.

Trypsin-like activity in elapid venom is low, while it is apparently absent from hydropheid venoms.

We will see later that some venoms contain thrombin-like enzymes, which are also peptidases. However, trypsin-like activity occurs in venoms which cannot convert fibrinogen to fibrin, so the two activities must represent separate enzymes.

Using modern chromatographic and electrophoretic methods, several different proteolytic enzyme fractions have been obtained from a variety of snake venoms. For example, five such enzymes have been reported from venom of *Trimeresurus flavoviridis*, three from *Crotalus atrox*, and two from *Bothrops jararaca*. Most appear to require Mg^{2+} or Ca^{2+} for activity, are inhibited by EDTA or heavy metal ions, but differ widely in pH optimum and substrate specificity.

The above enzymes act primarily on specific linkages in protein molecules. Peptidases acting on lower M.W. fragments have also been identified in many snake venoms. Leucine-aminopeptidases appear to be particularly widely distributed in all classes of venomous snakes.

Functions of peptidases and their contributions to venom toxicity are probably several. First, it is clear that some interfere with blood coagulation factors (see below), others act as "spreading factors" by aiding breakdown of proteins in tissues around the bite to allow more rapid penetration of toxins into the circulation. A more specific function is the release of vasoactive peptides, such as bradykinin, which are potent vasodilators and induce a marked hypotension. Crotalid venoms appear to be particularly rich in peptidases which catalyse the release of these kinins, which is probably the reason why shock and circulatory collapse are a feature of crotalid envenomation. Kinin-releasing factors have been isolated and purified from the venoms of several crotalid species. They are inhibited by aprotinin, a polypeptide derived from bovine lung tissue that is a potent inhibitor of kinin-releasing enzymes (e.g. kallidinogenase).

(10) *NADase:* It has been known for some time that various snake venoms are potent inhibitors *in vitro* of carbohydrate metabolism when tested on cell-free systems. A variety of possible explanations have been proposed, but one is the breakdown of NAD (and NADP), which is an essential cofactor for this process. A variety of different snake venoms have now been shown to catalyse the hydrolysis of NAD to niacinamide mononucleotide and adenosine-5′-phosphate. The enzyme responsible has not been purified, but is not inhibited by Zn^{2+}, which is an inhibitor for 5-nucleotidase. It appears that the enzyme attacking NAD will also act on coenzyme-A, with the formation of pantetheine-4′-phosphate and adenosine-3′,5′-diphosphate.

Again the role of this enzyme in venom toxicity is uncertain, but it could well act as a general cytotoxin by depriving cells of essential cofactors used in a variety of metabolic pathways.

(11) *ATPase:* Mention has already been made of the hydrolysis of ATP catalysed by the alkaline phosphatase of crotalid venoms, however, a more specific ATPase has been identified in venom from most groups of poisonous snakes.

The available evidence suggests that venom ATPase catalyses the hydrolysis of ATP to ADP and orthophosphate and that a separate enzyme then acts on ADP (see below). However, elapid venom may possess an ATPase which converts ATP to AMP and pyrophosphate.

(12) *ADPase:* It has proved possible to separate ATPase and ADPase activity from venoms of *Bitis gabonica* and *Bothrops jararaca*. A further species (*Bothrops cotiara*) has venom with ADPase, but not ATPase activities.

ADPase appears to be Zn^{2+} activated, while ATPase requires

Mg^{2+}. The former, but not the latter, is inhibited by α,α-dipyridyl, which is a chelating agent for Zn^{2+}. ADPase is also inhibited by Co^{2+}.

ATP and ADP play well-known roles in energy metabolism of all cells, so that their removal from tissues affected by snake-venom enzymes is likely to have serious consequences. It is also known that platelet activation in blood coagulation is initiated by ADP, so that depletion of ADP following envenomation is likely to impede this process. Finally, ADPase will further contribute to the release of adenosine from nucleic acids.

(13) *Blood-coagulation factors:* There is no doubt that many types of snake venom have a pronounced effect on blood coagulation. This is seen particularly with viperids and most crotalids, where haematomata develop at the site of a bite, and haematemesis and haematuria are common. The matter is complicated, however, by the discovery that manipulation of dose can induce either promotion or inhibition of coagulation. For example, venom of Russell's viper induces blood clotting both *in vivo* and *in vitro*, but very large doses may have the reverse effect. The active factor in the venom of this snake has been obtained in a relatively pure state. It is a protein of M.W. 105 000 with peptidase activity against gelatine and *p*-toluene-sulphonyl-arginine methyl ester (i.e. trypsin-like). However, it is not affected by soybean trypsin-inhibitor, while it also shows moderate activity against lysine ethyl ester.

Russell's viper venom is able to coagulate both normal and haemophilic blood and commercial preparations are available for use in the latter condition. Haemophilia of the classical variety is due to an inherited defect in Factor VIII of the blood-coagulation mechanism (Figure 4.1). Factor VIII participates in the activation of Factor X, which in turn participates in the conversion of Factor V into a prothrombin-activator. The Factor X of normal blood plasma has been isolated and shown to be a protein of M.W. 84 000. When incubated with the active factor of viperid venom, Factor X is broken into two smaller fractions, one of which is able to produce prothrombin-activator when added to a mixture of Factor V and phospholipid. It is likely, therefore, that the active factor of viperid venom is a proteolytic enzyme resembling Factor VIII of human blood.

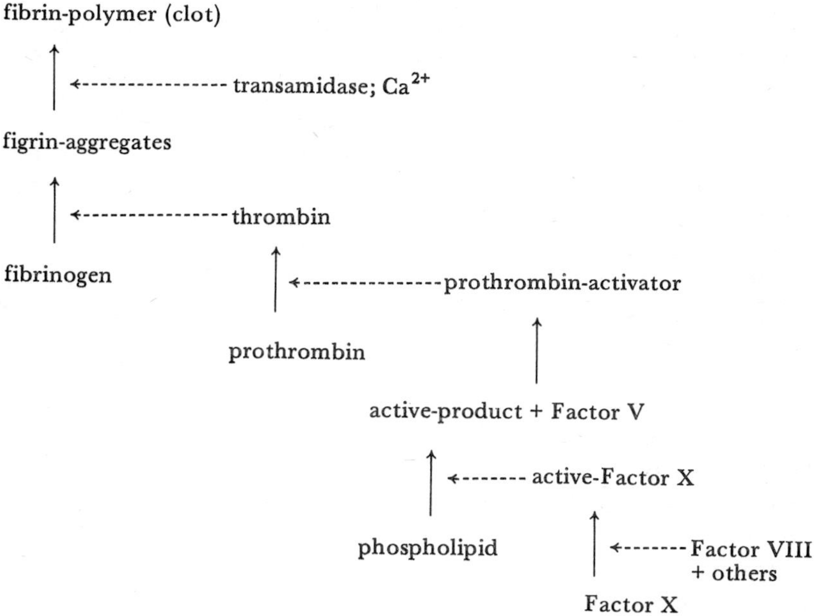

Figure 4.1. Blood coagulation mechanism (simplified).

Venoms of other snakes, particularly crotalids, contain enzymes which initiate blood coagulation by a direct effect on fibrinogen ("thrombin-like" activity).

Fibrinogen of human plasma is a dimer of M.W. 340 000. When acted upon by thrombin, fibrinogen is cleaved at two points, so that two fibrinopeptides are also formed. The amino acid sequences of these two short fibrinopeptides (A and B) are known (Figure 4.2). It is now known that the thrombin-like enzyme of snake venoms releases only fibrinopeptide-A from fibrinogen, though clotting still occurs. The peptide linkage cleaved by this enzyme is -Gly-Arg-. The consequence of this difference between human thrombin and the venom enzyme is that cross-linking of the fibrin-aggregates to the highly insoluble

Fibrinopeptide-A (release by thrombin and venom enzymes)
H.Ala.Asp.Ser.Gly.Glu.Gly.Asp.Phe.Leu.Ala.Glu.Gly.Gly.Gly.Val.Arg.OH
Fibrinopeptide-B (released only by thrombin)
Pyroglu.Gly.Val.Asn.Asp.Asn.Glu.Glu.Gly.Phe.Phe.Ser.Ala.Arg.OH

Figure 4.2. Fibrinopeptides released by thrombin from fibrinogen.

fibrin-polymer is inefficient, and the clot formed in response to crotalid venom is soluble in 10% urea, whereas the thrombin produced clot is not.

Several studies of venom "thrombin-like" enzyme have been undertaken. It appears to be relatively heat stable, has a pH optimum around 6.6, is inhibited by Cu^{2+}, Hg^{2+}, or glutathione, but not by many other divalent metal ions. Ca^{2+} appear to be required for full activity, while EDTA is an inhibitor.

It is noteworthy that some, though not all, crotalids and elapids possess both a thrombin-like enzyme and also a factor X- activator similar to that found in viperid venoms.

3. Absorption

In most cases snake venom is injected into the body through a bite, though a few snakes, such as the spitting-cobra, rely on absorption of venom through mucous membranes. The degree of skin penetration by teeth of a snake will vary considerably, depending on its species, age and the thickness of clothes at the site of attack. In some cases, venom will enter the blood stream directly, in others it may be deposited largely in sub-cutaneous fat. In the latter case, the presence or absence of spreading factors in the venom may be critical as to the time-interval before systemic effects of the venom appear.

The relative importance of venom distribution via the blood stream or via lymphatics has been investigated on a number of occasions in experimental animals immobilized and ligated so that blood circulation is normal but lymphatic drainage of a limb is blocked. In these cases, toxic symptoms usually develop much slower, so that transport in lymphatics may be important, particularly for toxins of high M.W. which are non-dialysable. However, water-soluble, dialysable toxins are probably transported mainly in blood.

Distribution of certain toxins within the body has been studied in animals following administration of ^{131}I-labelled venom. In many cases, accumulations of venom in lymph nodes, thymus and thyroid were detected, with lesser amounts in liver and kidneys.

4. Mode of Action

(i) Neuromuscular Junctions

Venom of most Proteroglyphae have a curare-like action, which can be demonstrated both *in vivo* and *in vitro*. Curare (i.e. d-tubocurarine and related compounds; see p. 319) induce a neuromuscular blockade by combining with postsynaptic receptors for acetylcholine. This effect develops rapidly, but can be reversed by cholinesterase inhibitors, which greatly increase the amount of acetylcholine available to compete with the drug for receptors. Investigations with snake venoms indicate that the effect develops slowly, while anticholinesterases are without effect on the blockade.

It seems likely that venom toxins combine with postsynaptic acetylcholine receptors in a non-reversible manner and cannot be displaced simply by increasing the amount of acetylcholine.

(ii) CNS

Respiratory depression induced by snake venom is largely a consequence of neuromuscular blockade. Administration of CO_2 to an envenomized animal to stimulate the respiratory centre has the same affect on phrenic nerve impulses as in a normal untreated animal. However, direct administration of venom into the brain has dramatic toxic actions on a number of higher centres. Ataxia, respiratory changes, agitation and muscular convulsions develop rapidly, and the pattern of signs and symptoms is unlike that seen following intravenous administration of the same dose of venom.

It is apparent that snake venom has the potential to disrupt CNS functions, but that this is unimportant except where venom is administered into the brain. This would be unlikely, even following bites to the head, unless the snake's teeth penetrated the skull.

(iii) Autonomic Effects

Some of the commonest effects of snake-bite in humans are alterations in respiratory and cardiac rates, together with increased salivation, diarrhoea, dilation of the pupils and muscular tremors. Studies on electrical activity of cervical sympathetic nerves following venom administration to experimental animals have revealed marked changes, so that direct effects of venom toxins on the autonomic nervous system seem very likely.

(iv) Direct Muscular Actions

We have discussed above the blocking effect of some snake venoms on neuromuscular junctions by combination with postsynaptic acetylcholine receptors. There is some evidence, however, for other, direct effects of venom toxins on muscle tissues.

Where a snake-bite penetrates muscles, a local necrosis often develops. Trismus, paresis and spastic paralysis of skeletal muscles is seen commonly in human snake-bite victims, and can be induced with ease in experimental animals. It has been suggested that some of these changes seen following purified toxins, such as crotamine, are inhibitors of glycolysis in muscle cells.

Smooth muscle may also be affected by kinins released by venom enzymes (see above).

(v) Cardiovascular Effects

Almost 40 years ago it was demonstrated that snake venoms in high dilution arrest the beating of an isolated frog heart. While this effect is seen with crotalid and viperid venoms, it may be more important for elapids, where cardiovascular collapse is the most prominent effect. A protein has been isolated from venom of *Naja naja* which will inhibit cardiac rhythm both *in vitro* and *in vivo*. This cardiotoxin has a M.W. about 46 000 and is non-dialysable. Other dialysable toxins occur in the same venom, but their chemical nature is still uncertain.

Hyperkalaemia may follow snake-bite, due to cell destruction. This also is likely to induce cardiac effects, some of which may be detected only by electrocardiogram, unless the potassium concentration rises markedly.

Changes in blood pressure following snake-bite may have

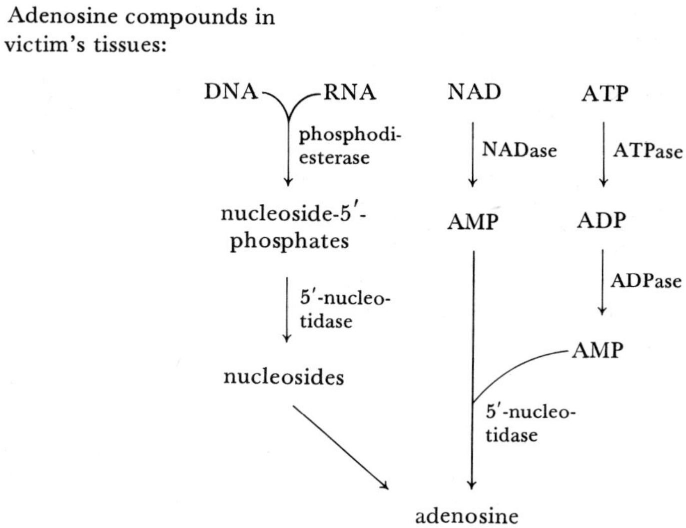

Figure 4.3. Co-ordinated effects of venom enzymes.

multiple causes. First, there are direct effects on the heart mentioned above, but the release of bradykinin and adenosine may also be important contributory factors. Figure 4.3 illustrates the way in which the various enzymes of snake venoms, particularly viperid or crotalid, augment to supply adenosine, which has a hypotensive action.

There is some evidence for increased release of histamine by mast cells following envenomation, and this also may be involved in the observed cardiovascular effects.

(iv) Haemolysis

Lysolecithin is a potent haemolytic agent, so its formation by phospholipase-A from lecithin of blood plasma and erythrocytes is an indirect mechanism for the haemolysis observed after some snake-bites. For elapid venoms, there is some evidence for direct haemolytic agents, which act on erythrocyte membranes. These agents appear not to be lysolecithin or phospholipases, but their chemical constitution is still uncertain. They have been obtained by fractionation of venom by chromatography and appear to be water-soluble basic compounds.

(vii) Other Actions

It is clear from the above discussion that most snake venoms are complex mixtures of toxins with wide-ranging effects on many different tissues. As essential enzyme cofactors (e.g. NAD, ATP, ADP, CoA) are reduced by venom enzymes, it is likely that tissues affected by venoms are likely to show changed metabolism. A number of investigations of specific metabolic processes have been undertaken on whole animals, tissue slices, homogenized tissues, and particular cell fractions. It has been found that many oxido-reductases (e.g. lactate dehydrogenase, β-hydroxybutyrate dehydrogenase, malate dehydrogenase) are much reduced following exposure to venoms containing NADase, presumably due to lack of their cofactor. The implication of this reduction in oxido-reductases is decreased metabolism of carbohydrates and lipids via glycolytic, β-oxidation and tricarboxylic acid cycle pathways. Apparently independant of this effect is an uncoupling of oxidative phosphorylation by mitochondria exposed to venoms in low concentration. This could be partly a consequence of reduced ADP supply, though a lytic action by lysolecithin, or other substance, could partially remove essential lipid cofactors for this process.

5. Antisera

The presence of a variety of toxic agents in snake venom gives rise to obvious difficulties in the development of antisera for the treatment of snake-bite victims, for it is likely that sera will contain antibodies effective against only one of the venom toxins. Nevertheless, antisera have been developed which are effective against the principal toxin of particular venoms. For example, antivenoms to elapid neurotoxins are of value in snake-bite by a member of this family, though ineffective against viperids or crotalids.

As it is often impossible to identify the snake causing a particular injury, polyvalent antisera have been developed which contain a mixture of antivenom effective against all groups of poisonous snakes. Unfortunately, considerable problems have arisen in the standardization of antivenom activities, and there is much doubt as to the most effective dose. It is also clear that efficacy of antivenoms decline rapidly with the interval between snake-bite and administration.

III. INSECT VENOMS

1. Classification and Distribution

Two groups of terrestrial arthropods form venoms toxic to humans: insects and arachnids. These venoms may be transmitted by stings (e.g. ants, wasps, bees, scorpions), bites (e.g. mosquitoes, spiders), or contact (e.g. blister-beetles, certain caterpillars). The average number of annual envenomations by insects is probably greater than the world population, for almost everyone is bitten or stung at least once a year. Few of these attacks require medical treatment, and of those that do fatalities are exceedingly rare. This is primarily due to the very small amount of venom transmitted by a single insect, rather than low toxicity. Mouse LD_{50} for snake venoms range from 0.1 to 3.0 mg/kg; comparable figures for insect venoms are shown in Table 4.7. The commonest venomous insects are the honey-bee (*Apis mellifera*), wasps (*Vespa vulgaris*), hornets (*Pollistes sp.*), ants (e.g. *Solenopsis soevissima richteri*), bed-bugs (*Cimex lectularis*), mosquitoes (e.g. *Anopheles sp.*), urticating caterpillars (e.g. *Megalopyge urens*), and blister-beetles (e.g. *Zonabris rubaica*).

Table 4.7. Mouse LD_{50} of some insect venoms

Species	LD_{50} (mg/kg dry venom)
Apis mellifera	3–4
Vespa vulgaris	2–4
Pollistes exclamans	2–4
Pogonomymex barbatus	1.2
Myrmecia pyriformis	2–5

Most investigations of insect venoms have been made on bees, wasps and ants. A little is known of vesicating venoms, and there is little information on venoms of such common biting insects as mosquitoes and fleas.

2. Chemical Composition

Table 4.8 lists the major toxins of some common insect venoms.

The major toxin of honey-bee venom is the polypeptide "mellitin", which contains 26 amino acid residues and has a

Table 4.8. Toxins of honey-bee venom

Substance	Approximate concentration (% dry venom)
Histamine	1.2
Histapeptide	1
Mellitin	50
Apamin	2
Apic acid	5
Phospholipase-A	3
Hyaluronidase	4
Unidentified peptides	15
Antigenic proteins and other enzymes	8

M.W. 2850. Its probable sequence is shown in Figure 4.4. As the LD_{50} of mellitin is close to the LD_{50} of dried whole venom, it is likely that the other components contribute little to toxicity. However, a structurally related polypeptide "apamin" (Figure 4.4) is also present and is a potent neurotoxin. The concentration ratio of mellitin to apamin in honey-bee venom is about 25

1. Histapeptide:

Ala.Gly.Ala.Gln.Histamine

2. Mellitin:

Gly.Ile.Gly.Ala.Val.Leu.Lys.Val.Leu.Thr.Thr.Gly.Leu.Pro.Ala.Leu.Ile.Ser-Trp.Ile.Lys.Arg.Lys.Arg.Gln.Gln.NH_2

3. Apamin:

Cys.Asn.Cys.Lys.Ala.Pro.Glu.Thr.Ala.Leu.Cys.Ala.Arg.Arg.Cys.Gln.Gln-His.NH_2

4. Mast-cell degranulation peptide:

Ile.Lys.Cys.Asn.Cys.Lys.Arg.His.Val.Ile.Lys.Pro.His.Ile.Cys.Gly.Lys.Asn-NH_2

Figure 4.4. Sequences of toxic peptides in honey-bee venom.

to 1, so that apamin may not be present in sufficient amounts to produce any marked effects.

The presence of phospholipase-A and hyaluronidase in bee venom probably contribute to the spread of the venom, as for snake venoms.

In contrast to honey-bee venom, wasp and hornet venoms both contain serotonin (5-hydroxytryptamine), in addition to free histamine. They also contain both phospholipase-A and phospholipase-B; the latter enzyme being absent from honey-bee venom. The amount of serotonin is about 0.3 mg/g dry venom for wasps, but up to 20 mg/g for dry venom of hornets. The reverse is true for histamine: dry venom of wasps contains about 16 mg/g, while dry venom of hornets usually contains less than 5 mg/g, though occasional specimens may be as high as 30 mg/g.

Pollistekinin 3

Pyrglu.Thr.Asn.Lys.Lys.Lys.Leu.Arg.Gly.Arg.Pro.Pro.Gly.Phe.Ser.Pro.Phe-Arg

Bradykinin
Arg.Pro.Pro.Gly.Phe.Ser.Pro.Phe.Arg

Figure 4.5. Wasp-venom kinin.

Hornet venom is unique among the hymenoptera in containing significant amounts of acetylcholine; up to 50 mg/g dry venom.

Peptides like mellitin and apamin appear to be absent from venoms of wasps and hornets, but their place is taken by other peptides with bradykinin-like activity. The structure of wasp-venom kinin is shown in Figure 4.5 contrasted with bradykinin. It will be seen that the peptide from venom possesses an additional 9 amino acid residues. The structure of hornet-venom kinase is similar, but has not been finally established.

There is no doubt that, at least in certain individuals, toxic reactions to bee-stings are dominated by allergic and anaphylactic responses. The same is probably true of responses to wasp or hornet venoms.

Surprisingly, hymenoptera venom has been used therapeutically; both by deliberately exposing patients to insects in circumstances under which a sting is likely, or applying extracts or preparations of venom by a variety of routes. One famous report describes 173 patients with rheumatic complaints who were "therapeutically" stung by bees a total of 39 000 times (Terc, F. (1888) *Wien. Med. Press.* 29, 1424). Despite enthusiasm of individual medical practitioners, the results of treatment were not particularly encouraging.

Many species of ant are venomous, but little is known of the

Table 4.9. Components of ant venoms

Component	Type of ant	Chemical structure
Limonene (80%*d* + 20%*l*)	*Myrmicaria*	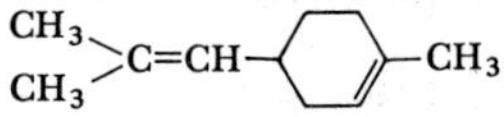
Isoiridomyrmecine and iridomyrmecine	*Iridomyrmex*	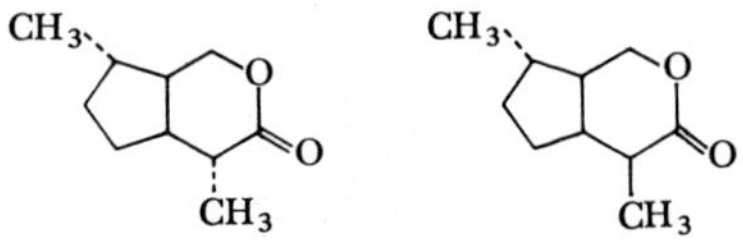
Dendrolasine	*Dendrolasius*	$(CH_3)_2C{=}CH.(CH_2)_2.C(CH_3){=}CH.CH_2CH_2$ 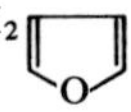
Iridodial	Dolichodernic ants	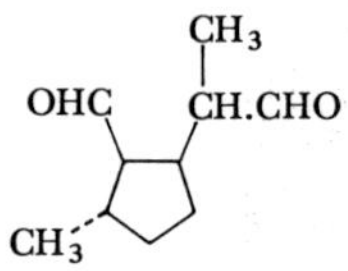
Dolichodial	Dolichodernic ants	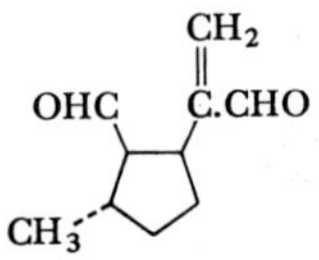
Methylheptenone	Dolichodernic ants	$CH_3.CO.(CH_2)_6.CH_3$
Methylhexanone	Dolichodernic ants	$CH_3.CO.(CH_2)_5.CH_3$
Propylisobutylketone	Dolichodernic ants	$CH_3.CH_2.CH_2.CO.CH_2.CH(CH_3)_2$
Fatty acids:	Many species	
formic acid		$H.COOH$
acetic acid		$CH_3.COOH$
isovaleric acid		$CH_3.CH(CH_2.CH_3).COOH$
propionic acid		$CH_3.CH_2.COOH$
isobutyric acid		$CH_3.CH(CH_3).COOH$
Histamine	Many species	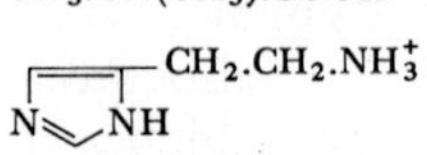

toxins present. Table 4.9 lists some of the compounds reported to be present and shows their chemical structures. However, the venom of some species is very toxic and none of these compounds possesses sufficiently poisonous characteristics to account for the properties of the venom. It is known, however, that the venom has kinin-like activity, histamine-releasing properties, spreading-factors, and cell-lytic substances, though the chemical nature of many of these remains to be established.

Cutaneous reactions develop in humans and animals coming into contact with certain lepidopterous larvae. The venom of these organisms is contained within poisonous spines which penetrate and break on contact with skin. The organization of the venom apparatus varies somewhat from family to family, but most urticating caterpillars possess either hollow hairs containing venom, or spines with a heavily chitinized cone at the tip that is readily detached.

Table 4.10. Toxic components of lepidoptera venoms

Substance	*Dirphia* sp.	*Megalopye* sp.
Histamine	0.02–0.04%	Absent
Acetylcholine	Absent	Absent
Serotonin	Absent (?)	Absent
Peptidase (against gelatin)	+	+
"Kinins"	+	+
"Allergens"	+	+

At least 8 families of Lepidoptera are known to include venomous species, and the distribution of urticating caterpillars is world-wide. However, venomous caterpillars of Europe and North America are rarely involved in serious human poisoning; the same cannot be said of urticating lepidoptera of Asia, Africa, or South America. The latter region is particularly abundant with toxic species.

Chemical studies of the nature of caterpillar venoms have been hindered by the very small amount present in each hair, and the fact that the venom is the product of a single-celled gland. Table 4.10 lists some of the toxic components so far identified in these venoms.

Poisoning by caterpillar venom is known as "erucism" (Latin *eruca*, caterpillar). However, a number of adult lepidoptera are also venomous and poisoning by these forms gives rise to a condition known as "lepidopterism". While early Greek and Roman writings contain accounts of poisoning by adult moths, it has proved impossible to identify the species involved, and at the present time lepidopterism appears to be confined to South America and the Caribbean islands, where the toxic animals are all members of the genus *Hylesia*. Members of a second genus, *Zygaena*, are toxic to certain animals, but not to humans.

Adult moths of *Hylesia* species are dull-coloured. The female is usually larger than the male and carries on her abdomen small, barbed chitinous "flechettes", which contain venom. Very little is known of the toxins present in this venom, but it is water-soluble, induces a marked urticarial response on contact with skin, and is said to contain "allergine": a group of chemically ill-defined basic organic amines. Many affected persons become sensitized to the venom, following an initial exposure.

While fatalities following contact with venomous caterpillars or adult lepidoptera do not appear to have been reported, the insects can induce painful and disfiguring lesions.

Finally, we must consider the most abundant venomous insects: mosquitoes and fleas. Some doubt has been cast on the true venomous nature of these species, but certainly they induce painful bites which sometimes are accompanied by systemic effects. Virtually nothing is known of the biochemistry of the venoms injected by these insects, but their effects are probably largely based on the antigenic properties of protein components. Some investigations on the venom of common bed-bugs (*Cimex lectularius*) have been reported. Hyaluronidase is present, together with histamine and a bradykinin-like peptide. It is likely that most biting insects possess venoms of similar constitution.

IV. ARACHNIDS

1. Scorpions

There are three orders of this class which are venomous: Scorpiones (scorpions), Araneida (spiders), Acari (ticks). Of these, spiders are the most serious public health problem.

Indeed, in some countries, spider envenomation is a greater problem than snake-bite. We will consider first scorpions and ticks, before turning to details of spider venoms.

Scorpions are well known in tropical regions, and administer their venom via a tail, which terminates in a telson with a prominent sting. Venom is secreted by a pair of glands contained by the telson, and is injected into a victim by rapid repeated strikes of the tail.

Scorpion venom is not as toxic as venom of snakes, spiders or some insects. The most poisonous species, which belong to the Buthidae family, possess venoms which show an LD_{50} in rodents of about 1 to 20 mg/kg. The most medically significant genera are: *Centruroides* (Central America), *Tityus* (South America), *Buthus* (North Africa and Asia), *Androctonus* (North Africa to India), *Parabutus* (South Africa). Most venomous scorpions induce very similar symptoms, so it is likely that scorpion venom varies little from species to species. However, detailed studies on venom biochemistry are lacking in most cases.

Scorpion stings are painful and result in systemic neurotoxic effects. While fatalities are recorded, particularly in children, they are rare compared to the probable frequency of scorpion envenomations (about 1% of adult cases). Prominent symptoms are excessive salivation and lachrymation, accompanied by strabismus and occasional convulsions. An initial hypotension is followed by hypertension and tachycardia. Hyperglycaemia and glycosuria may occur. These effects appear to be mediated via a multitude of neurological effects, including toxic changes to the CNS, spinal cord, and autonomic nervous systems. There is also evidence of an eserine-like effect at neuromuscular junctions (see p. 320), together with a direct stimulatory action on smooth muscles.

The amount of venom injected by a single scorpion sting is uncertain, and may vary considerably. Average yields of dry venom obtained from *Tityus serrulatus* and *T. Bahiensis* by electrical stimulation were 0.62 and 0.39 mg respectively, though occasional individuals yielded as much as 2.0 mg or as little as 0.03 mg.

Like most venoms, the toxic secretions of scorpions contain a complex mixture of substances, several of which contribute to the overall toxic effect of the venom. Histamine is usually absent, but about 1 to 2% is lysine. However, this is the only free amino acid present in significant amounts. There are traces of serotonin, but these low molecular weight compounds probably contribute little to the toxicity, which appears to be shared by several proteins. These can be fractionated into a number of components, each of which is toxic to experimental animals. Some of the proteins are enzymes: lecithinases are present in most venoms, but in relatively low amounts. Some also contain hyaluronidases, though peptidases are absent.

Table 4.11. Amino acid composition of scorpamines

Amino acid	Residues/mol Scorpamine I	 Scorpamine II
Aspartate	9	8
Threonine	2	3
Serine	6	2
Proline	6	3
Glycine	6	7
Alanine	1	3
Valine	4	4
$\frac{1}{2}$-Cystine	8	8
Isoleucine	2	1
Tyrosine	3	7
Phenylalanine	1	1
Lysine	6	5
Histidine	1	2
Arginine	2	3
Tryptophan	1	1
Glutamate	6	4

The major toxins of scorpion venom have been named "scorpamines", for they are basic proteins of relatively low molecular weight (6700 to 5200). Two well-defined components have so far been identified, but others may well be present in lower quantities. The LD_{50} of scorpamines I and II in mice are 19 and 10 μg/kg respectively. Amino acid compositions are given in Table 4.11. In both cases, the N-terminal amino acid is lysine, but for scorpamine I the C-terminal amino acid is threonine, whereas it is glycine in scorpamine II. Partial sequences are known, and both toxins show several similarities. It is likely that their mode of actions are identical, though full details are still unavailable. Both stimulate *in vitro* the release of acetylcholine from the ileum and *in vivo* the release of adrenal catecholamines.

2. Ticks

Despite their widespread distribution and the frequency of bites to both humans and animals, little is known of tick venoms. For most species, the bite is rarely noticed initially, though itching, pain, redness and swelling develop later. Exceptions are species such as *Ornithodoros coriaceus* of North and Central America, which give an immediately very painful bite and serious local reactions. Tick saliva is introduced into the victim, and sometimes contains sufficient neurotoxins to induce a serious spinal paralysis. The toxin can also block neuromuscular transmission.

Table 4.12. Venomous spiders dangerous to humans

Common name	Systematic name	Occurrence
Black-widow	*Lactrodectus mactans*	Widespread
Brown-widow	*L. geometricus*	Widespread
Malmignatte	*L. tredecimguttatus*	S. Europe
Shoe-button spider	*L. indistinctus*	S. Africa
Funnel-web spider	*Atrax robustus*	Australia
Brown spiders	*Loxosceles recusus*	N. America
Brown spiders	*L. laeta*	S. America
Wolf-spider	*Lycosa ornata*	S. America
Banana-spider	*Phoneutria nigriventer*	S. America

The seriousness of tick envenomation depends on the species of tick, together with the amount of toxin introduced during the bite. Many species can remain firmly attached to the skin for many hours while they feed, so that large amounts of venom can be transferred. Paralysis normally develops only after tick envenomation over a period of 5 to 6 days (by a female tick).

3. Spiders

At least 25 000 species, beloning to over 2000 genera, have been described. Most possess venomous saliva, but few are able to inflict bites on human skin, though they are poisonous to other animals. Of species dangerous to humans, only female spiders, which are more aggressive than males, are able to inflict fatal bites. In some tropical countries, spider envenomation is a significantly greater public health problem than snake-bite. Some of the more dangerous species are listed in Table 4.12. Most are relatively small; in contrast, the much feared large hairy species are relatively harmless, though are capable of inflicting painful bites if sufficiently aroused. The term "tarantula" is often loosely applied to any large hairy spider, though it is more properly reserved as a name for members of Orthognatha (a sub-order).

The venom apparatus of spiders consists of one pair of chelicerae and one pair of venom-secreting glands. Venom can be "milked" from many species if they are restrained and electrically stimulated. Maximum yields vary with both species and individual, but 1.2 to 6.5 mg dry venom can be obtained from orthognatha and 0.5 to 8.0 mg (usually 1 to 2 mg) from Labidognatha. Toxicity of spider venoms varies from about LD_{50} (mouse, s.c.) 0.25 mg/kg for black-widow venom to about 2.5 mg/kg for *Eupalaestrus* species (a non-aggressive brown tarantula with yellow-red bands on its legs). Perhaps the most poisonous species is *Phoneutria nigriventer* where the mouse LD_{50} is about 0.01 mg/kg. This is an aggressive South American species. One individual has been known to kill two children with successive bites.

Symptoms of spider envenomation vary with the species, but usually involve severe local pain at the site of the injury, followed by excessive salivation, priapism, prostration, distensive paralysis, and death in fatal cases. The condition induced in humans by spider envenomation is often described in terms of the species involved; thus envenomation by *Loxosceles laeta*, for example, would be named "loxoscelism", while bites of *Lactrodectus mactans* induce "lactrodectism".

Despite their obvious importance, surprisingly little is known of spider venom components. Unlike scorpions, where the results of envenomation are relatively independent of species, signs of spider envenomation vary considerably. *Lycosa* and *Loxosceles* species induce oedema and vesicular lesions, sometimes progressing to necrosis and ulceration. Renal and hepatic involvement in loxoscelism are well described, with haemoglobinuria and death resulting from internal haemorrhages. In contrast, *Lactrodectus* and *Phoneutria* induce conditions predominated by neurotoxic effects. This clinical evidence points to the existence of species-specific toxins with different modes of action, as is the case with snake venoms.

Extracts of *Lactrodecus* venom have been partially purified by paper chromatography and electrophoresis to yield a relatively potent hypertensive protein, which apparently acts by stimulating release of adrenaline.

Rather more is known of *Phoneutria* toxins. The intense pain produced by the spider's bite is known to be caused by a low molecular weight peptide which stimulates histamine release. Small amounts of histamine (0.06 to 1.0%) and serotonin (0.03 to 0.25%) are also present in the venom. There is a remarkably large amount of free glutamic acid (*c*. 24%), with much small amounts of free aspartic acid (*c*. 1%) and lysine (*c*. 0.2%). Two peptides which stimulate contraction of the ileum have also been isolated, while a further electrophoretic fraction induces flaccid paralysis when injected into mice. None of these components is inactivated by heating at 100°C for 20 minutes, though all are readily inactivated by proteolytic enzymes such as trypsin or pepsin.

Sephadex-gel fractionation of *Phoneutria* venom has allowed the isolation in relatively pure state of the following further toxins: a trypsin-like enyzme of high molecular weight, a peptide inducing hypertension, two peptides inducing paralysis (one induces flaccid paralysis, the second distensive paralysis). There is also a neurotoxic peptide, and perhaps yet another responsible for priapism. All these components appear to lie in the molecular weight range 5000 to 6000.

The presence of a proteolytic enzyme in this spider venom complicates investigation of its toxins, for incubation of the venom alone results in auto-inactivation.

V. VENOMS OF MARINE ANIMALS

1. Porifera

The sea is an abundant source of venomous species. We will deal with common poisonous members of the major groups.

There are about 5000 known species of sponges, which are

organized colonies of unicellular monads. Very few are toxic to humans, though they may be poisonous to various marine organisms. Human envenomation by sponges occurs when skin is abraided by sharp spicules of a poisonous sponge and toxins are deposited in the wounds. Symptoms are usually local (e.g. redness, oedema, swelling, pruritis, etc.), though a few cases with systemic involvement have been described. The latter cases display shock, malaise and skeletal muscle weakness. Table 4.13 lists some venomous species.

Table 4.13. Some venomous porifera

Species	Occurrence
Callyspongia vaginalis	West Indies
Microciona prolifera	North Atlantic (USA)
Pseudosuberites	Gulf of Mexico
Suberites domunculus	Europe
Tedania toxicalis	California
Calyx nicaeensis	Mediterranean
Spongilla lacustris	Widespread

Toxins of sponges have been poorly investigated. The venom appears to be located in gemmules, from which it can be extracted with water. These extracts are haemolytic *in vitro* and also inhibit contractions of a perfused heart. A variety of toxic

$$(CH_3)_3.N^+.CH_2.COOH$$

Betaine

$$HN{:}C(NH_2).NH.CH_2.COOH$$

Glycocyamine

$$H_2N.CH_2.(CH_2)_2.CH_2.NH_2$$

Putrescine

Also present: acetylcholine
histamine
dimethylhistamine
homarine (*N*-methylpilolinic acid)
protein allergens
various unidentified low MW organic compounds

Figure 4.6. Structures of some constituents of sponge venoms.

signs occur in laboratory animals following parenteral administration of sponge venom preparations. The LD_{50} in most species is about 10 mg/kg, so that individual toxins of the venom may well be highly poisonous compounds. Death from injected sponge extracts in mice or dogs occurs mainly by respiratory failure.

A large number of compounds have been detected in sponge extracts, but it is uncertain which are venom toxins. Figure 4.6 gives the structures of some components.

Sponge venom is heat-stable and the toxic components dissolve in both water and various organic solvents.

2. Cnidaria (Coelenterates: Sea-anemones, Corals and Jellyfish)

About 9000 species are known, but only about 70 possess nematocysts capable of penetrating human skin. Several species are poisonous if ingested. A few toxic species are listed in Table 4.14.

Symptoms of cnidarian envenomation are both local and systemic. There is local redness and pruritis at the site of a sting, which sometimes develops into painful pustules and vesicles. Weakness and vertigo may develop, accompanied by headache, muscle spasm and difficulty in breathing. A few fatalities have been recorded, and appear to be due to cardiac arrest, which occasionally develops within a few minutes of being stung.

Table 4.14. Venomous cnidaria

Species	Common name	Occurrence
Millepora alcicornis	Stinging-coral	Pacific, Indian Ocean, Caribbean
Physalia physalis	Portugese man-of-war	Atlantic, Mediterranean
Lytocarpus philippinus	Feather-hydroid	Widespread (tropics)
Carybdea alata	Sea-wasp	Pacific, Indian Ocean, Atlantic
Catostylus mosqicus	Sea-blubber	Pacific
Chrysaora quinquecirrha	Sea-nettle	Widespread (tropics)
Actinia equina	Sea-anenome	Caribbean
Sagartia elegans	Rosy-anenome	Atlantic, Mediterranean

Sensitivity to cnidarian venoms appears to vary markedly with the individual.

Nematocysts are abundant on the tentacles of cnidarians. They uncoil on contact and embed in skin. Venom passes down the spines into the victim. The mouse LD_{50} of most cnidarian venoms is about 2 ml/kg by the subcutaneous route. As the crude venoms contain about 95% water, this implies an LD_{50} of about 10 mg/kg for dry venom. A few species, however, are very much more poisonous. The LD_{50} of *Chironex fleckeri* venom in mice is about 250 μg/kg.

$(CH_3)_4.N^+OH^-$

Tetramine

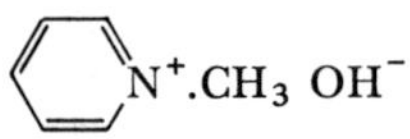

N-Methylpyridium hydroxide

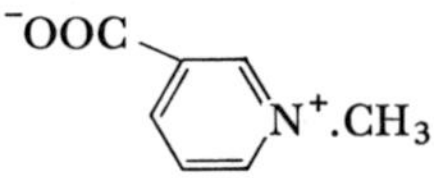

Trigonelline

Also present: serotonin
homarine
basic peptides

Figure 4.7. Structures of some constituents of cnidaria venoms.

Again many different substances have been identified in cnidarian venoms, so that the principal toxins are still uncertain. Figure 4.7 gives chemical structures of some components.

Serotonin is present in nematocyst fluid and is probably partly responsible for the initial pain of a cnidarian sting. Free histamine is also present. Four polypeptides separated by paper chromatography were found to account for 95% of the venom's toxity. Other polypeptides are also present, but are of lower toxicity. A number of enzymes have also been detected, including alkaline and acid phosphatases, 5-nucleotidase, and cholinesterase. Partially purified preparations of the toxic polypeptides have been obtained by gel filtration. The molecular weight range appears to be 10 000 to 15 000. The toxins, which are basic peptides, are inactivated by heat (10 minutes at 100°C), or by incubation with proteolytic enzymes. The mode of action appears to involve a direct cardiotoxic effect.

3. Echinodermata

About 6000 species of starfish, sea-urchins, and sea-cucumbers have been described, but only about 80 are venomous, though many others are probably poisonous if ingested. Venomous echinoderms possess spines associated with a toxic secretion which passes into the victim's tissues via the wounds caused by penetration of the spines. Anatomical arrangements of venom apparatus in these animals varies considerably with species.

Table 4.15. Venomous echinoderms

Species	Common name	Occurrence
Acanthaster planci *A. ellisi*	Venomous starfish	Pacific, Indian Ocean California, Galapagos
Diadema setosum	Long-spined or black sea-urchin	Pacific, Indian Ocean
Aesthenosoma varium	Leather-urchin	Indian Ocean, Indonesia
Toxopneustes pileolus	–	Indian Ocean, Pacific, Japan

There is immediate pain and swelling around the puncture wound caused by a venomous echinoderm. Systemic symptoms, such as weakness, syncope, muscular paralysis and respiratory distress, develop later. Fatalities are very rare, though have been documented following the eating of sea-cucumbers.

Table 4.15 lists the names and occurrence of some venomous echinoderms. Most envenomations by these animals arise from hand or foot contacts to swimmers in the sea.

Numerous toxins have been isolated from echinoderms, but their involvement in the signs and symptoms of human envenomations by poisonous starfish or sea-urchins is uncertain. The most interesting group of compounds are steroidal saponins (Figure 4.8), some of which are potent haemolytic agents. As intravascular haemolysis has not been described as a

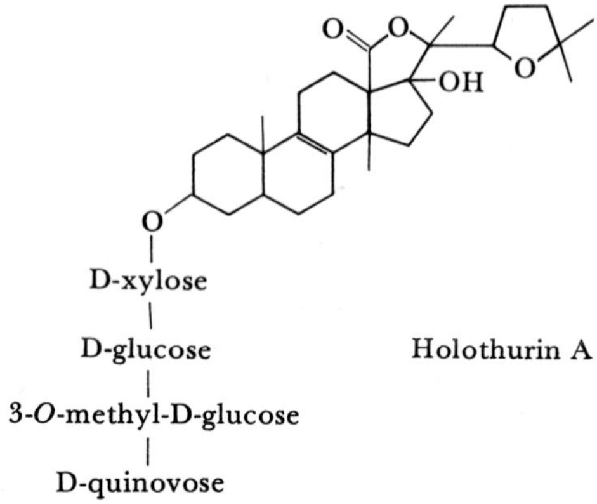

Holothurin A

Figure 4.8. Echinoderm saponins.

major feature of echinoderm envenomation, it is unclear whether these compounds are important to human victims. Table 4.16 lists some other toxic components of echinoderm venoms. Serotonin is apparently absent, but there are free peptides and low molecular weight proteins of unknown constitution.

Table 4.16. Components of echinoderm venoms

Holothurin A and related sterols
Homarine (*N*-methyl pilolinic acid)
Acetylcholine-like substance
Phosphagens
Polypeptides
Mucoprotein (anticoagulant)
Noradrenaline

4. Mollusca

Two classes of this phyllum are known to contain venomous species, though these account for only about 85 of the 80 000 molluscan species. These molluscs are truly venomous, in that they secrete a venom for injection into victims via wounds produced by specialized organs. Other species, however, while not venomous are capable of producing serious illness if eaten by humans. In most cases, the cause of this poisoning is toxins produced by micro-organisms (often Dinioflagellates) contaminating the molluscs.

Some species of cephalopod (octopus, nautilus, squid, etc.) produce a toxic secretion in their posterior salivary glands. This is administered to a victim by bites of powerful jaws or "beaks". In contrast, venomous gastropods (slugs, snails, etc.) possess needle-like teeth within their radula. These are thrust by the proboscis into prey in a sting-like action. A list of venomous molluscs is given in Table 4.17.

Table 4.17. Some venomous molluscs

Class	Species	Common name	Occurrence
Cephalopoda	*Hapalochlaena macalosa*	Blue-ringed octopus	Australia
	Octopus fitchi	–	California
Gastropoda	*Conus geographus*	Cone-shells	Pacific
	C. texile		
	C. omaria		

Signs and symptoms of envenomation by molluscs are difficult to define precisely, due to the relatively few cases which have been reliably reported, and to apparent individual variations, perhaps in both victims and molluscs. Octopus bites around the Australian coast appear to be more serious than those from other Pacific regions. Most of the former are probably due to a single species, *Hapalochlaena maculosa*, the bite of which causes little or no pain. This is followed by suppression of salivation, nausea and difficulty in swallowing. Finally, there is a general numbness and muscular paralysis that can be fatal. In contrast, octopus bites elsewhere are often described as painful and may bleed freely. Other effects are minor and systemic complications are rare. Cone-shell envenomations are similar, irrespective of species, though some appear to administer more venom than others. The sting is painful and affected areas become red and numb. This is followed by a spreading numbness and muscular paralysis, which can be fatal. The similarity of serious cone-shell and octopus envenomations suggest that their toxins may be similar.

Table 4.18 lists some compounds identified in saliva of

Table 4.18. Components of octopus venom

Tyramine
Octopine
Agmatine
Adrenaline
Noradrenaline
Histamine
Dopamine
Serotonin
L-*p*-hydroxyphenylethanolamine
Corticosteroids
Homarine

venomous cephalopods. Again it is uncertain which of these plays a significant role in toxicity. Investigations of peptide constituents have indicated that these components may be most important. A protein fraction known as "cephalotoxin" has been isolated from octopus salivary glands and shown to possess 4 electrophoretic fractions. A peptide with the sequence shown in Figure 4.9 has been obtained from other cephalopods. It is a potent hypotensive and vasodilating agent.

Pyr.Pro.Ser.Lys.Asp.Ala.Phe.Ile.Gly.Leu.Met.

Figure 4.9. Structure of eledoisin.

5. Pisces (Fish)

Many fish are poisonous if eaten by humans. This can be due to contaminating toxins of bacteria, or to accumulated industrial wastes. Some fish, however, are naturally poisonous, but others are truly venomous. About 200 species fall into this category, which includes string-rays, scorpion-fish, toad-fish, cat-fish, and some sharks.

Most venomous fish are slow-swimming, non-migratory species that live in shallow water. Many live in rocks or corals, or buried in sand or mud. Their venom is used as a defensive mechanism, rather than against prey. A list of some venomous fish is given in Table 4.19.

The many kinds of sting-ray have been known to be venomous since ancient times. They range in size from a few inches to 15 feet. The dorsum of the tail possesses a bilaterally

Table 4.19. Venomous fish

Systemic name (family)	Common name
Dasyatidae	Whiprays
Urolophidae	Round sting-rays
Myliobatoidae	Bat-rays and eagle-rays
Gymnuridae	Butterfly-rays
Potamotrygonidae	River-rays
Scorpaenidae	Scorpion-fish and stone-fish
Trachinidae	Weever-fish

serrated spine encased in an integumentary sheath, within which are specialized venom-secreting cells.

It has been estimated about 750 persons each year suffer sting-ray envenomation in North-America alone. While deaths are rare, the injury is painful and serious. The wound produced may be several inches long and require suturing. There is intense pain around the wound, which gradually spreads and may persist for up to 2 days. In most cases, symptoms are confined to the area of the sting, but some people develop systemic effects, including syncope, nausea, weakness, sweating, cardiac arrhythmias and respiratory distress. Rare cases develop muscle contractures simulating paralysis.

Little is known of sting-ray toxins, largely due to the difficulty of obtaining sufficient venom. Extracts of the sting contain a complex mixture of amines and amino acids, but the major toxin appears to be a protein of high molecular weight. At least three components have been separated by electrophoresis. Among the enzymes present are 5-nucleotidase and a phosphodiesterase, though peptidase and phospholipase activity are absent. Venom is inactivated by storage at room temperature for 4–12 hours, even at neutral pH.

Mode of action of the venom is unclear, but a direct cardiovascular effect has been demonstrated in laboratory animals. There is vasoconstriction of large arteries, arterioles, and veins. The mouse LD_{50} of sting-ray venom protein is about 3 mg/kg.

About 80 species of scorpion-fish are known, all of which are potentially venomous to humans. Envenomation by these animals is often more serious than by other venomous fish, with death and serious complications being more frequent.

Most scorpion-fish possess toxic spines, which are distributed in dorsal, anal and pelvic regions according to species. The spines are contained in integumentary sheaths with venom glands lying in their antero-lateral grooves. Spines are driven into a victim's skin, often a hand or foot. There is excruciating pain at the site of wounding, which gradually spreads to surrounding areas. The injured skin becomes red and swollen, with systemic complications becoming gradually more apparent. Nausea, pallor, vomiting, headache and syncope are common, while some victims develop cardiovascular and/or pulmonary symptoms that may lead to death.

Necrosis and sloughing around the wound often develop in a few days.

Extracts of scorpion-fish venom glands yield reddish solutions, pH *c.* 7, containing 10–15% protein. The toxic fraction is non-dialysable, and electrophoresis reveals 7–12 protein bands. Only one of the latter appears to be lethal.

The mode of action appears to be primarily a direct cardiovascular effect, with the induction of hypotension and myocardial ischemia. There is some action on skeletal muscle, perhaps mediated via neuromuscular synapses.

The final group of venomous fish that we will consider are the weevers which, unlike the other groups mentioned above, are confined to relatively small regions of the sea (eastern Atlantic and Mediterranean). Their name is thought to be a corruption of "viper" (Anglo-Saxon "wivre"). They possess venomous spines (2 opercular, 5–8 dorsal) within interumentary sheaths with associated venom-secreting cells.

Stings by these fish may be single or multiple and cause intense pain, which gradually increases in intensity for about one hour, and may persist for a day. The pain spreads to involve surrounding tissues and axillary involvement is common. There is some swelling, with blanching and local necrosis.

In severe cases, there is weakness, nausea, shock and respiratory distress, though fatalities are rare.

Weever-fish venom is rich in serotonin, histamine and catecholamines, but the major toxin is a series of high molecular weight proteins. From 8–15 fractions have been separated by electrophoresis, but only one or two appear to be toxic. Mouse LD_{50} appears to be about 5–10 mg/kg for intact dry venom and about 2 mg/kg for individual protein toxins.

As for other venomous fish, weever toxins appear to be predominantly cardiovascular poisons.

VI. VENOMOUS PLANTS

While venoms are usually considered to be confined to the animal kingdom, there is little doubt that plants such as stinging nettles (*Urtica sp.*) fulfil most criteria, while others, such as poison ivy (*Rhus radicans*) induce a marked dermatitis on contact that is in many ways similar to the action of some insect venoms.

Little is known of chemical constituents of plant venoms, but the fluid in hairs of stinging nettles contains about 0.1–0.5% histamine, 1% acetylcholine, and a bradykinin-like substance of uncertain composition. The constituent of poison-ivy which causes dermatitis is probably related to 3-pentadecyl-catechol, for this compound is used successfully to reduce sensitivity to the plant.

VII. SELECTED REFERENCES

1. Bücherl W., Buckley E. E. and Deulofeu V. (editors) (1968) *Venomous Animals & Their Venoms: Vol. 1—Vertebrates.* New York: Academic Press.
2. Bücherl W. and Buckley E. E. (editors) (1971) *Venomous Animals & Their Venoms: Vol 2—Vertebrates.* New York: Academic Press.
3. Bücherl W. and Buckley E. E. (editors) (1971) *Venomous Animals & Their Venoms: Vol 3—Invertebrates.* New York: Academic Press.
4. Buckley E. E. and Porges N. (editors) (1956) *Venoms.* Washington: Am. Assocn. Advn. Sci.
5. Raskora H. (editor) (1965) *Recent Advances in the Pharmacology of Toxins.* Oxford: Pergamon Press.
6. Raskova H. (editor) (1971) *Pharmacology and Toxicology of Naturally-Occurring Toxins.* Oxford: Pergamon Press (2 volumes).

INDEX